# of Arizona

Halka Chronic

MOUNTAIN PRESS PUBLISHING COMPANY
MISSOULA

February 1989
Sixth Printing

Library of Congress Cataloging in Publication Data

Chronic, Halka.
Roadside geology of Arizona.

1. Geology—Arizona—Guide-books. 2. Arizona—Description and travel—1981- —Guide-books.
I. Title.
QE85.C48 1983 557.91 83-2233
ISBN 0-87842-147-5

MOUNTAIN PRESS PUBLISHING COMPANY
P.O. Box 2399
Missoula, MT 59806
(406) 728-1900

Dedication

To my four daughters and those they love

Bold cliffs of dune-formed sandstone dwarf 800-year-old ruins at White House in Canyon de Chelly.

preface

Writing this book has been a revelation. It has taken me back to scenes of childhood, to early adventures on the desert, to hikes in untamed ranges, college field trips, and geologic work on the northern Arizona Plateau. To one brought up on the desert, its call is ever present. One's soul is in the wide spaces, the long vistas, the desert flowers, the soaring buzzards and blue sky, the hum of bees among golden palo verde blooms, the fragrance of wind after rain. And in the rocks, bright with color, hot with sunshine, crag and cliff and desert mountainside.

Man has moved into the desert in force now, altered it and changed it from what it was in my childhood, rarely, alas, for the better. In moving in, man has come into intimate interplay with the geologic forces that created the mountains and valleys and that control the water supply – of vital importance in this dry land.

The mountains of central Arizona and the high plateaus of the north are no less a land apart. Pines and mountain lakes, snow-fed streams, are to some a pleasant surprise. This is not the arid Arizona they had imagined! Throughout much of geologic time – and we have a record of about the last half of it in Arizona, the last 2 billion years – the rise of mountains and the spread of deserts have marked the geologic history of the state. Landscapes that we see here today are here because of landscapes of the past, and because of tremendous forces deep within the earth, forces that carry continents into collisions and then drag them apart again, forces of heat and pressure and the slow churning boil of the earth's interior. Landscape features result, too, from more comprehensible, more recent forces: the unending attack of water and wind and frost, the building of volcanoes, the short-term geologic happenings like landslides and rockfalls, earthquakes and floods, and a gopher digging a hole.

Material in this book comes largely from published geologic literature, particularly publications of the United States Geological Survey – the USGS – and the Arizona Bureau of Geology and Mineral Technology, and from field trip guidebooks of a number of geologic organizations, notably the Arizona Geological Society. I owe a debt of gratitude to the many geologists who have contributed to this body of literature. Other geologists helped me directly by discussing their work with me or by reading over and commenting on part or all of my manuscript: Edwin D. McKee, John Harshbarger, Don Peterson, Wesley Peirce, Troy Péwé, Harold Drewes, and Don Hyndman. I have Tad Nichols to thank for many of the photographs. Unless otherwise noted, other photos and artwork are my own.

Maps are derived from the Geologic Map of Arizona published by the USGS in 1969, with some updating from more recent geologic literature. Geologic maps show the age of the rock present at the surface or just below the soil layers. In many cases, especially in northern Arizona, the maps also identify specifically named **formations** *(recognizable rock units) or* **groups** *of formations. The same holds true of the sections across, along, or parallel to the highways. Be forewarned: for the sections, vertical scales are exaggerated, so mountains appear much too high, valleys too deep, and slopes too steep for an automobile!*

Geologic abbreviations used on the maps in this book are modified versions of abbreviations commonly used by geologists – a sort of geologic shorthand. Similarly, the symbols used are standard symbols suggestive of the rock types: dots (sand grains) for formations that are largely sandstone, dashed lines for formations that are largely shale, blocklike symbols for limestone, and so forth. A legend to these symbols and abbreviations is on the inside of the front cover. Geologic terms are defined where first used, and again in the glossary at the end of the book.

This book is designed especially (though not exclusively) for those with little or no geologic training. Students and professional geologists may seek here an introduction to Arizona that can be fleshed out later from more detailed literature. If you are excited by your own curiosity and if you want to understand your natural surroundings, geology is a good place to begin. Arizona has had a long and checkered past; here we follow the story of that past. Here we look at the foundation – the bare rock – and interpret its story, or sometimes several conflicting stories, with the conflicts in interpretation that are rungs on the ladder of knowledge.

I have tried to present here the main aspects of Arizona's geology, embellished by details that can be seen from the highways. But please, don't be satisfied with what you can see from your car window. Stop occasionally, and linger. Step off the pavement and onto the face of the land. Find a path or trail if you can – an easy task in national forests, parks and monuments.[1]

On Interstate highways stop at rest stops for a good look around. Look closely at the rocks themselves, handle them, register in your mind their texture and color, their seeming durability or weakness, and any other features you can identify. Notice contacts between one rock type and another. Poke at the soil, search for a crystal or a fossil, watch a "dust devil" in action, sift stream sand between your fingers, listen as a rain-swollen stream grinds rock against rock. Feel the weight of a cobble, and imagine the pressures deep within the earth, under miles of such rock. Then fit what you find into the context of this book.

Rocks in Arizona are well exposed – there is no doubt about that. And many can *be seen from the highway. An arid climate in particular is responsible for the lack of a soil mantle and the consequent lack of vegetation that makes large parts of Arizona a geologist's paradise. Precipitation ranges from 2 to 5 inches in the barren lowland deserts, from 20 to 30 inches in the high central and northern mountains and plateaus. There is variation in precipitation with altitude; snow mantles the high country for part or all of the winter. The main rainy season is in summer.*

---

[1]*A few words of warning: When you walk in the desert carry water, and if there is no trail keep your car or a familiar landmark in sight. Obey a simple rule, good for both cactus (very common) and rattlesnakes (quite rare): never put your hand or foot down without looking first. Stay clear of old mines – many are near collapse. In all parts of the state respect private, public, and Indian property. Fossil collecting is now prohibited or severely limited on public land.*

A glance at the map that precedes Chapter I will show you that the roadlogs in this book read in "every which" direction, apparently completely at random. A second glance will show that there is some *rhyme and reason: Many loop trips are possible, most of them starting at one of Arizona's three largest population centers: Phoenix, Tucson, or Flagstaff. In places it has been easy to use cultural and geographic names of towns, rivers, and mountain ranges for orientation and location. Elsewhere in this spaced-out land I've had to resort to mileposts. Regardless of the direction in which the mileposts read, "mile" in the text means the* mile following the milepost *in the direction for which the log is written. "Milepost" in the text means within a stone's throw of the mile marker itself.*

For finding your way around, use any good roadmap; the one issued by the Arizona Department of Transportation is just fine. Most roadmaps name streams, mountains, and other prominent geographic features; some give elevations as well. Except for I-19 from Tucson to Nogales, highways and roadmaps are marked in miles; I-19 has gone metric.

Using this book, read Chapter I first, and refer to it as often as you need to. It's a mini-course in geology. Then before you start off down the highway read the introduction to the chapter covering the area you'll be traveling in – the southern Basin and Range Province (Chapter II), the Central Highlands (Chapter III), and the Colorado Plateau Province of northern Arizona (Chapter IV). From there on, follow the individual roadlogs. National parks and monuments are discussed in Chapter V.

Contents

III. MAZATZAL LAND — the Central Highlands

IV. SCENIC WONDERLAND — the Colorado Plateau

V. THOSE SPECIAL PLACES
— National Parks and Monuments

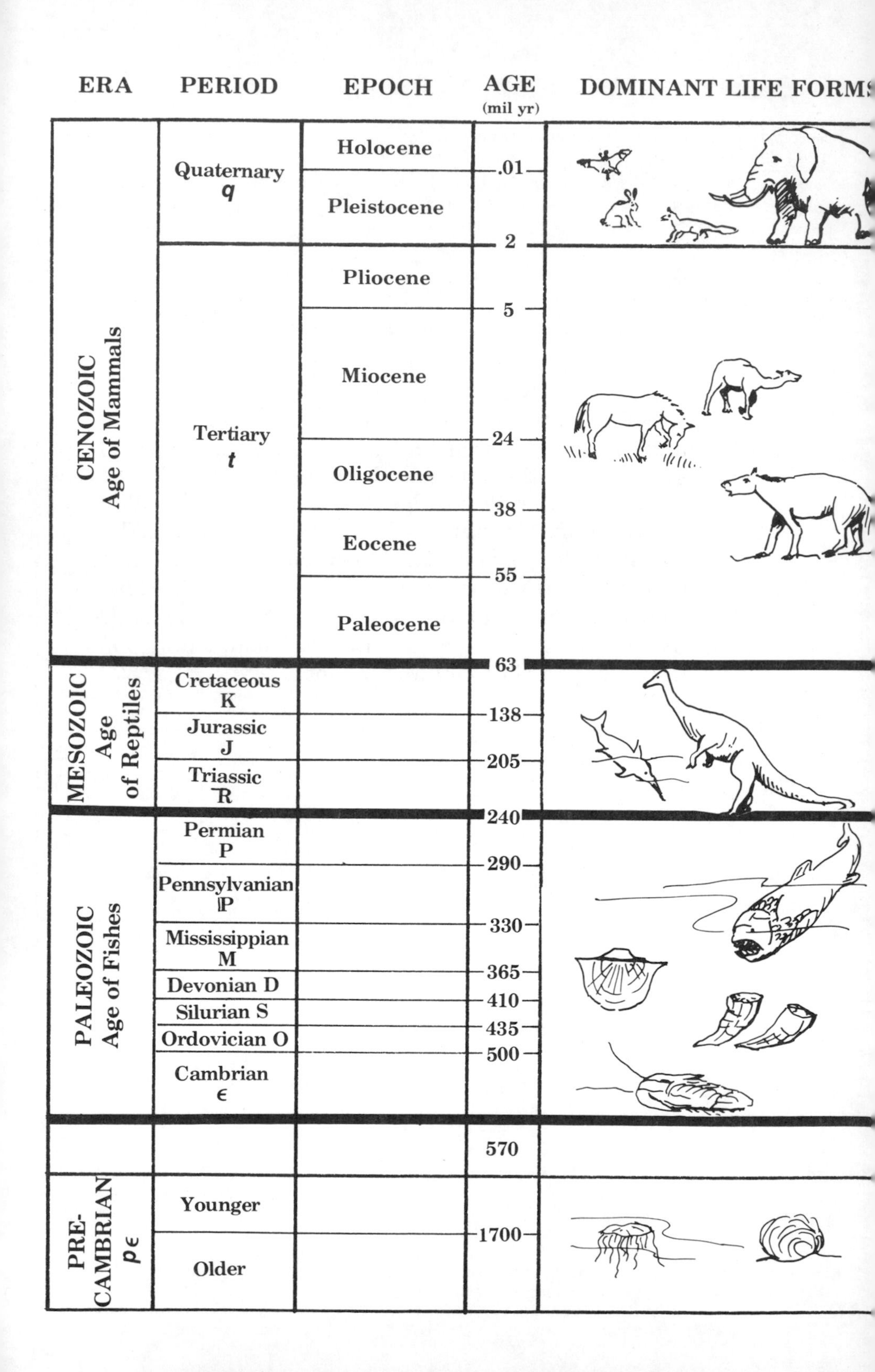

ERA
PERIOD
EPOCH
AGE (mil yr)
DOMINANT LIFE FORMS
CENOZOIC Age of Mammals
Quaternary q
Holocene
.01
Pleistocene
2
Tertiary t
Pliocene
5
Miocene
24
Oligocene
38
Eocene
55
Paleocene
63
MESOZOIC Age of Reptiles
Cretaceous K
138
Jurassic J
205
Triassic R
240
PALEOZOIC Age of Fishes
Permian P
290
Pennsylvanian P
330
Mississippian M
365
Devonian D
410
Silurian S
435
Ordovician O
500
Cambrian ϵ
570
PRE-CAMBRIAN pϵ
Younger
1700
Older

EVENTS IN ARIZONA

Present erosion cycle gouges Pleistocene and Tertiary deposits. Basalt volcanism continues near San Francisco Peaks and at a few other sites.
Regional uplift accelerates erosion; cyclic erosion creates terraces. Basalt volcanism occurs in several areas; San Francisco Peaks grow, collapse, and are glaciated. Colorado River flows through to Gulf of California. Pluvial lakes occupy some valleys.
Colorado River turns west, initiates canyon cutting on Colorado Plateau. Little Colorado reverses as recurrent movements lift plateaus. In south, basins fill with stream and lake deposits.
Basin and Range Orogeny 15 to 8 million years ago creates fault-block ranges with NW-SE grain. Basalt volcanism widespread.
Mid-Tertiary orogeny 30-20 million years ago pushes up mountains with NE-SW grain. Metamorphic core complexes form. Colorado Plateau rises; Colorado River flows south, east of Kaibab Arch. Downdropped Verde Valley intercepts northward drainage. Explosive volcanism common, with calderas in Chiricahua and Superstition Mountains.
Tension faulting in south is accompanied by volcanism and intrusion of dikes, stocks, laccoliths. Intermountain valleys fill with debris from mountains. Verde Valley begins to form.
Laramide Orogeny ends 50 million years ago, leaving undrained intermountain valleys, some with lakes. No volcanism or intrusions mark "Eocene magma gap." Northbound streams deposit rim gravels.
In south, Laramide Orogeny creates mountains with NE-SW trend: overthrusting may have occurred. Explosive volcanism occurs. Abundant small intrusions appear, some containing copper, silver, gold. In north, plateaus begin to form as large blocks are lifted or dropped.
Seas invade briefly from west and south; volcanism widespread. Laramide Orogeny begins 75 million years ago as west-drifting continent collides with outlying plates.
Deserts widespread; thick sand dune deposits in north. Explosive volcanism in south and west is followed by erosion.
Extensive coastal plain, delta, and dune deposits spread north from mountains in central and southern Arizona. Faulting, small intrusions, explosive volcanism occur in south.
Dunes form across northern Arizona, then a western sea invades briefly. Alternating marine and non-marine deposition in south and west.
Marine limestones deposited in south and south-central Arizona; floodplain and desert prevail in north.
Widespread deposition of fossil-bearing marine limestone is followed by emergence and development of karst topography with sinks and caves.
Marine deposits form, then are removed from many areas by erosion.
No record.
Brief marine invasion, then no record.
A western sea advances across denuded continent, depositing conglomerate and sandstone, then shale and limestone.
Great Unconformity — long erosion.
Several episodes of mountain-building and intrusions of sills and dikes are followed by marine and near-shore sedimentation, faulting, and uplift.
Sedimentary and volcanic rocks accumulate, then are compressed and altered into NE-SW-trending ranges extending beyond Arizona. 1.7 billion years ago granite batholiths intrude these older metamorphic rocks.

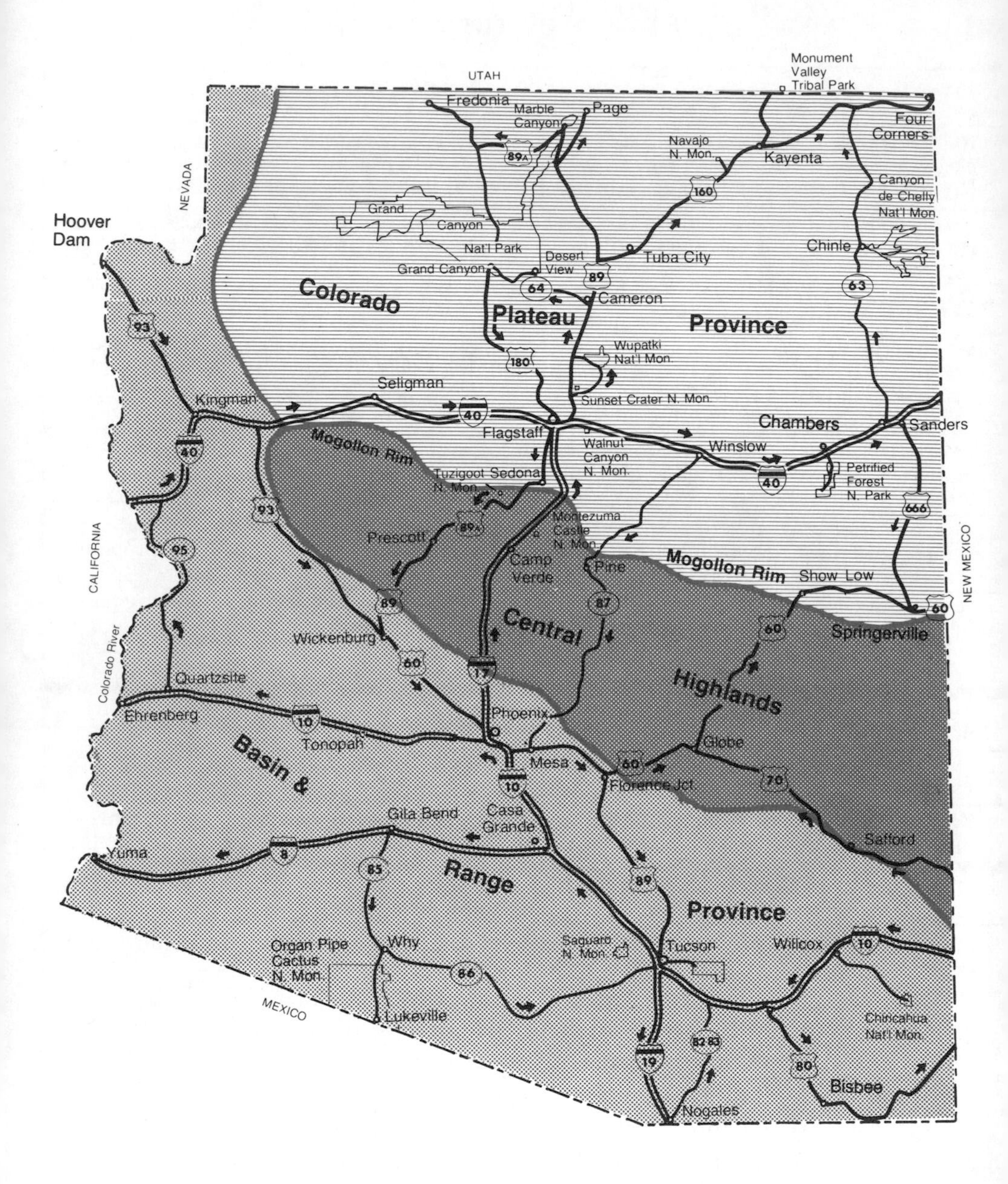

UTAH
Monument Valley Tribal Park
Fredonia
Marble Canyon
Page
Navajo N. Mon.
Kayenta
Four Corners
NEVADA
Hoover Dam
Grand Canyon Nat'l Park
Grand Canyon
Desert View
Tuba City
Canyon de Chelly Nat'l Mon.
Chinle
Cameron
Colorado
Plateau
Province
Wupatki Nat'l Mon.
Sunset Crater N. Mon.
Seligman
Kingman
Flagstaff
Chambers
Sanders
Mogollon Rim
Walnut Canyon N. Mon.
Winslow
Tuzigoot N. Mon.
Sedona
Petrified Forest N. Park
Montezuma Castle N. Mon.
Prescott
Camp Verde
Pine
Mogollon Rim
Show Low
CALIFORNIA
NEW MEXICO
Central
Springerville
Wickenburg
Colorado River
Quartzsite
Highlands
Ehrenberg
Phoenix
Tonopah
Basin &
Globe
Mesa
Florence Jct.
Gila Bend
Casa Grande
Yuma
Safford
Range
Province
Organ Pipe Cactus N. Mon.
Why
Saguaro N. Mon.
Tucson
Willcox
MEXICO
Lukeville
Chiricahua Nat'l Mon.
Bisbee
Nogales

I
Some Geology Basics

Arizona's dry climate and varied topography make it a geologic wonderland, an open textbook of geology. From desert lowland to barren mountaintop, from bent and broken rocks of the southern ranges to the layer-cake strata of the stable northern plateau, about 2 billion years of geologic "happenings" have left their traces for us to piece together into a coherent, albeit patchwork, history.

THREE PROVINCES

In this book, as in many other discussions of Arizona's geology, the state is divided into three regions or provinces: the Basin and Range deserts of southern and western Arizona (Chapter II), the mountainous Central Highlands (Chapter III), and in the north the Colorado Plateau (Chapter IV), named for the river which has so boldly and beautifully carved a canyon through it. In each province, geology plays the major role in governing the spacing and character of hills and mountains, canyons and valleys, cliffs and plains. By governing their habitats, geology has ruled over plant and animal life as well, and much more recently over the ways of man.

In the Basin and Range deserts each rocky mountain range at first seems different, out of character with its neighbor, the transitions between them buried deep beneath debris-filled desert valleys. Yet certain patterns emerge. Many of the ranges — particularly those with smoothly domed summits — are recognized now as having a common basic structural pattern. Many others display certain rocks in common, suggesting a single ancient mountain-building episode that created long-gone mountains to match today's Sierras. In yet others, older and much distorted rock exhibits a definite "fabric," a northeast-southwest (NE-SW) trend of cracking and jointing and mineral alignment, as well as of rock types, telling us of even earlier, even greater ranges of folded rock, mountains of Himalayan magnitude, crushed and forced upward by collision between continents.

Some of the rock types and trends of the Basin and Range Province continue into the Central Highlands, where we can learn from their continuity how these trends behave. The Central Highlands manifest also many features characteristic of the Colorado Plateau to the north, and therefore may be considered transitional between the Basin and Range Province

Mountain Ranges alternate with desert basins in the Basin and Range Province.

The Four Peaks form a well recognized landmark in the mountainous Central Highlands. Their rock is Precambrian quartzite.

and the Plateau. But in their tight-clustered ranges and narrower, shallower, and less numerous basins, the Highlands are distinct and deserve a chapter of their own.

The Colorado Plateau, on the other hand, bears little resemblance to the southern deserts. Ranging 4000 to 9000 feet above sea level, the Plateau is in reality a step-by-step series of flat-topped units separated from each other by cliffs or steep slopes — an unconventional landscape made more beautiful by the many and varied hues of bare rock. Seeming grotesque to some , this vast and lonely land is to many others imbued with haunting magic and singular charm.

PUSH-PULL GEOLOGY

Why the differences between Arizona's three provinces? Newest geologic thinking places them at the doorstep of **Plate Tectonics** — a now well-documented theory stating that portions of the earth's crust have drifted around on the surface of the earth throughout geologic time, breaking apart here, crunching together there, sliding past one another somewhere else. Arizona's southern deserts lie in what has been for the last

75 million years a particularly mobile region, where collisions and separations have frequently changed the tenor of the land. The Colorado Plateau is more stable; surprisingly little has happened to it in the last 600 to 700 million years. The Central Highlands are caught between the active, mobile Basin and Range Province and the firm, deep-rooted Plateau.

Let's look for a moment at the earth as a whole. At its center, we now know, is a large, spherical **core**, extremely hot, part liquid and part solid. Around the core is the thick **mantle**, also partly liquid (or at least plastic) and partly solid. Outside the mantle, floating on it, if you wish, is the earth's crust, a feeble 3 miles thick under the oceans and 20 to 25 miles thick under the continents. It is attached to the upper or outer part of the mantle; together they make up the **lithosphere**, a layer about 40 miles thick under the ocean, about 60 miles thick under the continents. Relative to the size of the earth the lithosphere is merely a thin film, one that varies in thickness and that, like a film of cooled fat on a bowl of chicken broth, can be rumpled and moved about by stirrings in the "liquid" below.

And stirrings there are: broad, strong, rolling, boiling convection currents as the mantle is heated by the great stove of the earth's interior. Because of the large size of the earth and the thickness and stickiness of the semi-molten mantle, the currents move agonizingly slowly by human standards — an inch a year or less. But with them they carry the film on the surface — the lithosphere — shifting it about, tearing it apart

In the Colorado Plateau Province, flat-lying rock layers are deeply dissected by streams and rivers.

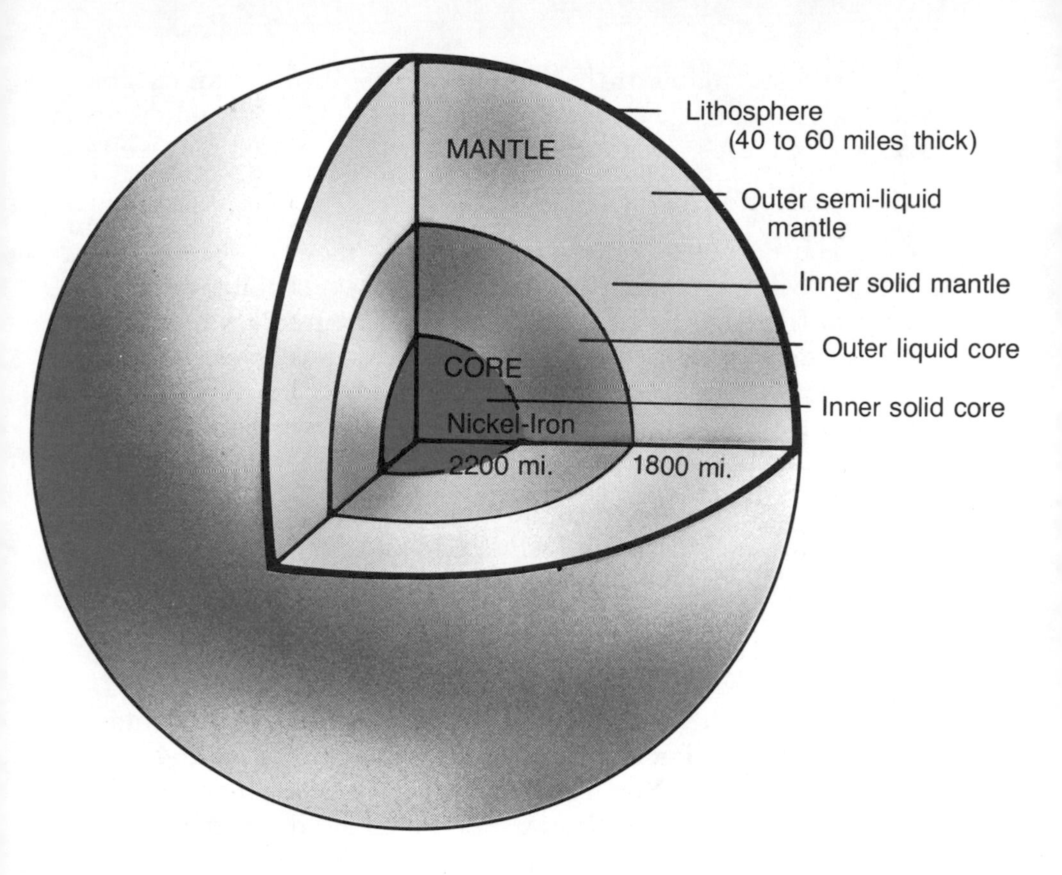

into separate plates, at times shoving the parts together. So the currents constantly, though gradually, reshape the face of the earth; they are responsible for the rising and falling, the folding and bending of rock layers in the earth's crust, as well as for cataclysmic geologic processes like earthquakes and volcanic eruptions.

As a result of all this motion the earth's lithosphere consists of a dozen large, relatively rigid **plates** and an as yet undetermined number of smaller ones. We've seen that under the oceans the crust is thin; there it consists of dark, heavy **basalt**, rich in iron and magnesium. Continental crust is more varied. It contains all the other kinds of rock, most of them lighter in color and weight than basalt because they contain much smaller amounts of iron and magnesium minerals. The continental lithosphere, with its thicker, lighter crust, floats high on the mantle, so that like an iceberg its upper surface rises above the level of the sea.

The crustal plates are bordered by either **mid-ocean ridges** or deep, arcuate **trenches** near the continents. At mid-ocean ridges the plates spread apart very slowly as molten rock wells up from the mantle to form new crust. At the trenches the plates collide; lighter continental lithosphere overrides the oceanic lithosphere, which is then pulled under — **subducted** — by downward-plunging currents. It eventually remelts, often with an admixture of light-colored continental rock that got dragged under too. So like giant conveyor belts operating in slow motion, oceanic plates form at mid-ocean ridges, cross their half of the ocean basin, and plunge again into the interior at the trenches.

North America is part of the North American Plate, a large plate which stretches from Alaska to the Caribbean and from the Mid-Atlantic Ridge to the Pacific shore. Like many others, this plate is part oceanic — the part beneath the deep Atlantic — and part continental — including both dry land and the broad continental shelf. The big plate is moving west at a rate of about an inch a year. On the west it abuts the large, almost entirely oceanic Pacific Plate. Where the two plates meet the thick edge of the North American Plate rides out over the Pacific Plate, which is drawn down under the continent in a typical oceanic-continental plate collision. Sometimes two continental plates collide, and neither goes under; their edges bend and crumple into magnificent mountains, as India and Asia have done, creating the Himalayas. There is evidence in Arizona that continental collisions of both these types — continental plate *vs* oceanic plate, and continental plate *vs* continental plate — have happened here in the past.

There is also evidence in Arizona that the crust was at times pulled apart rather than pushed together, giving us tensional as well as collisional **tectonics**. The word **tectonic** comes from the Greek *tekton* — carpenter or builder — and carries implications of how the earth, or any part of it, is put together.

What's the result of all this pushing and pulling, squeezing and stretching? Mountains, yes, and intermountain basins. But the present mountains and basins — the present landforms — are shaped as much by erosion and deposition as by tectonic events. The higher the uplift, the steeper the slope, the stronger the stream. And the stronger the stream (or river or

Differential weathering *and erosion of hard and soft, resistant and non-resistant layers of rock create the cliffs and ledges of Grand Canyon. Sandstone and limestone are normally more resistant than shale and mudstone.* Tad Nichols photo.

rivulet), the greater its ability to cut into the land. So uplift breeds downcutting — a strong geologic tenet. And what of basins? Down-dropping invites deposition — another geologic tenet — so the basins are gradually filled in. Some rocks are harder, stronger, more resistant to erosion than others, and wear into cliffs and ledges or into high-standing mountain peaks. Other rocks are weaker and end up as slopes or erode away altogether.

THREE DIVISIONS OF TIME

Arizona's pull-apart and push-together story can only be related in respect to time. Geologic time begins, of course, with the birth of *geos*, the earth, about 4.5 billion years ago. That's 4,500,000,000 years — a long, long, long time by any measure. Geologists divide that immensity of time into **eras**, **periods**, and **epochs**, somewhat as calendars divide a human lifetime into years, months, and days. But the geologic units are not as uniform as calendar units. The first era, the **Precambrian Era**, for instance, is longer than the other three put together. It lasted about 3.9 billion years, from 4.5 billion years ago when the earth formed until only about 600 million years ago. The oldest rocks in North America are 4 billion years old, the oldest in Arizona around 2 billion years old.

The next era, the **Paleozoic Era** (its name means "old life"), was a great deal shorter: about 330 million years. The third or **Mesozoic Era** ("middle life") lasted 180 million years, and the fourth or **Cenozoic Era** ("recent life") only about 60 million years. These figures, by the way, become more and more refined as techniques are perfected for measuring the age of rocks. More exact figures are given in the table at the beginning of this chapter.

Another way to think of the last three eras — the Paleozoic, Mesozoic, and Cenozoic Eras — is as the Age of Fishes, the Age of Reptiles, and the Age of Mammals. For a long time geologists thought there were no fossils in Precambrian rocks, and that no life existed in Precambrian time. However we know better now: Life did exist in younger parts of that great era, but shells and skeletons didn't, and it is shells and skeletons that are most likely to be preserved. A few fossils are now known from Precambrian rocks — cabbagelike mounds built by algae, imprints of jellyfish, and bacteria. Some of these occur in Arizona's Precambrian rocks.

Eras are divided into **periods**, which also vary in length, and periods are divided into **epochs**. I should say that the divisions of geologic time are not just haphazard, but represent significant and commonly worldwide breaks in the continuity of rock layers or in the fossil record, the earth's great diary of life. Eras begin and end at quite significant breaks, with major mountain-building and erosion and in some cases extinctions of whole groups of animal and plant life. Periods are separated by shorter and less significant breaks; epoch breaks are still less significant. Most of the time units were first recognized in Europe, and since they were defined research in other parts of the world has narrowed the gaps between units or even filled them in completely. Nevertheless the units stand us in good stead, enabling geologists around the world to speak the same language in regard to geologic time.

Names of the Paleozoic periods derive from specific locations where rocks of certain ages accumulated: Cambria (the Roman name for Wales), Perm (a province in Russia), and so forth. Mesozoic names vary: Triassic is three-layered (in Germany anyway), Jurassic is from the Jura Mountains in France, Cretaceous refers to the Chalk, the white limestone of England.

Cenozoic period names are relics of an early nomenclature that divided up rocks by how hard they were. Primary and Secondary are no longer used; we are left with Tertiary for poorly consolidated rocks and Quaternary for unconsolidated gravel, sand and clay. But let me warn you: These definitions do not always apply. Cenozoic epoch names, which all very nicely end in *-cene*, refer again to the development of life, particularly mammals. The Paleozoic and Mesozoic periods are divided into epochs, too, but I won't be using them in this book.

Though the Precambrian is longer than all the other eras put together, as we've seen, there is an opposite side to the coin: We know much more about the shorter, later eras than we do about the long Precambrian. This is partly because the older the rocks, the more often they have been eroded and bent into mountains and broken and squashed into more mountains and then eroded again, repeatedly; each time their story becomes harder to read. It's also partly because fossils — useful indicators of the age of rocks — are so few and far between in Precambrian rocks. And, too, it's because the older rocks have been covered over by younger rocks. So as we go back in time there are fewer and fewer pages in the earth's diary, and what pages there are are more and more faded and splotched and torn with age. The most recent rocks — those of the Recent or Holocene Epoch, lie right on the surface as sand and silt and gravel and soil, and as lava flows and new volcanic ash.

Rocks are dated by a variety of methods. For a long time only sedimentary rocks which contained recognizable fossils could be given a place in the time sequence; all other rocks had to be dated in terms of "older than" or "younger than" these sedimentary units. Today we can date volcanic rocks, including sedimentary rocks with a component of volcanic ash, by **radiometric** methods — by determining the relative amounts of certain radioactive elements and the daughter products of their gradual but very regular decay. Several isotopes of uranium and thorium, for instance, decay at invariable rates into isotopes of lead, so by measuring the quantities of isotopes of uranium, thorium, and lead in a rock we can determine the date at which the uranium decay began — the age of the rock. This method is most useful for Precambrian igneous and metamorphic rocks. The potassium-argon dating method, which works well for igneous rocks more than a million years old, works the same way, with measurements of radioactive potassium-40 and its daughter product argon-40. Carbon-14 decays quite rapidly and so is useful for sedimentary rocks laid down in the last 50,000 years or so, or for archeological remains — bone fragments and charcoal from early campfires — within the same range. A relatively new method of dating has developed since we have learned that the earth's magnetic field occasionally reverses, with the north and south magnetic poles switching their positive and negative polarities. Because the switches are irregular it has been possible to work out a yardstick of sorts, recognizable in both sedimentary and igneous rocks containing iron minerals that lined up with the earth's magnetic field as the rock formed.

Geologically the exact age of the rocks is in most cases not nearly as important as knowing their relative positions: which unit is above and therefore younger than another unit, or which is below and therefore older. It is a basic geologic "law" that layered rocks — including both sedimentary and volcanic layers — become younger upward. Geologic history therefore begins at the bottom, and one must read from the bottom of the page, or the end of the book, upward and forward. Look for a moment at the chart that precedes this chapter. If you want to know what happened in Arizona through geologic time, in the proper order, begin at the beginning with events of the Precambrian Era, and read upward toward Cambrian and Orovician and so forth, all the way up to the Recent Epoch. The rocks are arranged the same way except where they have been severely bent and broken by mountain-building processes.

MORE TERMS AND CONCEPTS

To handle time, as we've seen, we need names. To handle rocks and the landforms in which we see them, we need more names. Fortunately, many geologic terms are already in common use: silt, clay, sand, pebble, boulder, river, mountain, canyon, and so forth. But some terms are specific to geology or

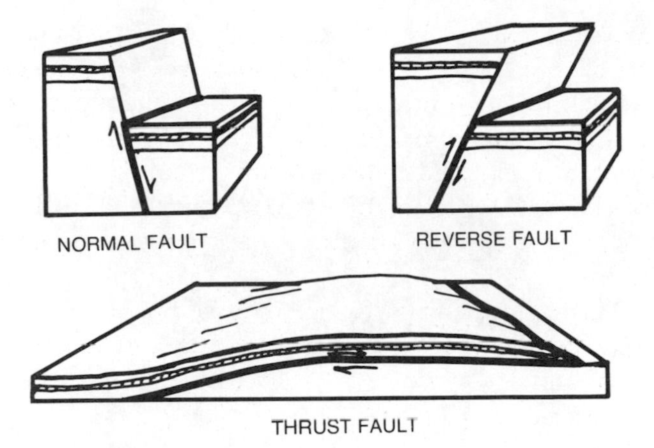

Geologists recognize three main kinds of faults. Half-arrows connote direction of movement. Normal faults *result from horizontal tension, while* reverse faults *and* thrust faults *are caused by horizontal compression.*

have specific geologic meanings: **formation**, **anticline**, **alluvial fan**, **pediment**, **fault**, **graben**. Still others — words commonly of Spanish origin — are pecular to southwestern geology: **mesa** and **cuesta**, **bajada**, **playa**, and **arroyo**. These and other terms are defined from place to place in the text, and also in the Glossary at the end of the book. The list of symbols inside the front cover lists the rock units — **formations** or **groups** — used in this guide.

Another useful geologic concept is an understanding that rocks and mountains and hills are *not* as eternal as poets would have us believe. High places like mountains and hills are worn down by erosion; low places like valleys and lake basins are filled up with debris worn off the high places. Without the work of those upwelling convection currents in the mantle, the earth's crust would long since have been flattened completely, and its entire surface would be covered with a sea of uniform depth.

But upwelling in the mantle *does* occur, and causes continents to drift and collide and crumple into mountains or rise into flat-topped plateaus. Geologists use the word **orogeny** for mountain-building, and give names to the various mountain-building episodes, orogenies, that have occurred through geologic time.

A highway cut exposes a two-layered thrust fault (arrows) where darker, somewhat more resistant older rocks have pushed up and over less well consolidated younger rocks.

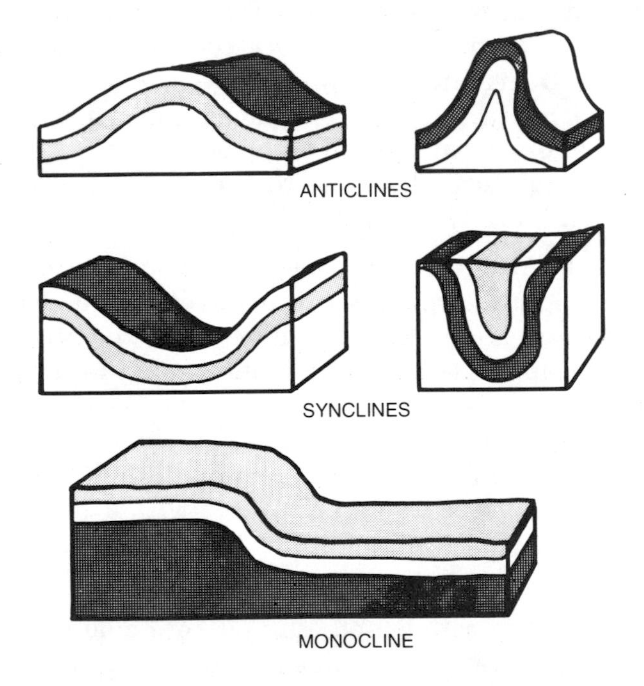

Three kinds of folds are easily recognized when they occur in layered (stratified) sedimentary or volcanic rocks.

This brings us to another basic geologic concept: Rocks can be broken, smoothly or jaggedly, or folded in a variety of ways. Some breaking takes place deep below the surface, where pressures and temperatures are tremendous. But both folds and breaks commonly show up on the surface, either because they extend through to the surface in the first place, as does the San Andreas Fault in California, or because erosion has cleaned off rocks that used to hide them. **Faults** are breaks along which displacement of the two sides has occurred. Breaks or cracks without relative displacement are called **joints**, and they are much more common than faults. There is hardly a place in Arizona where you can't see joints and faults and folds, or their surface expression in the form of mountains and cliffs and shattered rock.

THREE CLASSES OF ROCKS

By now you must be sure that geologists think in threes. They do. There are also three basic classes of rock, classed by their origin: igneous, sedimentary, and metamorphic rock.

• **Igneous rocks** originate from molten rock material or **magma** that comes from a source fairly well down in or below the earth's solid crust. Some magma pushes through to the surface and cools rapidly as **volcanic** or **extrusive igneous rock**. Extrusive rocks solidify so rapidly that crystals have little or no time to form, so these rocks are very fine-grained, with grains that are invisible to the naked eye. Common volcanic rocks are **lava**, **tuff** or **welded tuff** formed from volcanic ash, and **breccia**, a broken and recemented mixture of lava and tuff.

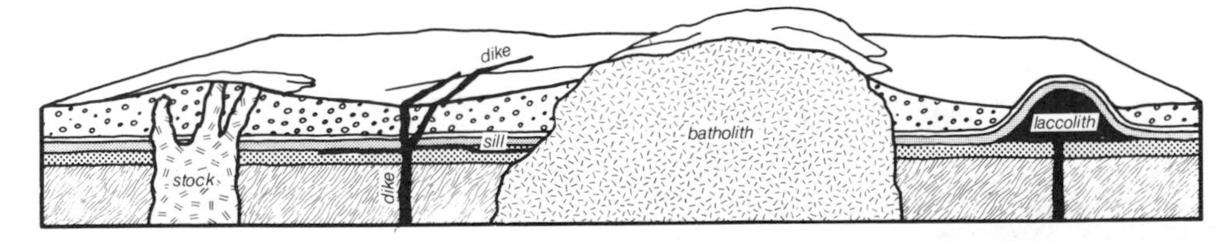

Shapes and sizes of intrusions *vary. They are named according to size and their position relative to stratified rocks.* Intrusive rocks *harden below the surface; those shown at the surface here have been bared by erosion.*

Some magma never makes it to the surface. Instead, it cools and hardens within the crust, becoming **intrusive igneous rock**. Intrusive rock crystallizes slowly and its individual crystals, forming over a long period of time, grow large enough to be visible to the naked eye. **Granite** is the most common intrusive igneous rock.

In Arizona a common and economically important intrusive rock falls between these two extremes. It cools in shallow intrusions, some of them probably just under former volcanoes where some of the same magma reached the surface and became volcanic rock. Some of the magma minerals had time to grow into large crystals, but others with different crystallization temperatures didn't. This kind of rock is called **porphyry**, and it is characterized by having large crystals, **phenocrysts**, scattered through a finer matrix.

Another thing about igneous rocks: Chemically and in terms of their mineral makeup they fall into two groups: dark ones of basic or **basaltic** composition, containing a high proportion of iron and magnesium, and light-colored, more acid, **silicic** ones with a high proportion of silica and a low proportion of dark iron and magnesium minerals. The dark rocks are thought to derive from the earth's mantle; light ones come from a shal-

Magma that pushes into vertical fissures may harden as dikes. *Later exposed by erosion, dikes commonly appear as dark, resistant walls cutting cross-country like this one near Tuba City.*

Tad Nichols photo.

lower source where continental-type rocks have melted, as they do where plates collide. When magma appears at the surface as volcanic outpourings, the dark basaltic lava is fairly runny; it flows easily and relatively quietly from volcanic vents; the light-colored silicic magma, thick and sticky, erupts as short, stubby flows or, plugging the volcano for a time, bursts out explosively, scattering volcanic ash far and wide. There are many examples of both basaltic and silicic volcanism along the highways of Arizona. Throughout the state most Tertiary volcanic rocks are silicic and most Quaternary ones are basaltic.

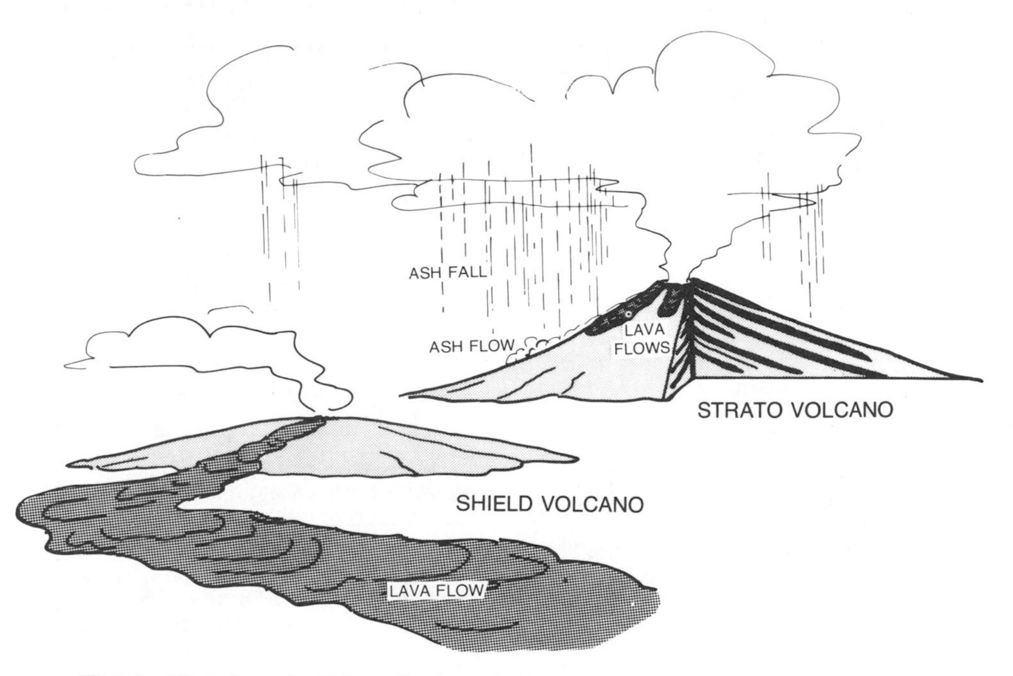

Thick silicic lava builds tall, graceful stratovolcanoes *in which stubby lava flows alternate with layers of volcanic ash. Basaltic lava, on the other hand, flows easily and forms low* shield volcanoes *or spreads out across the landscape in flat sheets.*

COMMON ROCKS OF ARIZONA

Class	Rock	Description
Sedimentary	Sandstone	Grains of sand cemented together
	Shale	Grains of silt and clay cemented together, usually breaking into flat slabs. When massive, called mudstone or siltstone.
	Conglomerate	Sand and pebbles deposited as gravel and then cemented together.
	Limestone	Composed mostly of calcite deposited as a limy mud. Usually white or gray, often containing fossils.
	Caliche	Impure limestone deposited close to the surface by groundwater evaporation.
Igneous Extrusive	Rhyolite	Light-colored, very fine-grained rock formed as lava or volcanic ash, the extrusive equivalent of granite.
	Dacite	Like rhyolite but with more alkali feldspar
	Andesite	Dark, fine-grained rock with abundant crystals of plagioclase feldspar.
	Basalt	Very fine-grained black or gray volcanic rock, often with visible gas-bubble holes or vesicles.
Igneous Intrusive	Granite	Common light-colored coarse-grained rock with visible quartz and feldspar crystals, usually peppered with black mica or hornblende.
	Monzonite	Medium-grained rock, often porphyritic, with feldspar predominating, some hornblende or quartz.
Metamorphic	Marble	Recrystalized limestone, commonly with visible calcite crystals.
	Quartzite	Sandstone and/or conglomerate so tightly cemented or welded that it breaks through individual sand grains and pebbles.
	Greenstone	Dark gray or green rock resulting from metamorphism of basalt or other dark igneous rock.
	Gneiss	Banded or streaky crystalline rock formed from older granite or sandstone.
	Schist	Medium-grained rock with mica grains lined up so that rock has streaky appearance and tends to split along parallel planes.

• **Sedimentary rocks** originate, as their name implies, from sediment — remains of other rocks broken up by weathering and erosion or dissolved by running water. They are deposited by water in river valleys, lakes, or seas, or by wind as dunes. Glacial ice also deposits sedimentary rocks, but Ice Age glaciers hardly touched Arizona so glacial deposits are of only minor importance here. Common sedimentary rocks are **limestone**, **shale**, **sandstone**, and **conglomerate**, the last full of pebbles or boulders. When not yet completely solidified or cemented, sedimentary deposits are called **sand** and **clay** and **gravel**.

Sedimentary rocks are characteristically **stratified**, arranged in **strata** or layers, so that even when they are later bent or broken they can be readily recognized. The sedimentary layers may be thick or thin; always their thickness is measured perpendicular to the layering or stratification. With rare exception sedimentary layers, or strata, were deposited in horizontal position.

• **Metamorphic rocks** also have their beginnings in other rocks, but are altered by long-term application of heat and pressure, usually deep below the surface. They may be only a little altered, as with **metasedimentary** or **metavolcanic** rocks, both of whose origins are clear, or they may be altered so severely that geologists are hard put to pin down their origins, as with **gneiss** and **schist**.

MINERALS

Rocks are made of minerals, natural substances with definite chemical makeup and very often with definite color and characteristic ways of crystallizing. Arizona has some particularly beautiful minerals, most notably ores of copper such as bright blue **azurite** and equally bright green **malachite**. Specimens of these and many other minerals are exhibited in the Geology Museum at the University of Arizona in Tucson, in museums in many old mining towns, and in rock shops and jewelry shops throughout the state. Common rock-forming minerals are listed below.

• **Quartz**: a clear glassy or milky white mineral so hard that it can't be scratched with a knife. Quartz comes in white, which is most common, pink (rose quartz), lavender (amethyst), or other pastel colors, tinted with minute amounts of other minerals. It is common as gray glassy grains in granite and in sandstone derived from granite.

• **Feldspar**: a family of translucent pinkish, grayish, or whitish minerals that *can* be scratched with a knife, and that normally break along flat cleavage faces that reflect sunlight. Feldspar crystals are the most abundant components of granite and many other intrusive and metamorphic rocks.

Gneiss may be highly contorted, as if it had been heated to a taffylike consistency. J. Gilluly photo, courtesy of USGS.

Desert varnish *and carbonaceous material streak the cliffs of Canyon de Chelly.*

• **Mica**: a group of black or silver minerals that separate into shiny, flat, paper-thin flakes. Mica can be scratched easily with a knife or even with a fingernail. Black mica is **biotite**, white mica is **muscovite** — the two most common varieties. Mica is common in granite and other intrusive rocks as well as in schist, a metamorphic rock.

• **Calcite**: a white or light gray mineral that makes up limestone. Sometimes transparent but in some varieties opaque, it can't be scratched with a fingermail but can be with a knife. When dilute acid is dropped on it, it fizzes.

• **Hematite**: an iron oxide easily recognized by its rust-red color. Adhering to quartz grains it lends its color, in all degrees from pale pink to deep purple-red, to sandstone, siltstone, and other rock types.

• **Limonite**: a dull, rusty yellow iron oxide that gives a mustard yellow or tan color to rocks in which it occurs.

• **Hornblende**: a black mineral that appears as rod-like crystals in dark igneous and metamorphic rocks.

• **Gypsum**: a translucent, fairly soft white or light gray mineral formed as an **evaporite** when sea water or salty ponds dry up. In red sedimentary rocks gypsum commonly appears as **veins** and **veinlets**.

Certain rocks can be recognized just by their color and texture. But the weathered surface of a rock may conceal its interior color, so we often give both colors: "Light gray limestone that weathers tan," for example. **Weathering** is just what it says: surface and near-surface changes due to sun and air and rain and, in some cases, plant and animal products. In Arizona dust may play a major role in weathering; recent research indicates that **desert varnish**, the dark, shiny coating common in rocks of the desert, may in part derive from ever-present desert dust.

TWO TOPICS OF INTEREST

Geologic topics of special interest in Arizona are water and copper. Here, water is in short supply, at least as far as man is concerned. Ways of increasing the water available to man revolve around geology. Ways to find more of it entail wells and geologic studies of water-bearing rock layers. Ways to transfer it from place to place entail knowledge of geologic features, including the **permeability** of rock over which, or through which, it must pass. And ways to keep it from misbehaving periodically, from damaging the works of man, entail dams and other flood control devices. You will see both dams and canals from some of Arizona's highways. A recent project to supplement Arizona's water supplies is the Central Arizona Project, a very large and intricate diversion system — dams, pumping systems, and canals — that will bring Colorado River water to the metropolitan and agricultural areas of the state.

Copper is of special interest because copper ores are abundant here; Arizona produces more than half the copper produced in United States, with the economic benefits and environmental detriments that go with mining of almost any kind. Southern Arizona is the great copper producer; towns like Ajo and Tombstone, Bisbee and Douglas, Clifton and Morenci, Globe and Miami are dependent on copper mining and smelting. Rich ores used to lure miners underground; most mines now are big open-pit affairs that move thousands of tons per day of **overburden** and low-grade ore. You are an invited guest at some mines, either at established viewpoints or with guided tours.

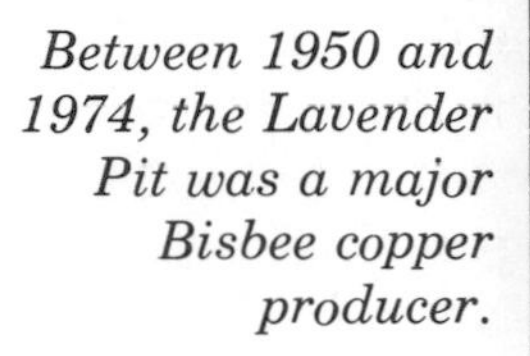

Between 1950 and 1974, the Lavender Pit was a major Bisbee copper producer.

ARIZONA THROUGH TIME

With this groundwork, let's take a very quick look at Arizona's place in the geologic history of North America. The record is fragmentary, as much of the evidence is deep down and hidden from view, or has been partly obscured by **metamorphism** or completely destroyed by erosion. Our earliest record comes from Precambrian gneiss and schist now exposed at the bottom of Grand Canyon, in the mountains of central Arizona, and in some of the desert ranges. It shows that around 2 billion years ago Arizona was placed much as it is now, along the southwest edge of a large continent having more or less the outline of the present North America. Layered sedimentary and volcanic rocks accumulated in offshore basins and along the edge of the continent, layers of sandstone and limestone and shale interspersed with lava flows and ash deposits. About 1.7 billion years ago these rocks were lifted into a great mountain range. Granitic magma pushed upward, in some cases following rock joints, and hardened into **batholiths** (large intrusive masses) and coarsely crystalline veins. The resulting ranges stretched up and down the edge of the continent much as the Andes of South America do today.

After many more cycles of deposition, emergence, mountain-building, and wearing down, over much of the world an unusually long period of erosion brought the Precambrian Era to a close. Except for a few scattered islands, Arizona and the rest of the proto-North America were beveled to a broad, flat, erosion surface near sea level.

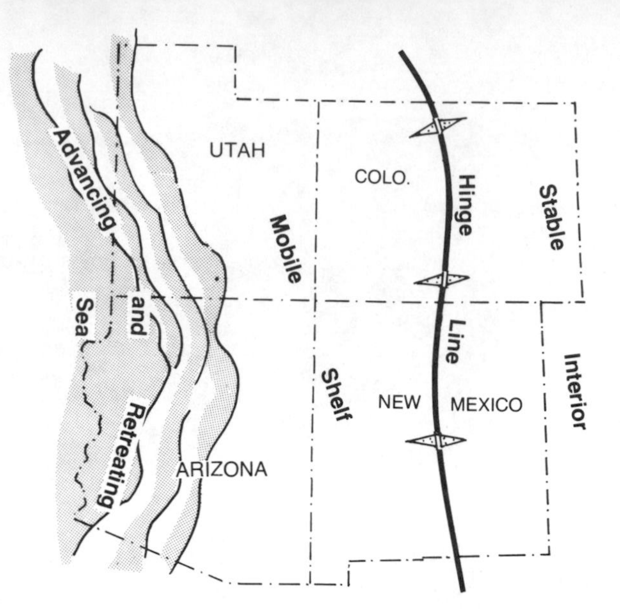

In Paleozoic time Arizona was part of a hinged shelf that flexed downward periodically, allowing the sea to advance across an almost flat land. As a result, Paleozoic strata thicken westward.

In Paleozoic time — the Age of Fishes — the nearly featureless continent was once more alternately elevated and submerged, tilting this way and that so that the sea periodically flooded across its margins. Especially along its mobile southwestern edge, which seemed hinged to the firmer, more stable central part of the continent, marine sedimentary rocks alternated with continental deposits that accumulated on river floodplains and deltas or as dunes on broad, sandy shores. At times widespread erosion removed previous deposits or prevented their deposition, so the Paleozoic record is not by any means complete. It's clear, though, that Paleozoic seas came from the west, lapping and then overlapping the west edge of the continent. At times rivers flowing from uplifts to the east and northeast swept silt and sand out over earlier marine sediments.

The Mesozoic Era — the Age of Reptiles — saw a good deal of **uplift**, with mountains in the central and eastern parts of the state. Erosional debris from the mountains spread northwestward and westward. And for a long time as the Sierra Nevada rose farther west, this area, cut off from sources of moisture, was a vast desert, with dunes blowing endlessly across an almost featureless landscape. Toward the end of the era, late in Cretaceous time, seas rolled one last time across the land, coming from the northeast. But incursion of the sea was short, and most Mesozoic deposits were of continental nature: sandstone and shale layers of dune and delta and floodplain, together indicating an environment not unlike that of northern Africa today, with its desert dunes and the great floodplain and delta of the Nile.

At the end of Mesozoic time North America broke away from Europe and began a westward and northwestward odyssey. As it collided with the Pacific Plate and overrode it, the bruised crust buckled and broke, mountains rose, and volcanoes spewed vast quantities of ash and lava. The newborn Rocky Mountains shed quantities of debris in the direction of Utah and Arizona; the Sierra Nevada was still contributing debris from the west. Collison and union with several offshore islands or microcontinents complicated the picture in southwestern Arizona. But through all the activity the Colorado Plateau area remained relatively stable and unaltered, with Paleozoic and Mesozoic sedimentary rocks still keeping their original horizontal positions.

Cenozoic time — the Age of Mammals — brought continued restlessness and uplift to all of western United States. Plateaus rose in northern Arizona and adjacent states, and in southern and western Arizona deep basins formed by faulting and downdropping between mountain blocks. Major rivers and streams cut downward through uplifted areas. The Colorado River shifted its course and brought about another period of increased erosion over all of its drainage area. And by late Quaternary time the landscape looked much as we know it

The sandy floor of a desert wash descends in a succession of steps separated by cobbles and boulders. Streamside vegetation forms a narrow, luxuriant border.

today, with deeply filled basins, some of them dissected anew by streams rejuvenated by uplift and boosted by meltwater during the Ice Ages.

* * *

With this glimpse of geologic processes let's go to Arizona's highways. The following logs are not strictly logs, but annotations of some of the things that can be seen as you drive along. Before starting off down the highway, read the introduction to the chapter covering the area you'll be travelling in — southern deserts, central mountains, or northern plateau. Beyond that introduction, each roadlog stands alone. Town names are used to locate specific stretches; highway mileposts serve to refine specific locations where there are no towns.

Like these strata near the confluence of the Colorado and Little Colorado Rivers, sedimentary rocks *are nearly always deposited as horizontal layers. Note the delta of the Little Colorado at lower right.*
Tad Nichols photo.

II
Where Heavens Are Bright —The Southern Deserts

Arizona's southern and western deserts, broad, sun-baked, low-elevation basins divided by gaunt mountain ranges, arc from the southeast corner to the far northwestern edge of the state. Geologically they are part of the Basin and Range Province of the Southwest. Linear ranges that trend north-south (N-S) or northwest-southeast (NW-SE), the mountains are remnants of faulted blocks of the earth's crust. They alternate with basins of various widths, dropped down along nearly vertical normal faults. Since they formed, vast amounts of gravel and sand have been chiseled from the mountains and carried down into the valleys, so the true amount of displacement, doubtless thousands of feet, commonly cannot be told.

Alluvial fans develop where streams issue from the mountains; they usually merge to form aprons of gravel and sand that engulf the bases of the mountain blocks, aprons for which southwestern geologists use the Spanish term **bajada** (ba-HA-da). Near the mountains a sloping surface may be eroded back into the mountain mass, in which case a lightly graveled **pediment** merges smoothly with the **valley fill**. Valley fill in many cases is thousands of feet thick; in some basins it includes thick masses of salt deposited in saline lakes before through drainage became established.

Flash floods turn quiet sandy washes – avenues for cattle and desert wildlife – into muddy torrents. Tad Nichols photos.

Light rainfall — what there is of it — is soaked up by the desert and goes to nourish sparse, highly adapted vegetation. Most desert streams — "washes" in the Southwest — flow only after rare torrential rains. Within a few moments a tree-shaded, sand-floored picnic spot may become a roiling, muddy cataract quite capable of pushing and rolling cows and cars downstream. When the rain stops, the streams as quickly dry — the water sinking into the valley fill and replenishing the **groundwater** supply. Rainwater that sheets across the desert during such cloudbursts augments the floods, and nourishes, too, the desert's plants. A week after a heavy soaking the desert wears a carpet of green; in two weeks the carpet may be spangled with wildflowers.

A more persistent performer when it comes to desert erosion is wind. On almost any warm day, tall, whirling columns of dust — the "dust devils" familiar to all Southwesterners — spiral against the sky. Larger dust storms akin to infamous North African *haboobs* are, fortunately, less frequent, though they sweep north from Mexico to the Phoenix area an average of 3.5 times per year. These storms are traffic hazards, engulfing cars in thick, blinding dust. Where the desert has been plowed, the loss of visibility is especially severe. If you encounter a dust storm, pull off the pavement to the right, shut off your engine, close your car windows, and watch geology in action. The dust storms seldom last long, but may be followed by rain. The billowing brown storms are caused by a combination of high surface temperatures and downdrafts from decaying thunderstorms.

Desert dust storms like this one on May 29, 1972, periodically sweep the Phoenix area, reducing visibility, blowing away precious topsoil, and finely sand-blasting exposed vehicles, window glass, and painted surfaces.

Nyle Leatham photo, courtesy of the Geological Society of America.

Dust devils and haboobs testify to wind's effectiveness in erosion, which is also displayed on a lesser scale by dusty everyday desert air. Winnowing dust and at times sand from the desert surface, wind leaves behind an armorlike concentration of pebbles known as **desert pavement**. With time the pebbles may darken with **desert varnish**, a thin, shiny coating of iron and manganese minerals now thought to accumulate in tiny increments over thousands of desert years from the dust itself, coupled with moisture from rain or dew. Desert varnish also darkens mountain ledges and cliffs, doing a particularly good job on volcanic rocks, some of which are likely sources for the iron amd manganese.

Caliche, another desert phenomenon, is a white or whitish deposit of calcium carbonate and other soluble minerals. It forms as soil water containing these minerals in solution dries up near the surface in the arid desert climate. Caliche (ca-LEE-chee) forms a "hardpan" right in the soil, destroying soil permeability, preventing root growth, and in general decreasing the productivity of farm land. Since calcium carbonate is present in most Arizona water (it is what makes water "hard") caliche can "grow" in any soil irrigated with that water. The whitish deposits can be seen at road or street level in many parts of Arizona; they also form in the bottom of teakettles.

Other agents, too, help to shape the desert landscape. High mountain crags are sharpened by frost that forms in rock

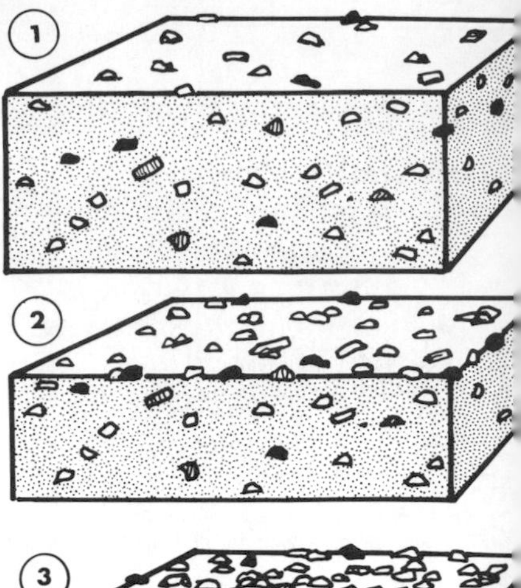

As wind removes sand, silt, and clay particles, pebbles concentrate as desert pavement. *Like a layer of armor they protect remaining soil.*
U.S. Geological Survey photo by James Gilluly

crevices, and rocks are pried apart by the roots of vegetation, to follow gravity's tug to rockpiles at the base of mountain cliffs. Many desert animals burrow into the ground to escape summer heat, throwing sand and silt out across the protective desert pavement. Massive rocks like granite suffer from slow disintegration that converts flat-faced, joint-edged blocks into rounded boulders — a process known as **spheroidal weathering**. Where rounded boulders produced in this way are exposed to sun, wind, rain, daytime heating, and nighttime cooling, thin curving sheets of rock may break off like leaves from a cabbage — a process called **sheeting** or **exfoliation**.

The mountain blocks themselves — a small part of the total landscape — are our only clues to the early history of the Basin and Range region. Fragmentary clues, to be sure. Assembling them is rather like trying to piece together a jigsaw puzzle with nine out of ten pieces missing. But let's take a look at them, paying particular attention to knowledge gained as many Arizona igneous rocks have been dated radiometrically. Many of the new dates differ from those deduced earlier from studies of rock strata, fossils, and the structural positions of the rocks.

Arizona's geologic history begins in early Precambrian time, more than 1700 million years ago, at a time when North America was drifting southeastward against another crustal plate. (As you will see, southern Arizona has a poor accident record, a history of one collision after another.) Mountains

Granite commonly weathers into oversized boulders surrounded by coarse, sandy debris. Under attack by wind, rain, and frost, thin sheets of rock flake off and slide to the ground, weathering there to coarse sand. Tad Nichols photo.

formed as the plates collided, their rocks crumpled and broken by the force of the impact, as the Himalayas have been raised by India's collision with Asia. These are the oldest rocks that can be identified in southern Arizona; they appear in more than half of the ranges of the Basin and Range region, as well as in the Central Highlands and Colorado Plateau: strong gneiss and schist with a NE-SW pattern to their crumpled structure, roots of an ancient mountain range that extended far beyond the confines of Arizona. Later these rocks were intruded by vast granite batholiths, the tops of giant magma chambers formed along the overheated collision zone, then cooled while still far underground.

Still later in Precambrian time the sea lapped across the land, beveled it, and laid upon it layers of sedimentary rock. Dark **sills** of magma squeezed between the layers and hardened there. Then for a billion years and more, the land lay quiet and was gradually beveled, once again, by erosion.

In Paleozoic time, more sedimentary rocks were deposited: layers of marine sandstone, siltstone, and limestone that can be correlated with Cambrian, Devonian, and Mississippian strata exposed in the Grand Canyon. Successive advances of the sea, lapping onto the continent, were interspersed with times of slight crustal bending, uplift, and erosion. Marine sediments gave way in late Pennsylvanian and Permian time to deposits of delta and desert, and then to new incursions of the sea. There seems to have been no volcanism at all in Paleozoic time, nor were any new intrusions formed. A quiet, flat world, where small changes in sea level brought great changes in the width of the submerged shelf, and where there was no greater violence than waves surging against the shelving shore.

During the middle of the Mesozoic Era, volcanism began again. Incredibly savage explosions blasted out great quantities of ash and broken volcanic rock, which today form tuff and volcanic breccia layers up to 3 miles thick. New granites pushed up as well.

But more significant changes came about close to the end of the era. About 75 million years ago, this area quivered and shook with the onset of another period of mountain-building and intrusion that was to last 25 million years — a series of events collectively known as the **Laramide Orogeny**. Caused as the continent drifted west and collided with and rode over a very active zone at the edge of the Pacific floor, this orogeny was definitely of the "push" type, with stresses directed toward the northeast. Involving tremendous pressure and high temperatures, the Laramide Orogeny reset the atomic calendars of many older intrusions, altered many Paleozoic and Mesozoic rocks, and created new, relatively shallow granite intrusions so closely matched chemically and mineralogically with the continental crust that we feel sure they formed from melted parts of the colliding plates. Intrusions as well as volcanic activity of Laramide time crept from west to east.

In places magma with a different chemistry rose almost to the surface, magma unusually rich in copper minerals that would later make Arizona the leading U.S. producer of copper ore. It is thought that the copper ores were the last substances to crystallize out of the hot magma of their associated intrusions, and that prior to crystallization the copper-bearing minerals were dissolved in the hot water that permeates most magma, and were thus able to penetrate narrow fissures and fractures and more open solution cavities in the channels of adjacent rock. They came out of solution particularly readily along folds and faults in Paleozoic limestone, itself quite soluble in water.

Porphyry copper deposits *of Arizona, Mexico, and New Mexico lie in a broad NW-SE belt. One theory suggests that they formed as the continent drifted northwest across a hot spot in the earth's mantle. A similar belt of these deposits extends from Montana and Idaho into Canada.*

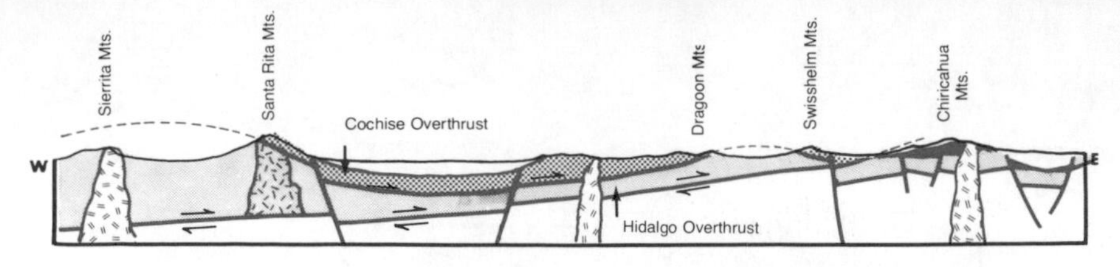

In Laramide time two thin overthrust sheets *may have moved horizontally across southeastern Arizona. Both sheets were later broken vertically by Basin and Range faulting. Much of the upper overthrust has now been lost to erosion.* (Adapted from H. Drewes)

There is some evidence that at this time two thin sheets of rock may have pushed northeastward along nearly horizontal **thrust faults**, in immense **overthrusts**, as the edge of the continent collided with several small islandlike plates that then became parts of the continent. Present-day patches of Precambrian, Paleozoic, and Mesozoic rocks, some of them in unorthodox older-above-younger sequence, seem to overlie zones of shattering, mineral alteration, and **foliation** (aligning of mineral grains in one direction by mechanical forces) that suggest horizontal movement of large overlying sheets. Because evidence of overthrusting is scarce and scattered among widely separated ranges, we run again into the jigsaw puzzle problem. North, northwest, and southeast of Arizona such thrusting is better documented, and overthrusts which moved on the order of 30 to 60 miles can hardly be denied. In Arizona it is postulated that overthrusting reached as least as far.

After 25 million years of mountain-building there seems to have been another relatively quiet time when erosion attacked and wore down the mountains formed during the Laramide Orogeny. During this time, 50 to 38 million years ago, streams and rivers drained north and northeast into the area that is now the Colorado Plateau region of Arizona and Utah. Fine sediment may have been carried as far north as large interior lakes that then existed in southwest Wyoming and northwest Colorado. Coarse sediment remained nearer its source in the southern ranges, forming alluvial fans that merged with sand and gravel of valley floors.

Another spate of mountain-building, the **Mid-Tertiary Orogeny**, occurred in Oligocene and early Miocene time 28 to 15 million years ago. Unlike the Laramide Orogeny, this mid-Tertiary episode involved tension, the "pull" type of mountain-building. It resulted in breaking and folding of the

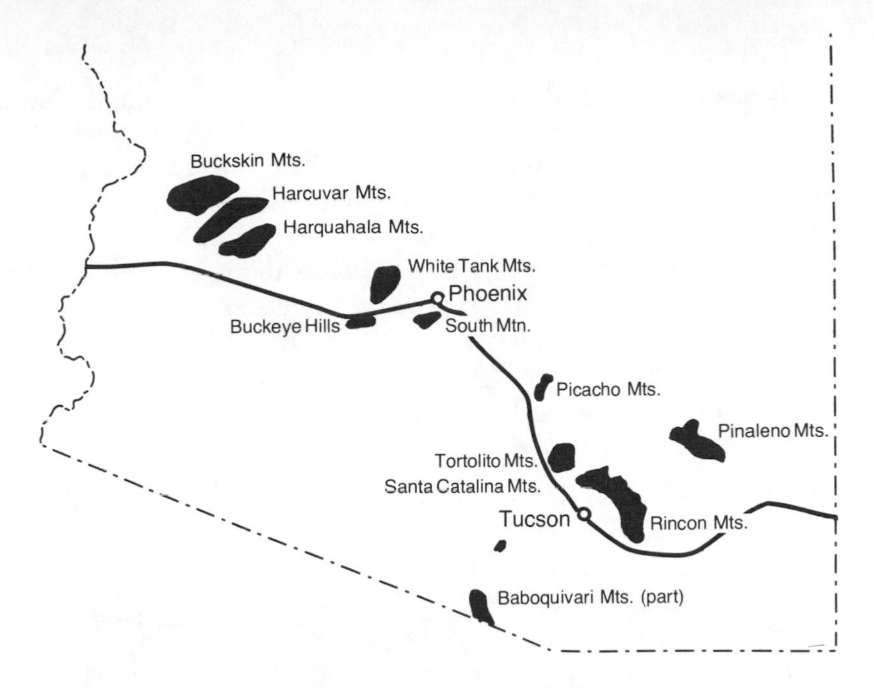

Interstate 10 weaves through an irregular line of mountains classed as metamorphic core complex *ranges. Most of these ranges trend NE-SW, across the trend of other Basin and Range mountains.*

crust and creation of several NE-SW mountain ranges at nearly right angles to the trend of Laramide ranges. Again there was volcanism — quite a bit of it, in violent explosive eruptions that spread thick layers of ash across southern Arizona. Lava flows reached into many intermountain basins. Once more there were intrusions of light-colored, granitic magma, accompanied by heating and further metamorphism of mountain cores. **Metamorphic core complexes**, domes of intrusive or metamorphic rock with outer shells of intensely stretched or sheared metamorphic rock, pushed toward the surface as well, by mechanisms that we are only beginning to understand.

In intermountain basins during this time, streams and rivers deposited coarse gravel and sand; fine mud and volcanic ash accumulated in lakes. These sediments were faulted and tilted during the dying stages of the Mid-Tertiary Orogeny, between 20 and 15 million years ago. At that time warping, thrust faulting, explosive volcanism, and intrusions of granite were replaced by near-vertical **block faulting** and much less violent basalt volcanism of the Basin and Range Orogeny.

Block faulting and basalt volcanism are both associated with "pull" tectonics, so the Basin and Range Orogeny produced a different type of mountain: long, narrow **fault block ranges** separated by deep basins. As the crust pulled apart, the tension faulting resulted not so much in mountain-building as in basin-sinking. Basins began to subside about 15 million years ago, and in most of southern and western Arizona continued to sink for about 8 million years. In far southeastern Arizona it is still going on. Only the highest parts of mountain blocks can be seen now; the faults that limit them are well hidden by the stream and lake deposits that fill the basins.*

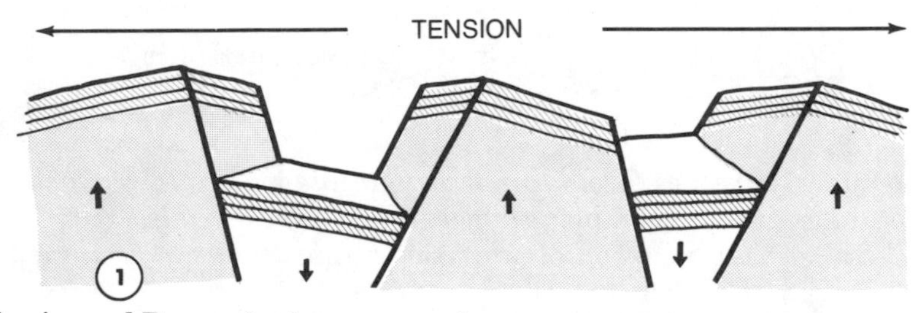

Basin and Range faulting created mountain blocks (color) separated by downfaulted basins. Structure within the blocks is simplified here.

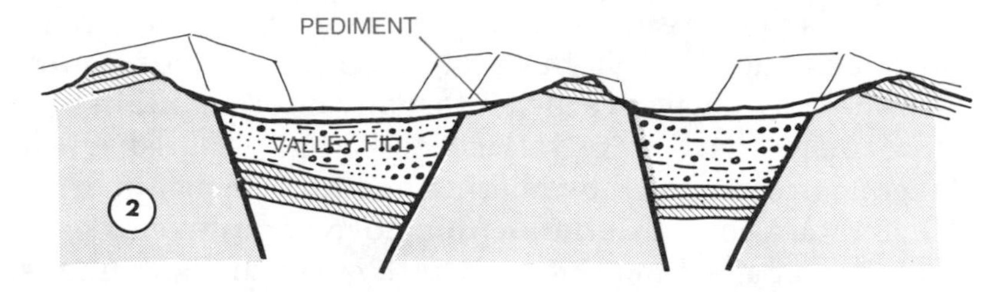

As debris from mountain blocks washed into basins, mountain fronts receded, leaving sloping pediments. The transition between eroded pediment and deposited valley fill is commonly obscured by a surface layer of gravel.

---

*Because most of these faults do not appear at the surface, they are not shown on geologic maps in this chapter. But they are there!

Many desert drainage patterns are dendritic, *branching like trees. Some are well marked by denser vegetation. As the large stream on the lower right deepens its channel, branching tributaries will cut headward into the valley rim.* Tad Nichols photo.

As subsidence isolated the basins, old drainage patterns were completely disrupted. New streams followed new courses, plunging down mountainsides into the basins, many of which were for a time occupied by lakes. In basins that had no external drainage at all, thick layers of salt accumulated in salt lakes.

Around 10 to 6 million years ago, most of the separate basins, each with its own drainage system, became so full of debris from the mountains that streams could flow from basin to basin. A few, however, such as Sulphur Springs Valley near Willcox, still have no external drainage. Most southern Arizona streams joined together as tributaries of the Gila River, which discovered a ready-made pathway across the state, as those who travel Interstate 8 will learn. From time to time volcanic outpourings dammed streams and temporarily isolated new lakes. Quite recently, less than 5 million years ago, the Colorado River became the master stream of this desert region, with the Gila (HEE-la) flowing into it near Yuma.

As through drainage was established, downcutting by rivers and streams increased. The Gila and its tributaries began to erode through the loose fill of the mountain basins. Within the last 2 million years, alternating rainy and dry cycles, which may have been in step with glacial and interglacial climates farther north, caused development of complicated arrays of terraces along the upper Gila River and its tributaries, and along the Colorado River farther west.

* * *

Let's go back a moment to the development of metamorphic core complexes in mid-Tertiary time — a hot topic among today's geologists. Since a line of such ranges extends clear across southern Arizona, we should look more closely at their structure.

Although no one seems to have come up with a completely logical explanation for all their characteristics, ranges described as metamorphic core complexes share certain features:

- Their centers are large domes of granite or gneiss, often but not always with the dome shape clearly recognizable in the shape of the mountain range.

- The granite domes are encrusted with severely stretched metamorphic rocks that seem to be derived from the core. These metamorphic rocks form a sort of shell, like a gigantic turtle's carapace, that distinctly tends to arch over the granite's domed surface.

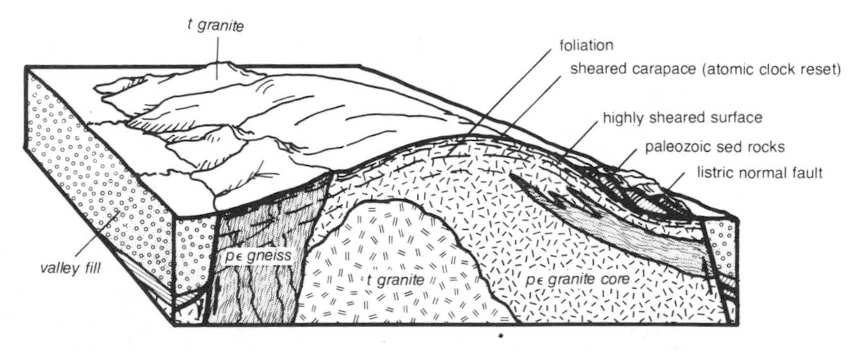

A typical metamorphic core complex consists of a core of Precambrian granite and metamorphic rock surfaced and interlayered with a sheared carapace or shell. Slices of Paleozoic and Mesozoic rocks may appear along the mountain flanks.

• In the carapace, streakiness or **foliation** in the rock, a result of flattening and shearing of individual mineral crystals, parallels the core surface also, and cuts across older banding inherited from the core.

• Large intrusions seem to have set the stage for development of most of the complexes, their heat softening surrounding rocks so that mechnical forces could act upon them, much as a blacksmith softens iron by heating it before bending or hammering it.

• Thin wedges of Paleozoic and Mesozoic metamorphic rocks appear to have become detached from the rising cores, perhaps dragged and broken as the cores pushed upward. A layer of intensely sheared rock now separates them from the core and carapace.

• The cores — whether Precambrian or younger — have had their atomic calendars reset. The new dates suggest that they finally cooled rather abruptly, geologically speaking, in mid-Tertiary time, about 25 million years ago.

• In Arizona most of the metamorphic core complex mountains trend SW-NE, transverse to the trend of most other desert ranges.

Tilted layers of Paleozoic sedimentary rocks mark many desert ranges. Some slid down the sides of rising metamorphic core complexes. Note the fold in the rock at lower left. Tad Nichols photo.

We should also look more closely at the origin of mineral deposits, and at some peculiar characteristics that may explain why they are common in this Basin and Range part of Arizona. Most mineral deposits are related to the rise of magma through the earth's crust. In terms of Plate Tectonics, **subduction** — the sinking of oceanic crust along the line of its collision with the continent — occurred here several times in the past, each time causing parts of the earth's crust to remelt at least to a slushy, porridge-like state — often at great depths and at some distance inland from the edge of the continent. Water, sea-floor basalt, and sea-floor sedimentary rocks were commonly involved in the melting. Mixing of the slush created a wide variety of magma types, which then rose through faults and fissures or melted their way up through the crust. Some of the magmas cooled into big granite intrusions we call **batholiths**. Others cooled in smaller intrusions such as **stocks** and **dikes**. Yet other magma burst through to the surface in volcanic eruptions.

As below-ground igneous masses cooled, minerals crystallized: quartz and feldspar, biotite and hornblende, and others. Some solutions remained — hot, highly concentrated liquids enriched with types of mineral ions that could come out of solution and crystallize into minerals only at lower pressures and temperatures. As they worked their way upward through rock fissures, these solutions reached regions of lower pressure, where they slowly cooled into today's ore deposits: veins and irregular masses of rock rich in copper, gold, silver, molybdenum, or other rock substances coveted by man. Changes in rock chemistry help ore deposits to form, so many of today's ore bodies lie in contact zones between shallow igneous intrusions and sedimentary rocks such as limestone.

The most common ores in Arizona are **porphyry copper deposits**. These deposits usually occur in fingers of intrusive rock that project above the roof of deeper batholiths, where they cooled at rates between the very slow cooling of granite in the batholiths and the quite rapid cooling of volcanic rocks. About 90 per cent of Arizona's copper comes from porphyry deposits. The rest comes from fingers of highly broken rocks called **breccia pipes**.

Silver and gold occur in narrow veins associated with vol-

An open pit copper mine produces large amounts of low-grade ore. At Pima Mine south of Tucson the 200-foot-thick overburden must be removed before ore can be mined. L. C. Huff photo, courtesy of USGS.

canic rocks, though porphyry and breccia copper deposits may contain small amounts of silver and gold associated with their copper ores. Veins are nothing more than crack fillings created as mineral-rich solutions seeped into narrow openings produced by mountain-building stresses or shrinkage of a cooling parent magma.

Most but not all of southern Arizona's mineral deposits are products of Laramide mountain-building. Thanks to its location near the southwest edge of the continent, this part of Arizona saw repeated folding and faulting, intrusion and volcanism, as we have already seen. But particularly during late Mesozoic and Cenozoic time, conditions seem to have been just right for ore formation, with the necessary heat from the earth's mantle, melting of sea-floor rocks saturated with sea water, and crustal folding and breaking that permitted mineral-rich solutions to rise toward the surface and solidify within reach of erosion and man.

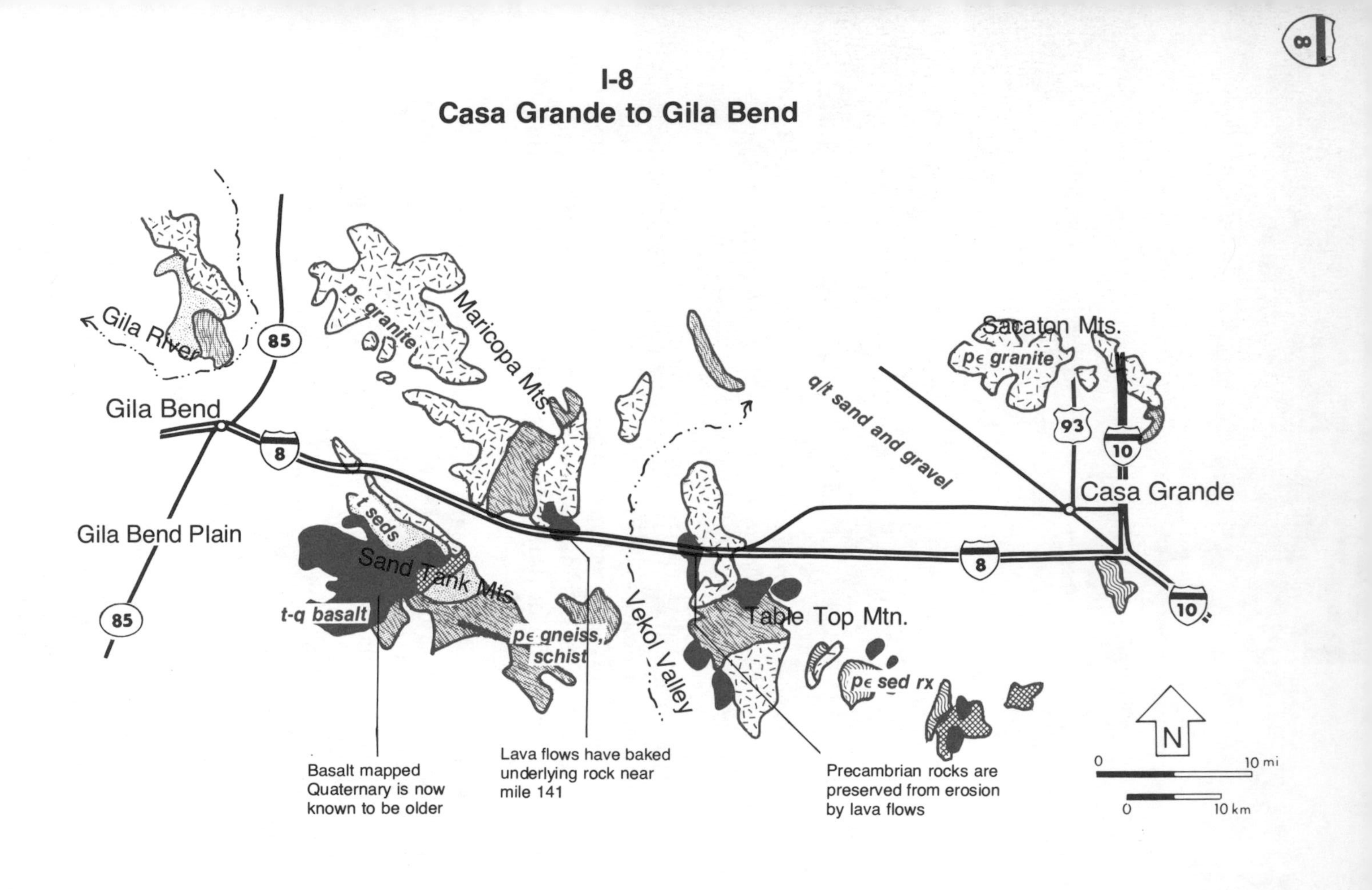
8
I-8
Casa Grande to Gila Bend
Gila River
85
Gila Bend
8
Gila Bend Plain
85
pє granite
Maricopa Mts.
t seds
Sand Tank Mts.
t-q basalt
pє gneiss,
schist
Vekol Valley
q/t sand and gravel
Sacaton Mts.
pє granite
93
10
Casa Grande
8
10
Table Top Mtn.
pє sed rx
N
0
10 mi
0
10 km
Basalt mapped
Quaternary is now
known to be older
Lava flows have baked
underlying rock near
mile 141
Precambrian rocks are
preserved from erosion
by lava flows

Abandoned mine buildings gaze empty-eyed across the Vekol Valley south of Interstate 8. Alkaline playa deposits mark low spots in the valley. Tad Nichols photo.

Interstate 8
Casa Grande — Gila Bend
(60 miles)

West of its junction with I-10, this highway crosses Quaternary sediments that floor the broad north end of the Santa Rosa Valley, heading toward the dark mass of Table Top Mountain. Although surface gravels in this and many other southern Arizona valleys are predominantly Pleistocene, exploratory drilling has revealed that they are underlain by thousands of feet of Tertiary stream and lake deposits, as well as by lava and volcanic ash layers by which these can be accurately dated.

Still farther west, beyond the irrigated farms of Casa Grande Valley, you may have a chance to observe wind as a major geologic force. In these open stretches, winds gain velocity and can pick up and carry sand and silt, sometimes as "dust devils" or whirlwinds and sometimes as billowing clouds. Loss of visibility is of course the great highway hazard. Health hazards are involved too — respiratory diseases like valley fever, carried by fungus spores. Sand grains and even gravel may blast structures and vehicles in the first few feet above the ground. Very fine dust is sometimes swirled quite high and picked up by high-altitude winds that transport it across the continent and even out over the Atlantic.

Desert washes in the area west of Casa Grande "flow" north toward the Gila River, about 20 miles away. The washes *do* occasionally flow, but their waters rarely reach the Gila; they sink instead into the loose gravel and sand of the valley floor, partly replenishing the groundwater supply. Par for the course in the southern deserts.

The upper part of the long, very gentle rise to Table Top Mountain is a pediment cut on granite bedrock, lightly covered with sand and silt. As its name might suggest, parts of Table Top Mountain are capped with lava flows that have not been bent or tilted but still lie in their original horizontal positions. The main parts of the range, however, are much older Precambrian granite and metamorphic rocks. Where the highway crosses these mountains, **outcrops** of weathered granite are preserved beneath the young basalts — the beginning and end of close to 1400 million years of geologic time. The granite is so deeply weathered and decomposed that mineral grains are no longer tightly joined; what once was granite is now just coarse, packed sand.

A few miles beyond, contrastingly dark lava, scarcely weathered at all, appears near the road. Though mapped as Quaternary, these flows may be older. Watch for reddish baked zones beneath the basalt flows.

The highway crosses Vekol Wash at milepost 151. Beyond Vekol Valley it slips between the Maricopa and Sand Tank Mountains, again composed of Precambrian gneiss and granite capped by Tertiary-Quaternary lava flows. In parts of the Sand Tank Mountains there are tilted and faulted Tertiary sedimentary rocks that were deposited before the last phases of Basin and Range faulting took place, and were caught up in the activity. Some of these sediments — hardly well enough consolidated to call rocks — can be seen as foothills at the north end of the range.

In general, you have been passing through a region where the overall grain or fabric of the land is NW-SE—sharply defined, ridge-like mountain ranges paralleling long, open valleys. Near Gila Bend we find evidence of a change of scene. For most of its journey from Phoenix to the Colorado River, the Gila River flows along a northeast-southwest trough that seems boldly to ignore the "lay of the land." **Seismic studies**, which analyze hidden subsurface features by bouncing man-made shock waves off underground reflective layers, show a **graben** 100 miles long and 10 miles wide, a down-dropped trough edged on each side by nearly vertical faults. The lava flows of the Gila Bend Mountains deflected the Gila River from this southwestward route, so that it bends south around both lava flows and Precambrian rock, then northwest to rejoin its former channel.

In the Gila Bend Mountains, Tertiary and Quaternary volcanic rocks overlie Precambrian granite and metamorphic rocks. Both are deeply eroded and darkened with desert varnish. New research on the dark brown varnish suggests that it forms from clay minerals in wind-blown desert dust deposited on rocks and then wetted by the rain that often follows dust storms. Over a few thousand years, rocks exposed to the dust and rain and to the alkaline desert environment, a necessary catalyst, gradually acquire this hard, shiny, iron-manganese coating.

Interstate 8
Gila Bend — Yuma
(120 miles)

Northwest of Gila Bend, the Gila River bends northward through a narrow channel between the Painted Rocks and Gila Bend Mountains and into the graben described above, the Gila Trough. The almost horizontal floor of Gila Bend Plain is now heavily cultivated; the main crops are cotton and cattle feed. Ranges surrounding the plain are almost all volcanic; lava flows make up all but the easternmost part of the Gila Bend Mountains. Don't confuse this range with the Gila Mountains near Yuma or with other Gila Mountains on the east side of the state near Safford. Flows of the Painted Rock Mountains are older — Cretaceous to early Tertiary.

About three miles south of the highway near milepost 102, a small shield volcano is an outlier of the Sentinel volcanic field. The highway soon rises onto the basalt flows of this field. Here, in one of the youngest displays of volcanism in Arizona, individual flows are thin; the basalt lava was fluid and erupted quietly, spreading in sheets and shallow lava ponds. Because of the ease with which it flowed and the apparent low gas content of the lava, there is little buildup around the **volcanic vents**, and they are hard to identify. Sentinel Peak — one of the vents — is hardly a peak at all. The lava can be seen in more detail at the rest stop at milepost 85. All these flows are less than 2 million years old, having erupted early in Pleistocene time. The

I-8
Gila Bend to Yuma

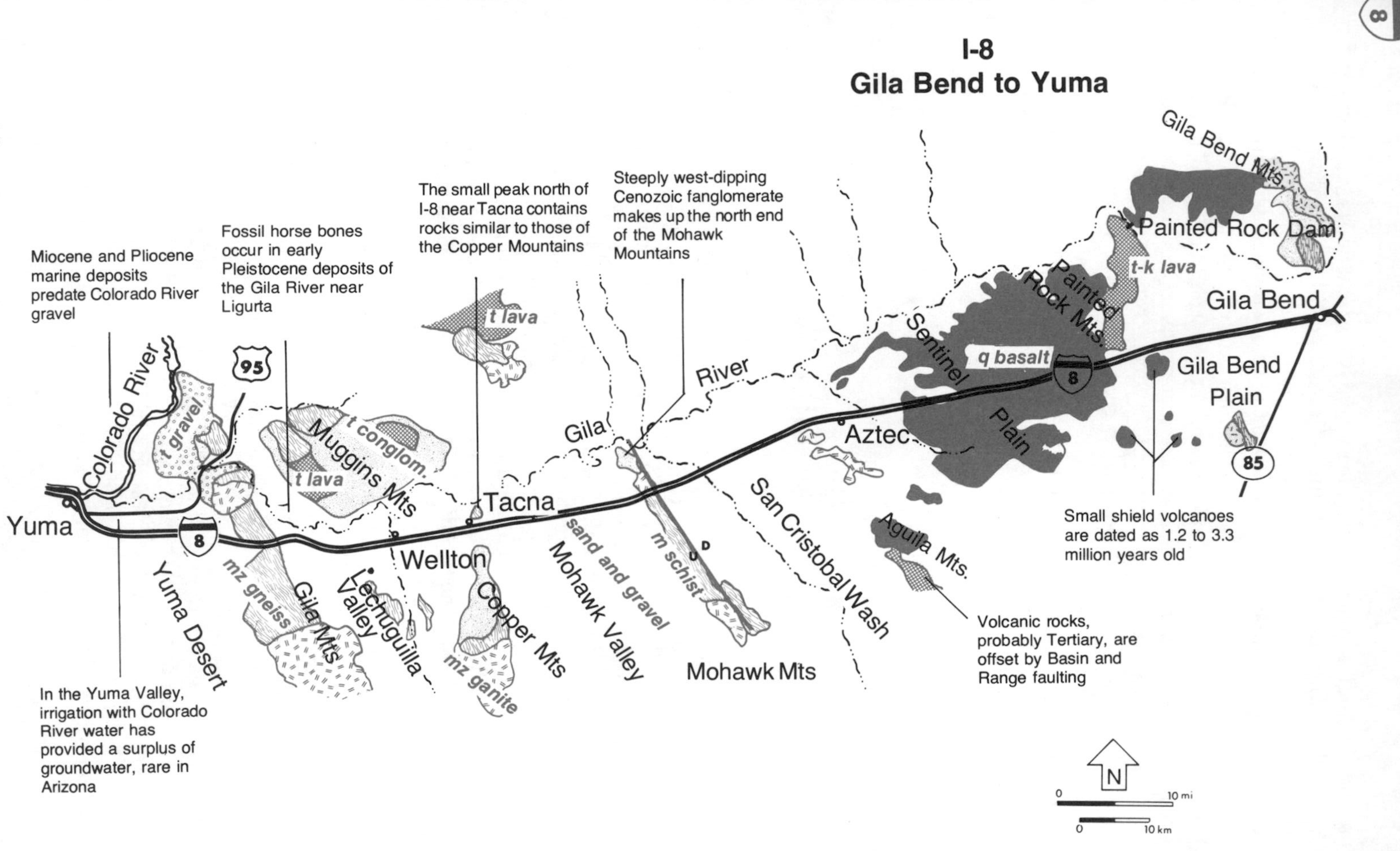

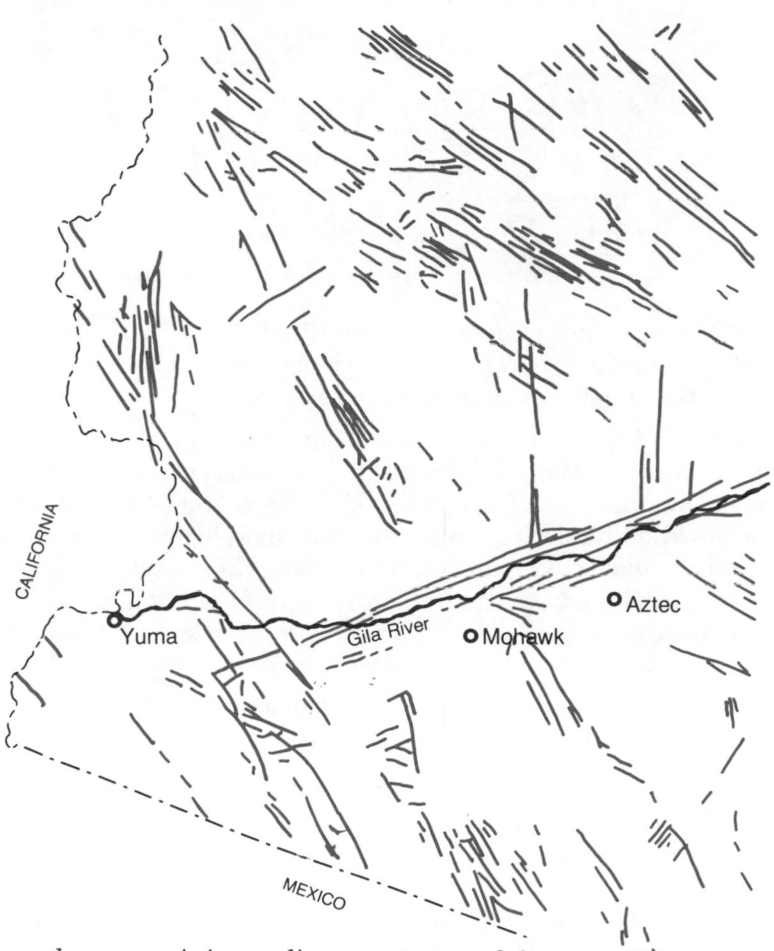

In southwestern Arizona, lineaments traced from satellite photographs give tantalizing clues to Basin and Range structure. Most lineaments run NW to SE. Often obscure at the surface, the lines may be faults, joints, folds, or contacts between different kinds of rocks.
Adapted from L. K. Lepley.

original surface of the uppermost flow has had time to weather and break up, leaving scattered blocks of basalt strewn about on the surface.

Sentinel lavas lie over Gila River sediments — fine, well sorted sand and silt — suggesting that the river, after being deflected southward around the Gila Bend Mountains, for some time flowed south of its present course. The Sentinel lava flows may have been the deciding factor in its return to the Gila Trough. West of these lavas the river and the trough gradually converge with the highway.

West of the Sentinel lavas watch for several small **cinder cones**. The Aguila Mountains, visible farther south, are capped by thin

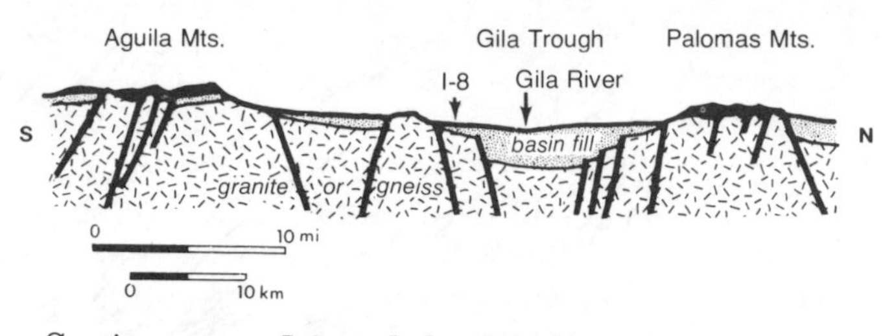

Section across I-8 and the Gila Trough near Aztec

basalt flows. Unlike those of the Sentinel Plain, these are slightly tilted and broken by faults, and are thought to date back to sometime before Basin and Range mountain-building.

Beyond Aztec the route crosses San Cristobal Valley and then approaches the Mohawk Mountains. A long, narrow backbone of a range, reminiscent of a "mohawk" haircut, these mountains are composed of Tertiary granite and metamorphic rocks, products of both Laramide and Basin and Range mountain-building.Gneiss and schist are well exposed near the highway. Looking along the fairly straight eastern flank of the mountains, you can well get a feeling for the approximate line of their bordering fault.

Dust devils — whirling columns of dust and sand — are frequent along this stretch of highway.

The next two ranges south of the highway, the Copper and Gila Mountains, are structurally similar to the Mohawks — slim, long, upward faulted slivers of Mesozoic granite and gneiss.

Muggins Mountains, to the north, consist largely of Miocene river and lake sediments, as do the light-colored foothills that edge the Copper Mountains. These sediments may once have sheeted across this entire area, but if so they wore away in Pliocene and Pleistocene time as the new-routed Colorado River cut rapidly downward, steepening the gradient and therefore the erosive strength of all the streams draining into it. Muggins Mountains contain some Tertiary volcanic rocks, too. Some of the Tertiary sediments contain uranium, and gravels at the mountain bases contain **placer gold**.

Not far upstream from Yuma, old Colorado River deposits appear as terraces, the highest 70 to 80 feet above the present river floodplain.

F. H. Olmstead photo, courtesy of USGS.

The Lechuguilla Desert between the Copper and Gila Mountains is now in part irrigated farm land. Desert soil, however parched and poor it may look, is quite fertile where it is irrigated, and orange and other citrus trees thrive in the frost-free climate.

I-8 cuts right through the Gila Mountains, though they deflect the Gila River northward. Spectacular roadcuts reveal the finely fractured black and white banded gneiss of which these mountains — yet another fault block — are made. Elsewhere, dark brown desert varnish hides their wiggling, taffy-like patterns.

The Yuma Desert between the Gila Mountains and the Colorado is desert indeed, especially farther south where it is not used for agriculture. Barren, less than 200 feet above sea level, it is almost lifeless in the shimmering heat of a long summer. Blowing sand and whirling dust erode the faces of the mountains, as well as man-made objects near the highway.

The town of Yuma lies on a slight rise, Yuma Mesa, overlooking the Colorado River. This once mighty river used to be navigable well north of Yuma. It also used to flood severely. A large proportion of its waters are now held back for flood prevention, irrigation, power generation, recreation, and the myriad other uses to which water can be put in a desert land. Water that returns to the river along natural aquifers is higher in dissolved minerals than it was before removal,

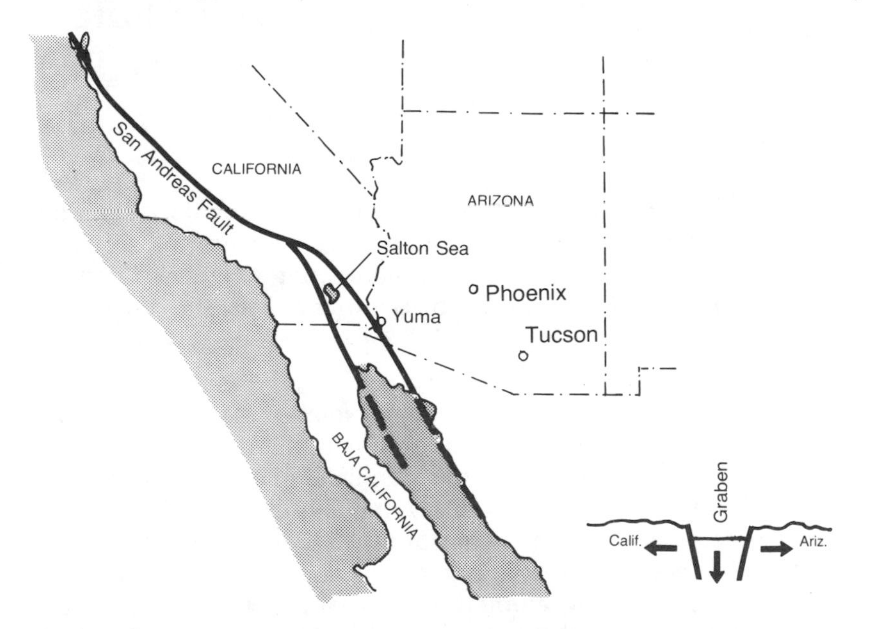

A site of crustal spreading that may signal the birth of a new ocean basin, the gulf of California-Salton Sea graben cuts across Arizona's southwest corner.

and can no longer serve the domestic and agricultural needs of the strip of Mexico that lies between here and the Gulf of California — a bone of contention between nations. Water for Mexico is now desalted in a large Bureau of Reclamation plant just east of Yuma, and put back into the river to complete its southward journey.

Yuma is very near the eastern margin of the Salton Trough, a northward extension of the rift that creates the Gulf of California. Formerly part of a mid-ocean ridge, this rift is here overridden by the continent. However, it keeps its character as the site of crustal spreading. In the last 20 million years it has opened up the gulf and the pronounced graben that extends northwest from Yuma, now partly occupied by the Salton Sea. In the Yuma area the trough is filled to a depth of several miles with marine limestone, Colorado River sediments, and interlayered volcanic rocks. The fault that forms the northeast edge of the trough cuts diagonally across the southwestern corner of Arizona between I-8 and the Mexican border. Northwestward it is virtually continuous with the San Andreas fault of California, and movement along it is related to the largely horizontal, shearing movement of the San Andreas fault.

Interstate 10
New Mexico — Willcox
(78 miles)

Interstate 10 enters Arizona just south of the Peloncillo Mountains, a volcanic range whose lava flows, tuff and breccia are almost continuous with a vast volcanic region that stretches north and northeast beyond Springerville and Show Low, 140 miles away. Some of the light-colored volcanic rocks appear in roadcuts east of San Simon, near the state line.

San Simon Valley just west of these mountains occupies a deep graben edged by northwest-trending Basin and Range faults which are concealed beneath broad alluvial fans that merge to form bajadas. The San Simon River, usually dry, is a tributary of the Gila River, a watercourse that cuts across southern Arizona in a predominantly westward direction. Flat-lying limestone and clay within San Simon Valley's deep layers of sediment show that periodically in Pliocene and Pleistocene time this valley held mountain-girt lakes. Sloping layers of coarse, porous **alluvial** (stream-deposited) sand and gravel converge near the center line of the valley. Near the mountains,

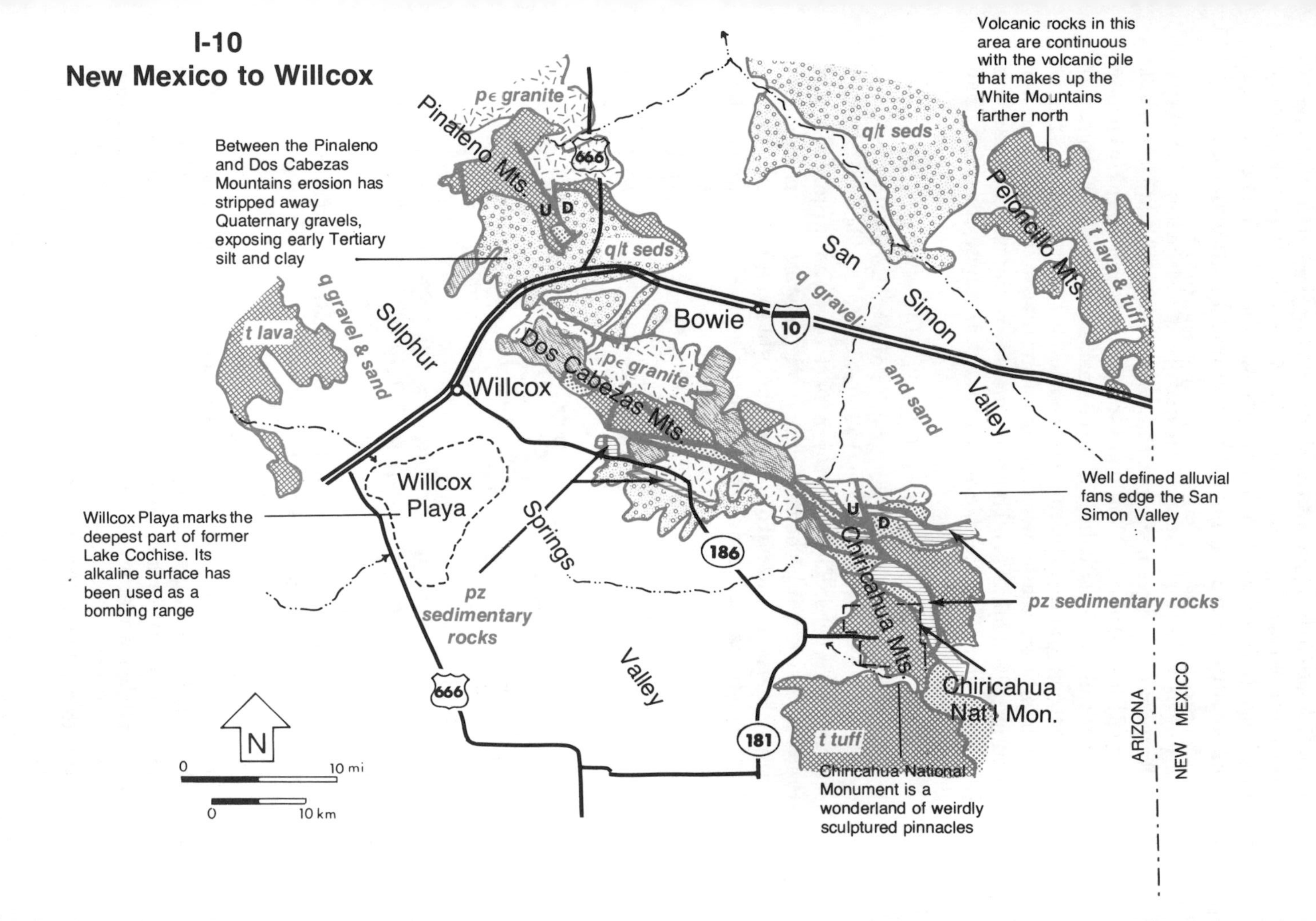
I-10
New Mexico to Willcox
Volcanic rocks in this area are continuous with the volcanic pile that makes up the White Mountains farther north
Between the Pinaleno and Dos Cabezas Mountains erosion has stripped away Quaternary gravels, exposing early Tertiary silt and clay
Willcox Playa marks the deepest part of former Lake Cochise. Its alkaline surface has been used as a bombing range
Well defined alluvial fans edge the San Simon Valley
Chiricahua National Monument is a wonderland of weirdly sculptured pinnacles
pє granite
Pinaleno Mts.
666
U D
q/t seds
t lava
q gravel & sand
Sulphur
Willcox
Bowie
10
San
Simon
Valley
q gravel
and sand
Peloncillo Mts.
t lava & tuff
Dos Cabezas Mts.
pє granite
Willcox
Playa
Springs
Valley
pz
sedimentary
rocks
186
181
Chiricahua Mts.
pz sedimentary rocks
Chiricahua
Nat'l Mon.
t tuff
ARIZONA
NEW MEXICO
N
0
10 mi
0
10 km

storm-fed streams sink into porous sand and gravel layers and flow down along them, trapped by overlying impermeable clay layers. Wells in the center of the valley tap this water. Because the well heads are lower than the level at which water enters the aquifers, well water rises to the surface without pumping. The valley's agriculture depends on the **artesian** water for irrigation.

To the west, between the Dos Cabezas (Two Heads) Mountains to the southwest and the Pinaleno Mountains north of the highway route, young layers of valley fill have worn away, revealing underlying Pliocene and Pleistocene gravel and clay and sand, most of it distinctively orange-brown because of gradual oxidation of minor amounts of iron minerals. These sediments are well exposed near the rest stop at milepost 358. More of them extend northward along San Simon Valley.

The Dos Cabezas Mountains are highly faulted northwest-trending slices of Precambrian granite and metamorphic rock and Cretaceous volcanic rocks, with a few thin slices of Paleozoic sedimentary rocks. Their northern part contains large masses of Precambrian granite. The Cretaceous volcanic rocks — mostly breccia or **volcanic agglomerate** — are thought to represent the unusually large throat, or conduit, of a volcano — one of the largest such known. There are many Tertiary dike swarms in these mountains as well.

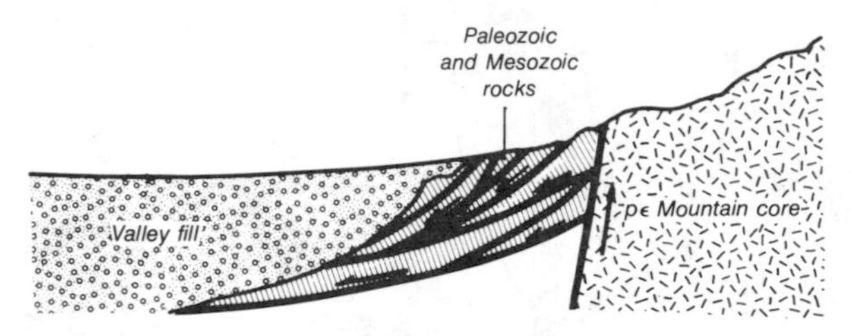

Curving wedges of Paleozoic and Mesozoic rocks rest against Precambrian rocks in a fanlike pattern known as listric normal faulting.

Though the fault slices of the Dos Cabezas Range extend southeast into the north end of the Chiricahuas, the main mass of the Chiricahuas is volcanic, layers and layers of volcanic ash (welded tuff) accumulated during a wild orgy of explosive volcanism around 30 million years ago. For more information on this interesting range, see Chiricahua National Monument in Chapter V. The coarse pink tuff shows a well developed pattern of vertical fractures, or joints. In parts of the range, erosion has shaped the fractured columns into forests of weird pinnacles; Chiricahua National Monument is one such area. One of the northern summits of the range, less highly fractured, forms a mountain you'll have no trouble recognizing from the highway: Cochise Head. It is separated from the rest of the Chiricahuas by a canyon eroded into sliced-up wedges of Paleozoic sedimentary rocks.

Like the Dos Cabezas and Chiricahua Mountains, the Pinaleno Range north of the highway is one of the many fault block ranges of the southern deserts. Here in southeastern Arizona the ranges are higher and closer together than those farther west — perhaps only because the valleys between them are less completely filled in. The Pinalenos, with Mount Graham (10,717 feet) as their highest point, are cored with a single large mass of Precambrian gneiss, flanked on the southwest by Precambrian granite. At the south end of the core both granite and gneiss are cut by faults that have also been dated as Precambrian, about 1.4 billion years old, and that line up with the general pattern of faulting and folding extending northwest across Arizona and south and southeast into Mexico and Texas, marking what must have been a Precambrian version of the Himalayas. At the southern end of the Pinalenos are some Cretaceous volcanic rocks about 100 million years old.

Despite their barren appearance, Sulphur Springs Valley and its adjoining ranges have yielded their share of Arizona's mineral riches. Surrounding ranges contain ores of lead, copper, zinc, silver, and gold, as well as good quality turquoise. From Willcox you can look southwest to the Dragoon Mountains (last stronghold of Apache Chief Cochise), southeast to the Dos Cabezas and Chiricahua Ranges, and over the shimmering salt-white floor of Willcox Playa. There is no outlet to this valley; during sporadic but sometimes torrential rain all streams drain into Willcox Playa, the lowest part of the valley floor. In Pleistocene time when the climate was much wetter, a 20-mile-long lake lay here. The **playa**'s glaring surface is a crust of alkali, a mixture of calcium, sodium, and potassium carbonates concentrated by evaporation as the lake dried up. After heavy rains, a new and temporary lake may form, just a few inches deep, with new deposits as it, too, evaporates.

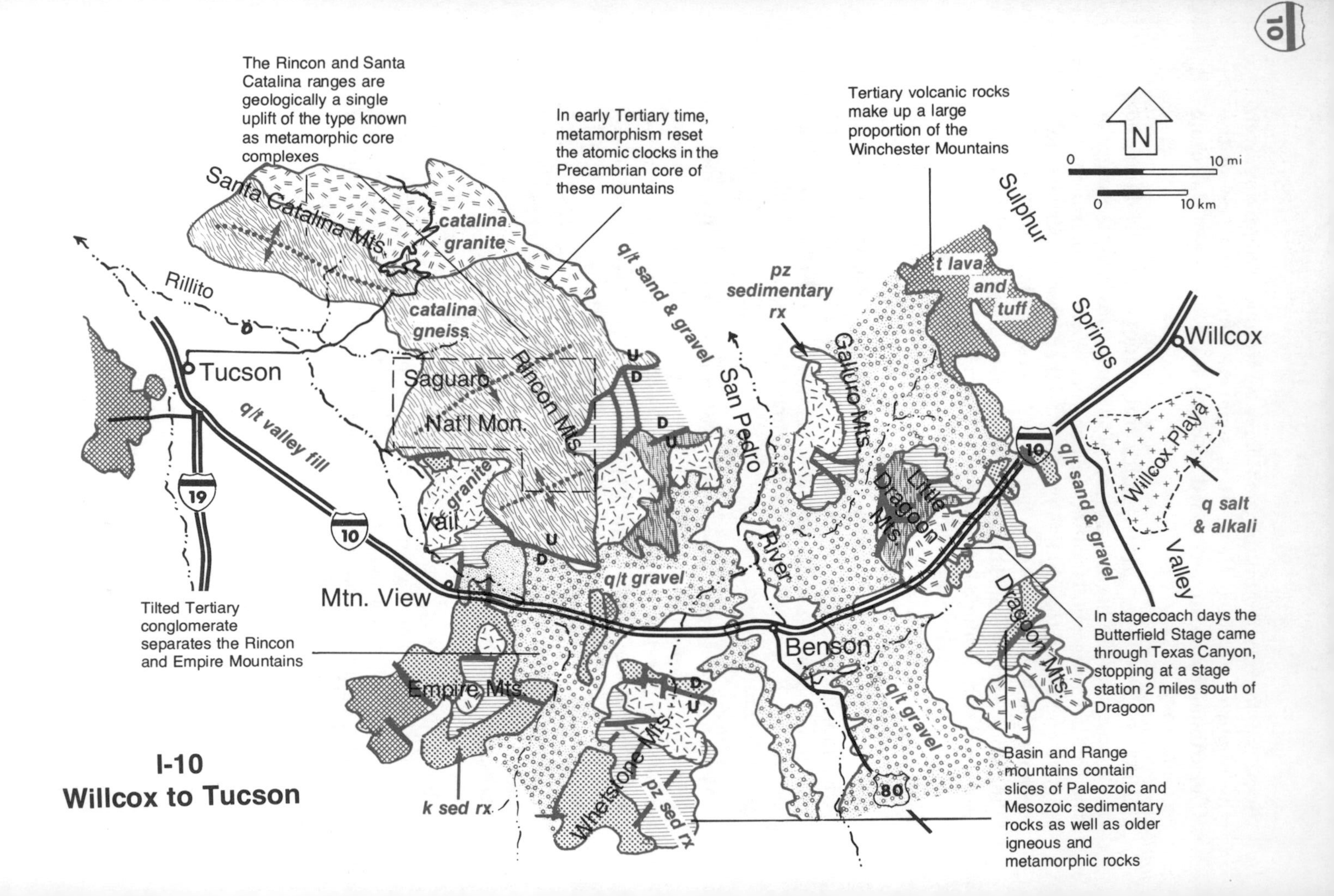

10
The Rincon and Santa Catalina ranges are geologically a single uplift of the type known as metamorphic core complexes
In early Tertiary time, metamorphism reset the atomic clocks in the Precambrian core of these mountains
Tertiary volcanic rocks make up a large proportion of the Winchester Mountains
N
0
10 mi
0
10 km
Santa Catalina Mts.
catalina granite
catalina gneiss
Rillito
Tucson
Saguaro Nat'l Mon.
Rincon Mts.
q/t sand & gravel
pz sedimentary rx
San Pedro
Galiuro Mts.
Sulphur
Springs
t lava and tuff
Willcox
Willcox Playa
q salt & alkali
Valley
q/t sand & gravel
Little Dragoon Mts.
q/t valley fill
19
10
granite
Vail
U
D
q/t gravel
River
Mtn. View
Benson
Dragoon Mts.
Tilted Tertiary conglomerate separates the Rincon and Empire Mountains
In stagecoach days the Butterfield Stage came through Texas Canyon, stopping at a stage station 2 miles south of Dragoon
Empire Mts.
Whetstone Mts.
pz sed rx
q/t gravel
80
k sed rx.
I-10
Willcox to Tucson
Basin and Range mountains contain slices of Paleozoic and Mesozoic sedimentary rocks as well as older igneous and metamorphic rocks

In Texas Canyon, spheroidal weathering rounds huge joint-edged blocks of porphyry. The pink-orange blush on the boulders results from breakdown of iron-bearing minerals; freshly broken surfaces are light gray.

Interstate 10
Willcox to Tucson
(90 miles)

For a brief discussion of the Sulphur Springs Valley and Willcox Playa, see the preceding section. Dunes now mark the ancient lake shore near Willcox. Valley fill here is about 2000 feet thick — lake deposits and sloping layers of silt, clay, sand, and gravel washed from the surrounding mountains: the Pinalenos, Dos Cabezas, and Chiricahuas on the east, the Galiuros and Dragoons on the west.

East of Willcox the highway climbs across a broad, undulating alluvial surface which merges imperceptibly with eroded pediments at the base of the Little Dragoon (north of the highway) and Dragoon Mountains. At the north end of the Dragoons, in faulted and metamorphosed Paleozoic rocks, marble has been quarried. The Dragoons also contain many old mines that once produced copper, gold, zinc, lead, and silver. Ghost towns and abandoned shafts are scattered through this range, once the hideout of the legendary Cochise.

The Little Dragoons just north of the highway are also rich in copper-lead-zinc ores. The mine on their east flank, near the contact between Paleozoic rocks and pink granite porphyry of the main mountain, has been in operation off and on since 1881. On their west flank, the Tungsten King mine produced tungsten ores from a mass of coarse granite porphyry.

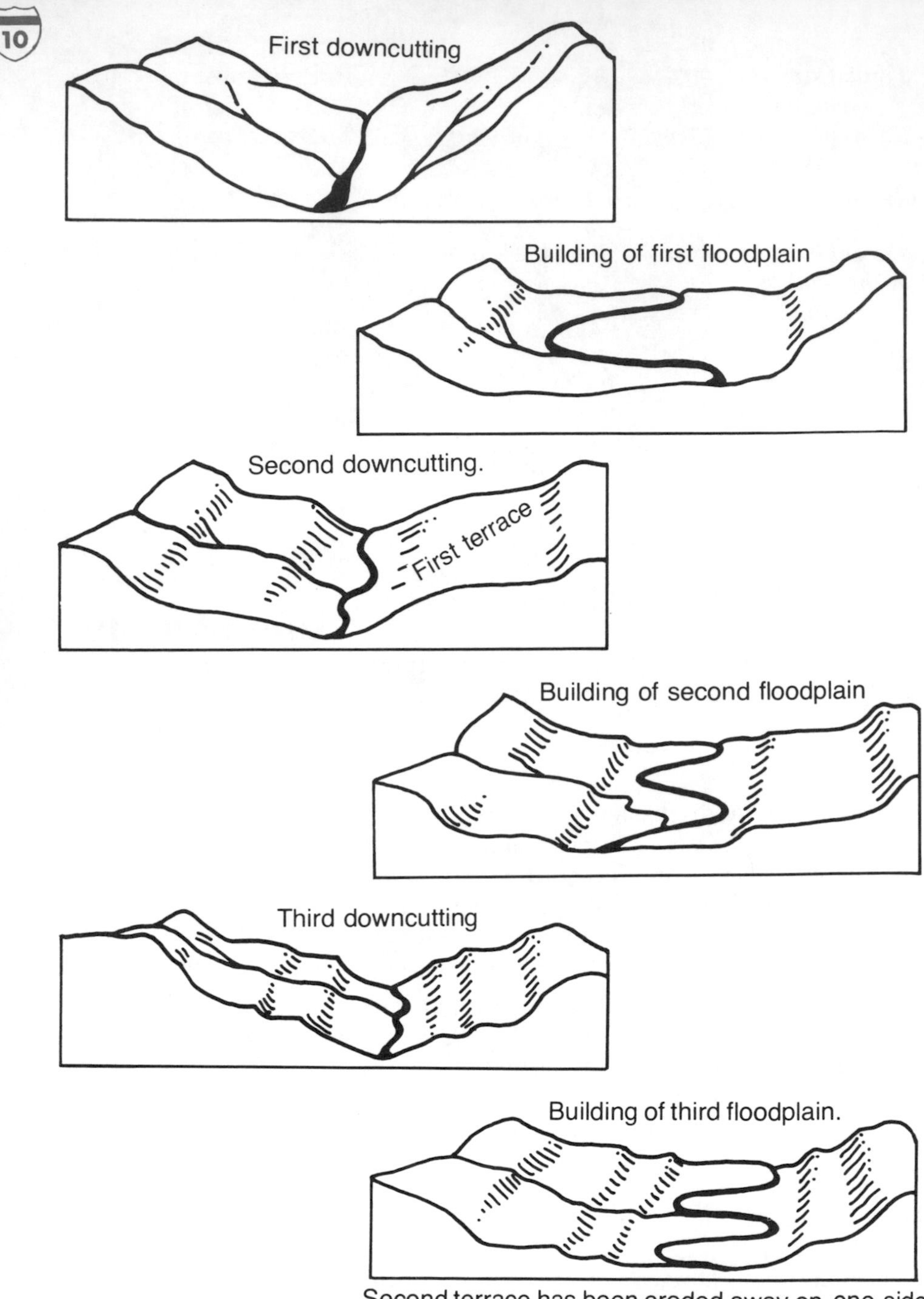

As many as three terrace levels border the San Pedro River. Each represents a period during which an overburdened stream widened its channel and filled it part way with rock debris, followed by a period of downcutting.

The highway continues through the pass between the Little Dragoon Mountains and the Dragoons themselves. Both these ranges are fault blocks, and both contain intricately faulted arrays of Precambrian schist and Paleozoic and Mesozoic sedimentary rocks surrounding Laramide intrusions.

Texas Canyon, west of the pass, penetrates one of these intrusions, and west of milepost 319 the highway travels through a wonderland of rocks that you may have seen before in western films or TV shows. Huge monoliths of quartz monzonite porphyry, looking much like granite but with large crystals of feldspar poking out like warts on a toad, are shaped by processes of jointing and weathering. Intersecting joints divide the rock into rectangular blocks, and then, even before they are exposed, weathering processes widen the joints and round the corners of the blocks. Further rounding of the separated boulders is accomplished by **spalling** or **sheeting**, also called **exfoliation**. Both processes are still taking place.

Beyond Texas Pass the highway descends gradually into the San Pedro Valley. Watch the roadcuts and the walls of deep ravines that parallel the road. Try to pick out the transition between the eroded mountain pediment beveled on partly decomposed granite and the bouldery gravels of the valley fill. Despite the differences in processes — erosion *versus* deposition — the desert surface crosses smoothly from pediment to valley fill near mile 316-315.

The San Pedro Valley, a deep graben downfaulted between the adjacent ranges, was lake-filled in Pliocene time; silty pink lake deposits are eroded into **badlands** west of Benson. In them have been discovered a rich variety of Pliocene and early Pleistocene fossil vertebrates: fish, turtles, lizards, snakes, birds, squirrels, rabbits, gophers, mice and rats, wolves, horses, peccaries, llamas, and camels! Such an abundance of animals confirms what the lake deposits tell us: that the climate here was once quite a bit wetter than at present. After the lake's demise, the deposits were blanketed with coarse gravel and cobbles, and then within the last 700,000 years dissected again by the river. The present river channel cuts through some of the former floodplain deposits. Along a terraced river valley such as this, the highest terraces are the oldest, the lowest are the youngest.

To the right as the highway climbs out of the San Pedro Valley are hills marked by low gray ridges of Precambrian and Paleozoic sedimentary rocks. They lie against 1.6 billion-year-old Precambrian schist and granite, southern Arizona's equivalent of the dark rocks of Grand Canyon's Inner Gorge, rocks that seem to be the foundations of this part of the continent and that are often referred to by geologists as the Precambrian **basement**. Here they extend out onto the eroded upper portions of the Rincon Mountain pediment, but they are gravel-covered and hard to see at highway speeds.

Both the Empire and the Whetstone Mountains south of the highway are made up of relatively undeformed but quite steeply tilted Paleozoic and Mesozoic sedimentary rocks. Low hills of Precambrian granite mark the north end of the Whetstones. The mustard-colored soils near the highway are derived from volcanic ash that is probably Oligocene in age, about 25-30 million years old.

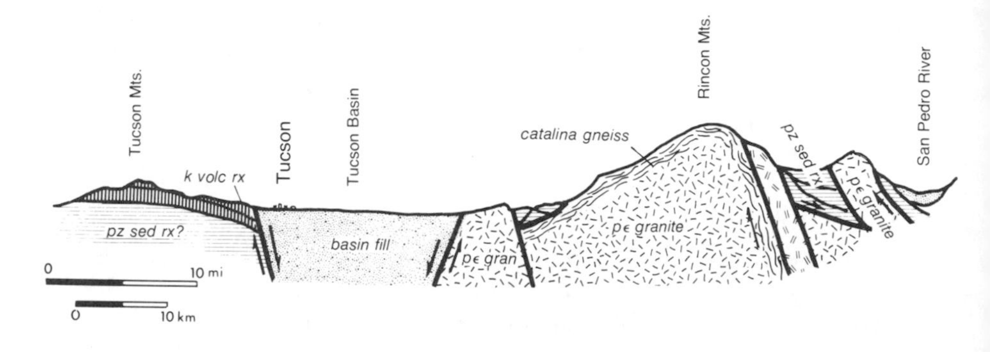

Section north of I-10, Benson to Tucson

As you come between the Empire Mountains and the Rincons, you can see that a large ridge stands away from the main Rincon Mountain mass. Edged with faults, this ridge dropped about 5000 feet relative to the Rincons, possibly before the rise of the great metamorphic core complex of the Rincons. The sliding surface — thought by some geologists to be the **sole** of a much more extensive overthrust fault — extends south across the highway at mile 295. Movement along this fault took place sometime between 14 and 10 million years ago.

Other detached wedges of Precambrian gneiss and slices of Paleozoic and Mesozoic rocks border the south and west sides of the Rincon's bulbous granite core, which is encased in a carapace of highly sheared rock — a typical metamorphic core complex. Near Vail, one of the Paleozoic slices holds Colossal Cave, a limestone solution cavern decorated with stalactites, stalagmites, and other types of cave ornaments (see next section). East of Mountain View, Mesozoic sedimentary rocks extend across the highway and are visible in highway cuts as tilted layers sliced by numerous faults. Without these exposures, one would be hard put to discern the faulting here!

Entering Tucson's wide valley, I-10 drops gradually across the broad, gentle slopes that surround the Rincon and Empire Mountains. Thousands of feet of valley fill — about 7000 feet under Tucson — conceal the real bases of the desert ranges. Roadcuts near mile 284-283 show some of the fill quite well, with typical stony Pliocene-Pleistocene gravel overlying an irregular, hilly surface cut in light colored, tilted volcanic rocks, reddish clay, and gravel.

Because there is no central river draining it, Tucson Basin seems unusually flat-floored. Its eastern half is drained by Pantano Wash, its north edge by the Rillito (Little River), and its western portion by the Santa Cruz River. The town originally sprang up on the banks of the Santa Cruz, which a century ago flowed all year around. Tucson has grown to fill the entire northern end of the basin. All of its water now comes from deep wells, and the water table has dropped markedly over the years. The rivers now flow only after heavy rains or snowmelt.

Open pit copper mines south of Tucson are in the Pima Mining District, discussed under I-19 Tucson to Nogales. Clockwise around the Tucson Basin from these mines are the Tucson, Santa Catalina, and Rincon Mountains, discussed under Saguaro National Monument (Chapter V) and Tucson and Vicinity, and the Empire and Santa Rita Mountains, described under I-19 Tucson to Nogales and AZ 82-83 Nogales to I-10.

The forerange of the Santa Catalinas, an anticline of banded gneiss, hides the domelike granite core of a large metamorphic core complex.
Tad Nichols photo.

Tucson and Vicinity

The mountains around Tucson offer several interesting excursions: to Sabino and Bear Canyons and Mt. Lemmon in the Santa Catalina Mountains, to the Tucson Mountains and the Arizona Sonora Desert Museum, to Saguaro National Monument (discussed in Chapter V), and to Colossal Cave at the base of the Rincon Mountains. Longer trips are also possible: to Oracle and the Pinal Pioneer Parkway (see US 89 Florence to Tucson); to Nogales (see I-19 and AZ 82-83); and to Kitt Peak (see AZ 86).

The Museum of Geology at the University of Arizona displays many beautiful minerals from Arizona mines, as well as other geologic specimens. A major rock and mineral show takes place in Tucson each February.

SABINO AND BEAR CANYONS

These two canyons, both now city parks, slice through the Catalina Gneiss that makes up the forerange of the Catalinas — part of the great Rincon-Catalina-Tortolita metamorphic core complex. The gneiss seems to lean up against the dark, forested dome of Mt. Lemmon, high point of the mountains (9157 feet), but it is actually part of an anticline separated from the dome by a deep valley in part occupied by the east and west forks of Sabino Creek and by Sycamore Creek, a tributary of Bear Creek. With some severely deformed Precambrian and Paleozoic sedimentary rocks (still exposed near the summit of Mt. Lemmon), the gneiss seen in Sabino and Bear canyons was altered by immense pressures, high temperatures, and the upward thrust of the mountain core. It formed from the granite of the summit dome and probably also, in part, from sedimentary rocks that once lay above that granite.

Tumbled boulders of Catalina Gneiss add to Sabino Canyon's beauty. They fell from the canyon walls during an 1887 earthquake.

The gray Catalina Gneiss is attractively banded, with white quartz veins of all sizes. It is a particularly resistant rock except along joint planes, where crushing and signs of movement are present.

Streams in these canyons, only partly fed by springs near the summit and in the valley behind the forerange, do not always flow. Their canyons appear basically V-shaped from a distance, but from close up are seen to be modified by the cliff-forming tendency of several strong, vertically fractured bands of gneiss. Lichens and desert varnish on the cliffs conceal the true gray color of the rock, which does however show up well on stream-washed boulders.

Both Sabino and Bear Canyons give access by trail to the valley behind the forerange and the main dome beyond it. At Seven Falls up Bear Canyon, Sycamore Creek tumbles into deep pools, super sized **potholes** formed as swirling water scoured solid rock with sand and boulders. Smaller potholes can be found in both canyons.

CATALINA HIGHWAY TO MT. LEMMON

Visible from Tucson as a rounded granite dome, Mt. Lemmon rises to 9157 feet. Weather up there is 20 degrees cooler than in the desert around Tucson. Rainfall is higher, and a conifer forest darkens the summit and upper slopes. In front of the dome is the rocky ridge of the Santa Catalina forerange, separated from the dome by a deep east-west cleft that to a great extent controls the direction of streams that pass through the forerange onto the desert floor. Together the dome and the forerange make up the Catalina part of the huge Rincon-Catalina metamorphic core complex, the largest in the Basin and Range region.

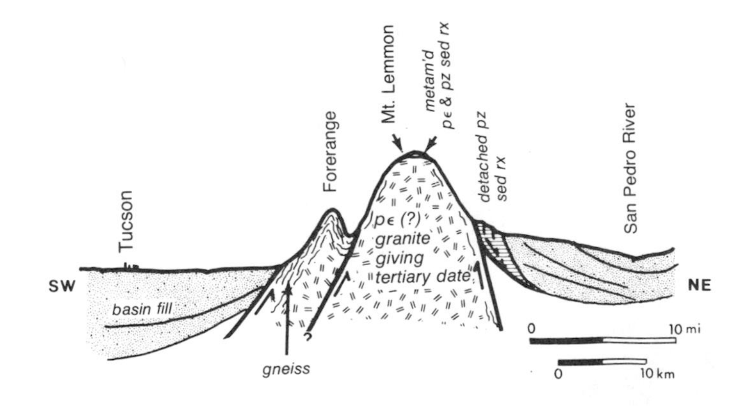

Section across the Santa Catalinas Mountains

The Catalina Highway, branching from East Tanque Verde Road, crosses the hidden fault at the base of the Santa Catalinas and climbs through banded dark and light gray Catalina Gneiss, well exposed in roadcuts and natural outcrops, that makes up the forerange. Some of these rocks may have formed from Precambrian granite, which appears in its original form only at the north end of the range. Others are banded in ways that suggest derivation from sedimentary rock layers that once overlay the granite. The forerange gneiss forms a steep anticline along the face of the mountains, with **dips** — the direction and angle at which the layers slope downward — expressed in layering or foliation.

As the highway climbs through the forerange the crest or axis of this anticline can be seen; beyond milepost 2 the dip is reversed. The gray gneiss, with dark and light veins, shows up well in cliffs visible from milepost 2 or mile 4-5.

The central core of the Catalinas is massive granite, probably Laramide in age but so altered by heat, pressure and movement during mountain-building stresses that its atomic calendar was reset to give a post-Laramide age. At the same time a faint layering was created, so that the rock could now be called a **gneissic granite**. The highway reaches some of the granite (we'll call it that) near milepost 6, just beyond Molino Basin Picnic Area, and then comes to its main mass, much less fractured, between miles 10 and 11. It is well exposed near another picnic area at mile 12-13. Chunky crystals of glassy quartz and feldspar, sparkly flakes of black and white mica, and occasional dark brown crystals of **garnet** make up this rock. In places it is banded by **pegmatite veins** containing larger crystals of the same minerals. Everywhere it is cut with several sets of joints; along some, crushed white sand shows that movement has taken place.

Along this road there are many fine examples of weathering processes that gradually turn the granite into loose sand. Along joints and faults where water can penetrate the rock, the rock begins to decompose, with mica crystals changing chemically to clay, and with

With a fanciful hand erosion has carved strange figures in granite along the Catalina Highway.

the bonding of quartz and feldspar crystals gradually loosening. In the fine fissures, water freezing on wintry nights forces the rock apart; on a small scale it separates individual crystals. Tree roots pry apart boulders, and plants release acids that further break down crystal bonds. Near and above mile 14 such weathering has carved strange pillars or **hoodoos** that tower above the road and march down the mountainside.

Below the ground surface, similar weathering takes place, and chemical decomposition turns the granite into loose sand. Rounded, half-buried knobs are very typical of weathered granite terrains.

On the north side of the Catalinas, and visible from the vista point at milepost 18, are dipping Precambrian and Palcozoic sedimentary rocks fringing the north side of the mountain core. Elsewhere, particularly near the summit, these rocks were caught up in the metamorphism that affected the core of the range. Near Summerhaven and along the summit ridge from mile 20-25 some of the same rocks, severely folded and metamorphosed, have been recognized as part of the carapace of the metamorphic core complex. The northern part of the summit is cut by many dikes, some of dark **diabase**, some of lighter, coarsely crystalline pegmatite.

Many small streams begin life as springs and little ponds up here near the top of the mountain, where rainfall is several times heavier than on the desert below. Carving V-shaped canyons down through the granite, the streams join others that flow parallel to the mountain front, behind the forerange, between the granite and its gneiss carapace. They then cut through the forerange as Soldier Creek, Sabino Creek, Bear Creek, and other delightful canyons.

TUCSON MOUNTAINS

(see also Saguaro National Monument in Chapter V)

This small range is composed of Tertiary intrusive and volcanic rocks bordered by faulted, folded Paleozoic and Cretaceous sedimentary rock. Flat-lying basalt flows make up "A" Mountain, while Safford Peak at the north end of the range is a **volcanic neck**. Cat Mountain Rhyolite — partly volcanic ash hardened into tuff — forms Cat Mountain and other parts of the range near Ajo Road (AZ 86). Dikes cut across and along the range. Small, light-colored hills on both sides of the range are Paleozoic limestone. The southern part of the range is a confused mixture of lava flows, layers and patches of volcanic ash, volcanic breccia, and blocks of Paleozoic limestone — such a hodge-podge that it is defined as the Tucson Mountain Chaos!

On the west side of the Tucson Mountains, in the Avra Valley, the Arizona Sonora Desert Museum (and zoo) tells the tale of life on the desert; recently added earth science exhibits explain the geologic history of this part of Arizona. Nearby is the western section of Saguaro National Monument, as well as "Old Tucson," a one-time movie set maintained as an outdoor museum of the old west.

COLOSSAL CAVE

Dissolved in Mississippian rock — the Escabrosa Limestone — of a detached block of Paleozoic sedimentary rocks on the west side of the Rincon Mountains, Colossal Cave is a typical small solution cavern, well decorated with stalactities, stalagmites, and other cave ornaments.

From the parking lot near the entrance to the cave, one can look out at this **detachment**, which with others makes up the foothills of the Rincons. Above the cave entrance massive layers of Escabrosa Limestone form a steep slope; younger, thinner limestone layers lie above, and parts of the sequence are repeated below, with large and small folds complicating the picture. Whether they were dragged along the side of an uplift under the weight of a mile or more of overburden, or were moved into position by other means (we don't yet know the answer to this knotty problem), the detached blocks did not retain their shape but squeezed and folded, rumpled or broke, depending on their resistance to the stresses involved.

To the southwest and west, bordering a small valley, are reddish brown slopes of the Pantano Formation, parts of which are mined for clay. Granite over the hill to the north is Precambrian, whereas the Catalina Gneiss covering the main mountain mass has had its atomic calendar reset to about 30 million years.

Rincon Peak is one of two parallel anticline ridges (shown on the map of I-10 Wilcox to Tucson), with the foliation of the gneiss arching up over them. A syncline comes between them at Rincon Valley.

Colossal Cave itself, limited in extent by the size of the block of Paleozoic sedimentary rocks, was dissolved by groundwater seeping through joints in the Escabrosa Limestone, probably early in Pliocene time. (The tour guide may tell a different story.) Groundwater solution occurred most readily right at the surface of the groundwater, where acidified rain water trickling through joints in the limestone most readily dissolved it. As the land rose, lower levels of the cave were excavated. Sometime after solution of the cave, the water drained away. From then on, rainwater trickling through

In limestone caverns like this one in the Santa Catalina Mountains, water dripping from the ceiling builds slender, icicle-like stalactites *and stubbier* stalagmites. Flowstone *(background) forms as water seeps across cavern walls.* Tad Nichols photo.

overlying limestones, dissolving calcium carbonate from them, began to decorate the cave. **Dripstone** and **flowstone** ornaments built up where this trickling water came into the cave, dripping from the ceiling or flowing thinly down the walls. Drop by drop, crystal by crystal, as calcium carbonate came out of solution, **stalactites** formed — thin "soda straws" at first, later thickening and building up from the outside. Sturdy, splatter-topped **stalagmites** below them grew as water dripped from the stalactites. Occasionally stalactite and stalagmite met, forming a **column**. Thin **draperies** built where flowing water trickled down the walls.

With increasing aridity at the end of the Pleistocene rainy spells, the cave dried up. No water drops or flows there now, and the ornaments bear a thin coating of dust.

As you leave the cave, look southward from the portico to the high, flat level between the distant mountains. This level, too, is Pleistocene — an old gravel and sand deposit that partly surrounds the Santa Rita, Empire, and Whetstone Mountains. It is discussed under AZ 82-83.

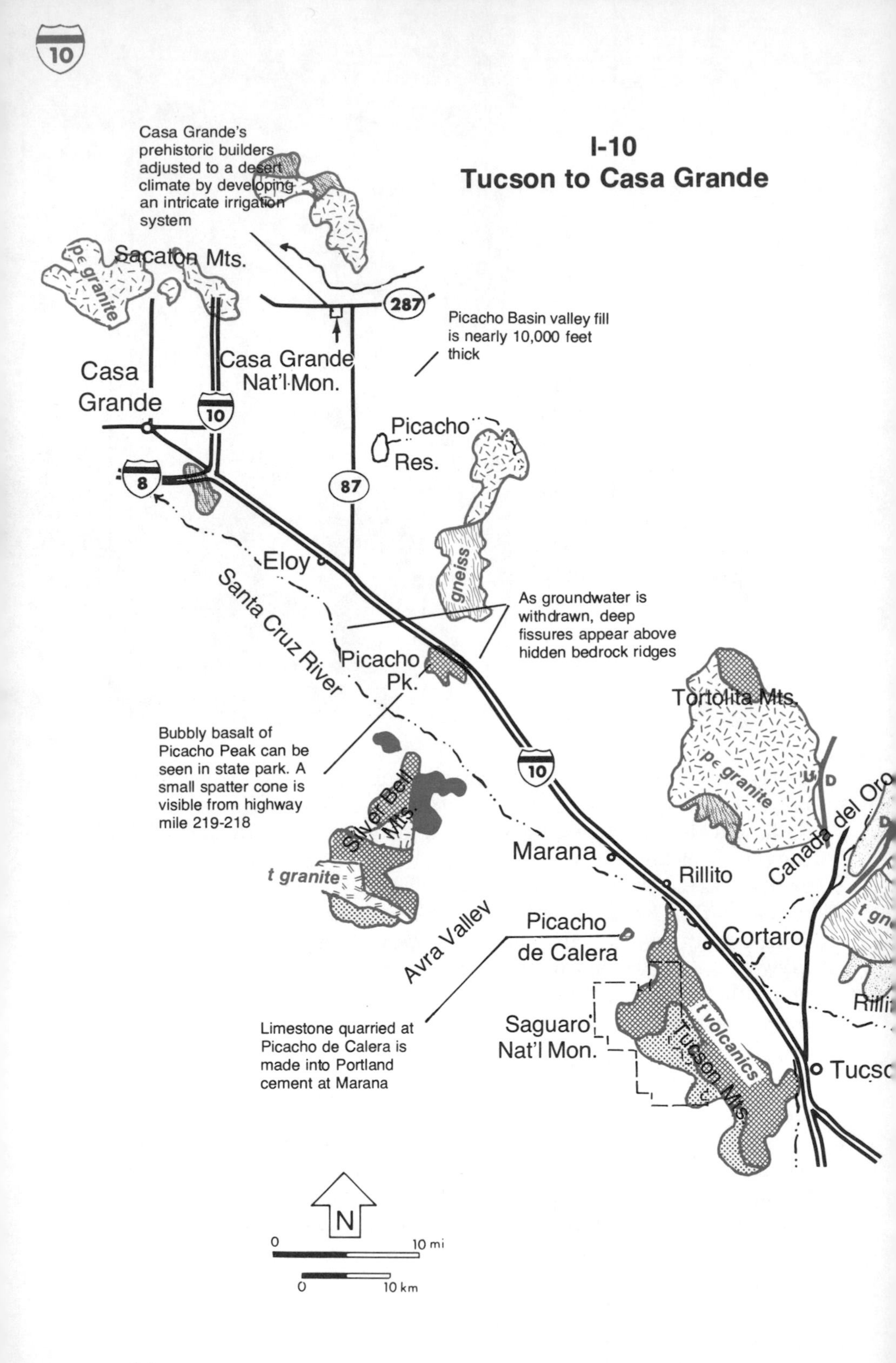
10
I-10
Tucson to Casa Grande
Casa Grande's prehistoric builders adjusted to a desert climate by developing an intricate irrigation system
Sacaton Mts.
pє granite
287
Picacho Basin valley fill is nearly 10,000 feet thick
Casa Grande
Casa Grande Nat'l Mon.
10
Picacho Res.
8
87
gneiss
Eloy
Santa Cruz River
As groundwater is withdrawn, deep fissures appear above hidden bedrock ridges
Picacho Pk.
Tortolita Mts.
Bubbly basalt of Picacho Peak can be seen in state park. A small spatter cone is visible from highway mile 219-218
10
pє granite
U D
Silver Bell Mts.
Canada del Oro
Marana
t granite
Rillito
Avra Valley
Picacho de Calera
Cortaro
Limestone quarried at Picacho de Calera is made into Portland cement at Marana
Saguaro Nat'l Mon.
t volcanics
Tucson Mts.
N
0
10 mi
0
10 km

Nourished by water from the Salt River, green fields supplant desert near Picacho Peak.

Interstate 10
Tucson — Casa Grande
(59 miles)

Northwest of Tucson I-10 follows fairly closely the course of the Santa Cruz River. Quite dry most of the time, or with a diminutive trickle across the sand, this desert wash "drains" the western part of the Tucson Basin. The eastern part is drained by Pantano Wash and the Rillito, the latter flowing into the Santa Cruz near mile 251. Another stream, Canada del Oro (Canyon of Gold!) drains the northwest side of the Santa Catalina Mountains. Gold *can* be washed from its sands, but adequate stream flow is infrequent. During or after heavy rains these streams may flow bank to bank; even so, the water is likely to sink into porous sands of the Tucson and Avra Valleys, rarely if ever reaching the Gila River.

Looking back at the Santa Catalina Mountains from about mile 248, you can see the sloping sides of the forerange anticline, fronting the great dome of granite that makes up the highest part of the range, the core of the metamorphic core complex. In the other direction are clustered small intrusions and lava flows of the Tucson Mountains. Both these ranges are described in the preceding section. The prominent peaks in this northern part of the Tucsons are small intrusions, probably the hardened contents of volcanic conduits.

North of the highway and northwest of the Catalinas rise the Tortolitas, geologically a single large mass of light-colored granite. In the Tortolitas as in most fault block ranges the edges of the upfaulted blocks have eroded back, creating a pediment flush with surrounding valley deposits. Looking between the Tortolitas and the Catalinas you can see a high platform of Tertiary sediments that fills much of the downfaulted graben between the two ranges.

An industrial complex at Rillito manufactures cement from Paleozoic limestone quarried from two small hills known as Twin Peaks or Picacho de Calera. One of several sets of Twin Peaks in Arizona, these are about 3 miles southwest of Rillito. Much farther south along the Avra Valley are the Baboquivari and Quinlan Mountains, their northernmost summit bearing the white observatory towers of Kitt Peak National Observatory.

Silver Bell Mountains west of Marana draw their name from Silver Bell Mine, an old underground mine now reactivated as an open pit. The little range is an odd mixture of Tertiary and Quaternary volcanic rocks around both Precambrian and Tertiary granite cores.

Marana is an agricultural center; in every direction from it are cotton and alfalfa fields irrigated with well water drawn from deep **aquifers**, water-bearing layers of sand and gravel.

In the broad desert valleys of the Basin and Range Province, wind erosion is every bit as important as erosion by water. Blowing winds lift immense amounts of soil skyward, especially where the desert surface has been plowed for agricultural use. Dust devils, whirlwinds that create tall, slender columns of dust, scour the surface and transport sand and even gravel for sizeable distances. The soil here is tan, without the darkening vegetal matter that characterizes soils of wetter climates. In places the surface of plowed and planted soil is coated with white alkali salts from water used for irrigation. High mineral concentrations retard or prevent plant growth, and eventually soils may become too highly mineralized for further use.

Except for an enigmatic block of granite near its summit, Picacho Peak is composed of volcanic rock.

Courtesy of Picacho State Park

To the north now, projecting against the sky and looking very much like a volcanic neck, rises the prominent "Ship of the Desert," Picacho Peak. Recent studies show that it is not a volcanic neck at all, but the faulted, tilted and eroded remains of a sequence of lava flows. The rest of the sequence, faulted and separated from this part, lies under several thousand feet of valley fill. The summit of Picacho, however, contains a single large block of Precambrian granite that must have been ripped from the wall of a conduit and carried up toward the surface.

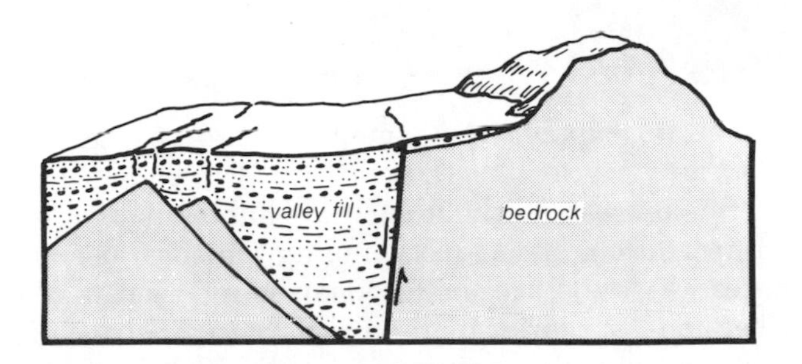

When withdrawal of groundwater causes subsidence, earth fissures *develop over buried ridges and the edges of pediments where, because the valley fill is thinner, total compaction is not so great.*

Light-colored hills directly across the highway from Picacho are the Picacho Mountains, their southern end gneiss, their northern end granite. Like that of the Tortolitas their granite is similar to that of the northwest dome of the Santa Catalinas.

In the Picacho area rapid increases in groundwater pumping during the last 30 years have caused severe lowering of the water table. The Santa Cruz Valley as a whole has been sinking as water is withdrawn, and near Picacho and the Picacho Mountains many deep, irregular cracks — a result of this sinking — cross the desert floor. Because the fissures have all developed since 1951, when deep irrigation wells began to appear, they are blamed on compaction of valley sediments as water is removed. The fissures develop above bedrock ridges (located by precise gravity measurements) because the degree of compaction is proportional to the thickness of the sediments. The cracked zone crosses both the railroad and the highway, necessitating frequent repairs.

Groundwater is now being pumped from southern Arizona valleys at rates far exceeding the rates of normal, natural recharge. Depletion in the Tucson and Phoenix Basins is four times as great as natural recharge. Geologists call this process — taking water out faster than it is replenished — "mining" the water, depleting the

Earth fissues such as this are caused by man-induced lowering of the groundwater surface.

T. L. Holzer photo, courtesy of USGS.

total remaining supply. Only a major climate change, for instance a **pluvial** (rainy) cycle like those in Ice Age time, could naturally reverse this process. Without such a climate change, conservation, water importation, and reduction of agricultural use are necessary alternates. As wells are drilled deeper and deeper the irrigation water becomes, of course, more and more expensive. It is getting to the point now where pumping water to farm the desert is no longer cost-effective. Some farms have already been abandoned.

Picacho Basin north of the Picacho Mountains is, like other Basin and Range valleys, underlain by several thousand feet of sand and gravel, and by thick layers of salt and other evaporites. As their collective name implies, such evaporites form by evaporation, here in saline lakes rather than from the sea. Almost all desert river water is somewhat salty, and if it flows into undrained basins in a dry climate, such as that which has prevailed here off and on for several million years, the basins become just so many giant evaporating pans. It's clear that this and a number of other basins once held briny lakes like Great Salt Lake in Utah and the many smaller saline lakes in Utah, Nevada, and the southern desert of Calfornia.

The history of irrigation in the Casa Grande Valley is a long one. The Hohokam people who lived here long before the coming of the white man — the people who built the Casa Grande, now a national monument — developed an intricate irrigation system to water their corn, beans, and cotton. However, they developed no water storage dams and dug no wells. They lived with vagaries of weather and climate and eventually had to give up their desert home, probably because their agricultural lands became waterlogged and unproductive.

At mile 199, I-8 branches off I-10 to the west. The town of Casa Grande is in the angle between the highways, several miles from either interstate.

Interstate 10
Casa Grande — Tonopah
(100 miles)

The broad valley that contains both the town and the ruins of Casa Grande is one of Arizona's richest agricultural areas. Nourished by a spiderweb of irrigation canals bringing water from the Salt River, as well as by subsurface water, farms raise alfalfa, cotton, barley and other grains, and garden crops.

Now a national monument, the Casa Grande ruins — site of earlier agricultural endeavors — can be reached from exit 185. The prehistoric apartment house, built by agricultural people known today as the Hohokam, rises above fields then planted to four staple crops: corn, beans, squash, and cotton.

Blowing dust can be a hazard along this stretch of highway. No doubt dust storms troubled the Hohokam people long before the coming of the white man. Much of the desert's surface dust, as well as fine sand, had already been removed by the wind, leaving a moderately stable, pebbly desert surface — desert pavement — which itself protected finer soil beneath. But man's activities — farming, ranching, building, and driving vehicles (particularly in the last few years off-the-road vehicles) — have destroyed or plowed under the protective pebble pavement, and millions of tons of newly loosened topsoil, including the part that characteristically holds soil moisture and nutrients, can be blown away by a single violent windstorm.

Between mileposts 190 and 180, the highway goes through a small granite range, the Sacaton Mountains, and past Santan Mountain, partly granite, partly Tertiary volcanic rocks, and partly gneiss. Between the two ranges flows the Gila River. Coming from headwaters in western New Mexico, the Gila crosses the state of Arizona to join the Colorado River near Yuma. Upstream it has carved a broad, fertile pathway that parallels the edge of Arizona's Central Highlands. Here in the Casa Grande and Phoenix Basins much of its water goes for agricultural use. Downstream, both natural and man-made dams further retard its progress, but by and large its course is con-

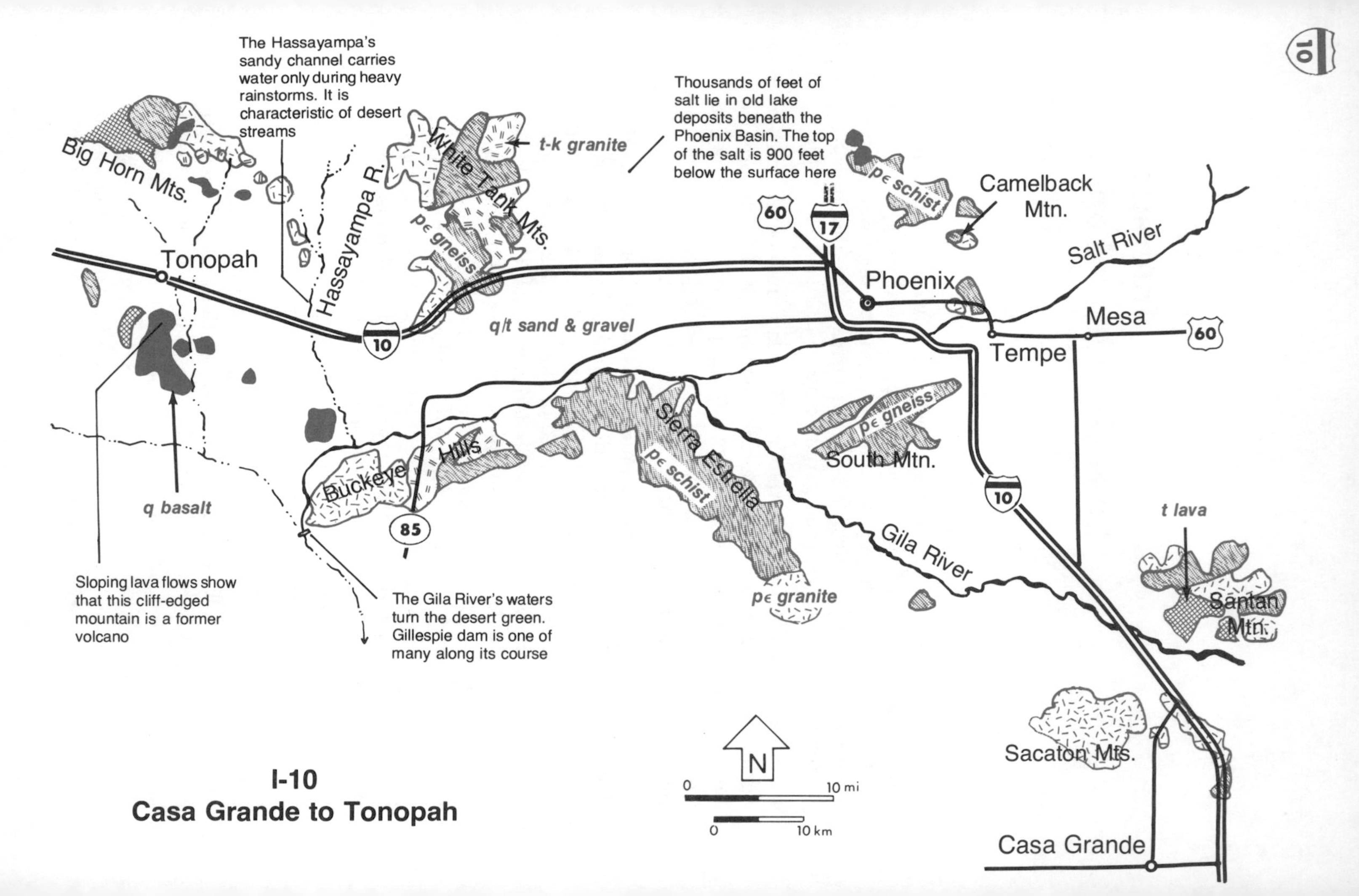

I-10
Casa Grande to Tonopah

trolled by a NE-SW-trending **graben** or downfaulted block that guides it directly across the NW-SE grain of the surrounding basins and mountain ranges.

Phoenix Basin, like the Casa Grande Valley, is almost flat. Deep sediments beneath it include lakebeds. Both southeast and west of Phoenix the valley fill also includes thick salt deposits formed in saline lakes.

South Mountain, northwest of mile 160, is a small but typical metamorphic core complex, centered with light-colored granite and surfaced with a sheared, arched carapace of metamorphic rock.

The highway crosses the Salt River at mile 151-150. Though dammed upstream, this river brings disaster as well as irrigation benefits to this part of Arizona. Major floods in 1978 and 1980 inflicted heavy damage in the Salt River Valley and the Phoenix agricultural and metropolitan areas. The storm that led to the 1978 flood is considered the most costly in Arizona's history, with homes destroyed, roads and bridges heavily damaged, and 12 fatalities. The 1980 storm closed eight of the area's ten bridges of the Salt River. Water from several dams farther up the river and from deep wells irrigates citrus orchards, cotton and bean fields, and other croplands in the Phoenix area. Pumping of groundwater in the Salt River Valley has led to land subsidence and fissuring.

It's quite apparent that in the deserts of Arizona, water tells people where to live and where to farm. The abundant waters of the Salt River dictated the location and prosperity of Arizona's largest city and the communities that surround it — Mesa, Tempe, and others. But growth of the great metropolitan area, which now contains half the state's population, is outstripping the available water supply — both the river water and water pumped from wells. Eighty per cent of this water goes to farming, even though much farmland has been or is

Irrigated land around Phoenix is nourished by the Salt River, dammed upstream to furnish irrigation water and hydroelectric power.
W. T. Lee photo, courtesy of USGS.

being converted to urban use. However, the high cost of energy for pumping water has led to abandonment of some farmland. Now well along in construction, the Central Arizona Project will bring Colorado River water to this area.

At this writing, Interstate 10 was not completed within western Phoenix. Its route turns west at about mile 145, across the Salt River Valley toward the White Tank Mountains and Buckeye Hills. West of Phoenix near Luke Air Force Base, the top of a huge mass of salt (actually **gypsum**, **anhydrite**, and **halite**) rises to within about 900 feet of the surface. This salt accumulated in salt lakes soon after Basin and Range mountain-building, and later was covered with sand and gravel brought in by streams and rivers. It is now being mined commercially.

Sierra Estrella, south of the highway about 10 miles away, is a long northwest-southeast ridge of Precambrian gneiss, schist, and granitic rocks. The same rock types make up the eastern end of Buckeye Hills.

The White Tank Mountains north of mile 115 make up a more typical metamorphic core complex, with Precambrian gneiss and granite forming their core and their western slope, and with Tertiary gneiss (or Precambrian rock altered to gneiss that comes up with a Tertiary radiometric date) on the eastern side. The embankment between these mountains and the highway is part of a flood-retarding system designed to collect and store water from desert downpours. This concept is fairly new in Arizona, and embankments in the White Tanks area are the pilot project for others east of Phoenix and farther west near the Harquahala Mountains.

Well south of the interstate, and south also of the Gila River, the Buckeye Hills make up a less well known, less well developed or more deeply eroded metamorphic core complex. Dark hills farther west, nearer to Tonopah, are made up of Tertiary volcanic rocks and a few horizontal Quaternary lava flows. Because a nuclear reactor is sited southeast of Tonopah, the geology of this region has been well and thoroughly studied!

On the arid wasteland of the Tonopah desert the creosote or greasewood bush is almost the only plant that will grow. And it has a few things to tell us about the amount of soil moisture it finds here. These bushes, well adapted to life on the desert, secrete a poisonous substance from their roots, preventing the growth of other plants — of their own or another species — close enough to steal their water. The less the rainfall, the farther the roots of each plant must go to find nourishment, and the farther out the ring of poison. So the farther apart the bushes are on the desert, the less the rainfall. Where they

Ephemeral streams trace irregular dendritic patterns across the desert floor near Phoenix. Vegetation is closely spaced along stream channels (washes), sparse on arid flats.
Tad Nichols photo.

grow near the road the shrubs are large and much closer together than on the desert flat, for they enjoy the benefit of runoff from the highway. In late March and April the highway may be edged with wildflowers sharing this runoff.

Along desert washes, including Hassayampa Wash at mile 104, other plants find enough soil moisture to grow; palo verde, mesquite, and varieties of cactus.

Interstate 10
Tonopah to Ehrenberg
(94 miles)

The dark color of the Big Horn Mountains northwest of Tonopah is distinct from the much lighter tones (despite the desert varnish) of granite ranges farther east. The Big Horns mark the northeast limit of a region where rocks lifted in mountain blocks are largely volcanic. True, the volcanic rocks may rest on Precambrian gneiss, schist, and granite that peek out at the margins of some of the ranges, but these are the exceptions rather than the rule.

Most of the volcanic rocks in west central Arizona have been radiometrically dated as early Tertiary. Short, sloping flows may be interspersed with breccia and light tan layers of volcanic ash such as those which slope away from the central **volcanic neck** or **plug** that

I—10
Tonopah to Ehrenberg

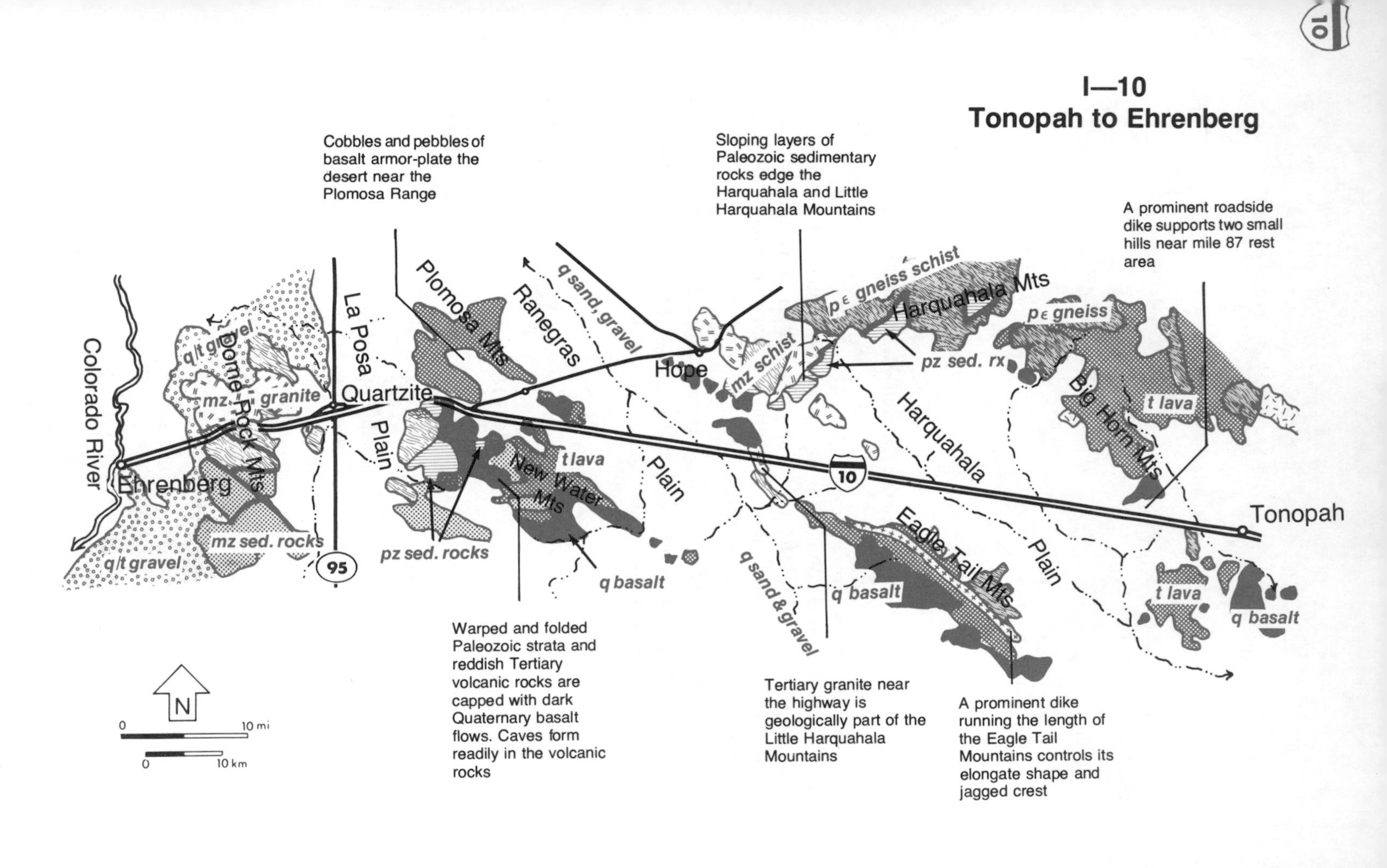

projects above the Big Horn Mountains. Many do not show such well established relationships with their conduits. Some younger volcanic rocks, mostly basalt, appear both as lava flows and as dikes.

The Harquahala Plain west of the Big Horns is another of the broad intermountain basins in this open, less crowded part of the Basin and Range Province, where ranges are smaller and farther from each other. The desert here is surfaced with desert pavement caused by and yet giving a measure of protection from erosion by wind and rain. In warm weather, dust devils spiral skyward, removing what remaining dust they can find. There are not many stream courses in this flat land. Creosote bushes, their root systems poisoning any rivals, are widely spaced except at the edge of the highway.

On the west side of the Harquahala Plain are some flood control embankments which hold back floodwater draining from the nearby mountains. As you continue westward you'll see similar embankments bordering canals of the Central Arizona Project, a diversion system that is to bring Colorado River water to the metropolitan and agricultural areas of central Arizona.

The Harquahala Mountains north of the highway, and north also of the volcanic ranges, trend SW-NE across the upper end of the Harquahala Plain. Like the Harcuvar, Buckskin, and Rawhide Mountains farther northwest, they run across the general "grain" of this part of Arizona. These transverse ranges all fit into the metamorphic core complex pattern. All are cored with Precambrian gneiss, schist, or granite; most are associated with Tertiary intrusions; some are flanked with Paleozoic or Mesozoic sedimentary rocks. The Little Harquahala Mountains contain major thrust faults that shuffle their Paleozoic, Mesozoic, and Precambrian rocks.

South of the highway and converging with it are the Eagle Tail Mountains — again in the volcanic category — with a prominent dike running through them lengthwise like a jagged backbone. The highway crosses a little Tertiary intrusion at the northwest end of the range, in sight of flat-lying Tertiary lava flows, and then drops into another desert valley, the Ranegras Plain.

Both the New Water Mountains and the Plomosa Mountains are, like the Eagle Tails, mostly volcanic. Lava and ash flows in these ranges are mid-Tertiary. As a rule of thumb, the ages of the volcanic rocks can be distinguished by their color and their position or **attitude**. If they are reddish in color and tilted steeply, they are older than 17 million years; if they are black and horizontal they are younger than about 15 million years. Also, in these mountains, the reddish Tertiary volcanic rocks show the irregular flow structure of thick, viscous types of lava such as rhyolite. In contrast, the thin, parallel, gray or black flows form from fluid basalt lava. Many black basalt boulders are coated with white caliche deposited by evaporation of the soil water.

Other rocks occur here, too; patches of Paleozoic sedimentary rocks similar to some of those in Grand Canyon. They include the strata that border the highway near the little town of Brenda.

La Posa Plain is the last intermountain valley before the valley of the Colorado River. Not far north of the highway there are thick layers of gravel deposited by the Colorado River before it incised its present channel. Some of these deposits contain vertebrate fossils.

Beyond La Posa Plain, between it and the Colorado River, is one more mountain range: the Dome Rock Mountains. Where I-10 crosses it the range consists of Mesozoic granite and volcanic rocks. A thick, south-tilted faulted sequence of Mesozoic sedimentary rocks — as much as 15,000 feet of them — occurs farther south. The sedimentary rocks record the onset of uplift and volcanic activity in an interesting way: Lower layers consist of conglomerate and sandstone containing claystone and limestone fragments clearly derived from sedimentary rocks; those higher up and therefore younger include sandstone with grains made largely of volcanic rocks, and conglomerate containing fragments of volcanic rock and granite bared by uplift and erosion. Mineral veins in the Dome Rock Mountains have been mined for gold, silver, copper, lead, zinc, molybdenum, and mercury.

West of the mountains I-10 crosses a wide two-level terrace of river deposits formed when the Colorado River carried much more rock material than it does now. The terrace gravels, seen to advantage near miles 4 and 3 where they have been dissected by tributary streams, date from Pliocene and Pleistocene time. Except where mountain ranges cut obliquely through them, these pebbly deposits border the river from Hoover Dam to Yuma, and contribute to our knowledge of the evolution of this greatest of southwestern rivers.

The Colorado is one of the most highly controlled rivers in the world, with portions of its waters assigned to Colorado, Utah, California, Arizona, Nevada, and Mexico. Upstream from Yuma a string of dams controls its flow, preventing floods and providing irrigation water, electricity, and recreation sites. The river no longer floods across its former floodplain, 20 feet or so above the present river level.

Picturesque San Xavier Mine produced copper ores from underground workings. Availability of heavy machinery and new processing techniques made large open pit mines more practical.
Tad Nichols photo

Interstate 19
Tucson — Nogales
106 kilometers (66 miles)

This highway is marked with metric units. One kilometer (km) equals 0.622 miles.

The southern end of the Tucson Mountains can be seen to advantage from this route. Here the range consists of faulted, tilted blocks of Cat Mountain Rhyolite, volcanic rock dated as less than 70 million years old. South of the intersection with Arizona 86, at kilometer 100, the hills near I-19 are composed of an odd, somewhat controversial rock unit named the Tucson Mountain Chaos — a plum pudding mixture of broken rock fragments and volcanic materials that may have resulted from explosive eruptions similar to the 1980 eruption of Mt. St. Helens.

From kilometer 92, San Xavier Mission shows up to the west against dark mid-Tertiary and younger volcanic rocks of Black

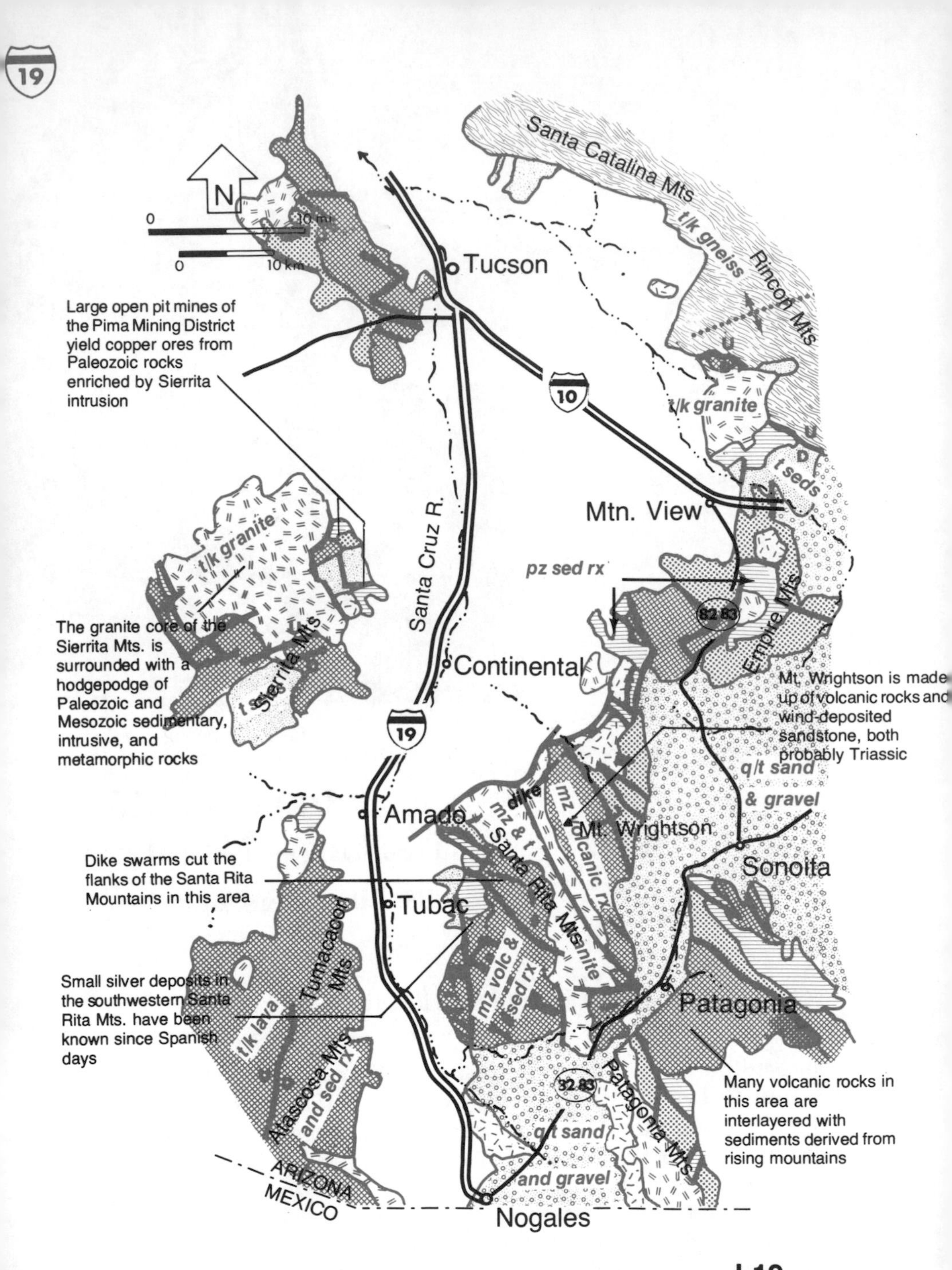

I-19
Tucson to Nogales
AZ 82/83
Nogales to I-10

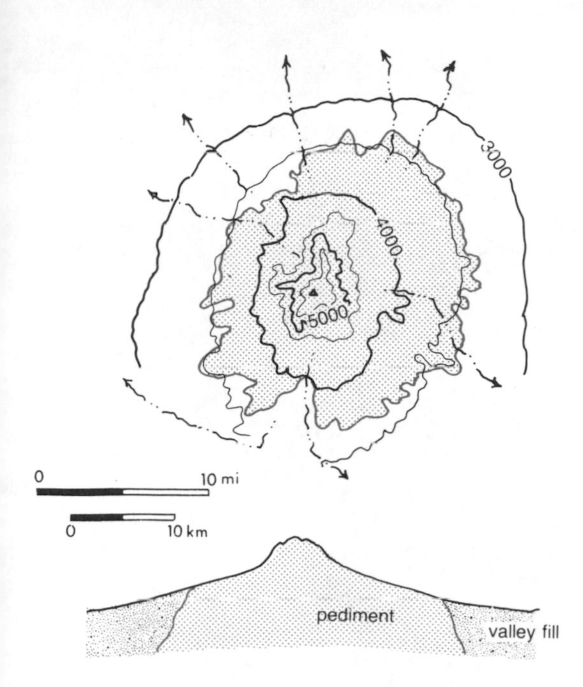

The Sierrita Range is surrounded by a broad, evenly sloping pediment that blends with no change in angle with surrounding valley fill. Drainage is radial. Large mine tailings east of the range now interrupt the symmetry shown here.

Mountain. Built between 1783 and 1797, this mission succeeds an earlier one constructed around 1700 by Jesuit priest Father Kino, one of the first white men to explore and describe this area.

The Santa Cruz River west of the highway flows north and then northwest toward the Gila River. Its present channel, cutting through soft, fine floodplain deposits, postdates 1880, when a cycle of **arroyo** or gully cutting began in the southwestern deserts. Prior to that time the river meandered broadly on its floodplain; early reports say that it flowed all year. The downcutting, which may or may not have been triggered by overgrazing, continues to this day.

The area south and east of San Xavier overlies the deepest part of the Tucson Basin, with about 7000 feet of valley fill.

Southwest of kilometers 90 to 80, dumps of several large open pit copper mines mark the slopes below the Sierrita Mountains. The Sierrita region — the Pima Mining District — is endowed with some of the largest low-grade copper deposits in the world. From 1908 to 1975 this district produced, in order of abundance:

copper	2,842,498 tons
zinc	more than 100,000 tons
molybdenum	91,592 tons
lead	43,587 tons
silver	1,260 tons
gold	2 tons

Open pit copper mines south of Tucson penetrate copper deposits associated with a Tertiary intrusion. Both the Rincons (background, right) and the Santa Catalina Mountains (left) are metamorphic core complexes. Tad Nichols photo.

The bulk of these metals came from great open pit mines that began operations in the 1960s near sites of earlier underground workings. The ore bodies fit into the pattern of **porphyry copper** deposits, where shallow Tertiary intrusions of granite porphyry penetrated, altered, and enriched Paleozoic limestone and quartzite. The principal ore mineral is chalcopyrite.

The granite intrusions, as well as the blocks of Precambrian, Paleozoic, and Mesozoic rocks, were lifted during Basin and Range faulting. Some of the sedimentary rocks may represent the upper slab of a broad thrust sheet now many miles northeast of its original position. The possibility of such thrusting leads as you might expect to a good deal of speculation as to the whereabouts of the original base of the intrusion and of the mineral deposits!

To see one of the big open pit mines, turn west on Duval Mine Road at exit 69, go west about 3 miles, and follow the signs to the visitor viewpoint. The depth of the pit can be calculated by counting the benches: each bench is 40 feet high. Trucks down in the pit have wheels ten feet in diameter and carry 75 to 130 tons of rock. The ore is greenish, and was originally covered by several hundred feet of overburden. Ore is carried out of the pit and to a crushing mill by conveyor belt. Crushed and ground to powder, it is mixed with water and chemical frothing agents that bring copper-bearing particles to the top of the mixture, where they are skimmed off. One ton of ore yields about 40 pounds of the skimmed-off concentrate, a third of which is copper. Molybdenum is recovered from the concentrate by a similar process, and then the concentrate is shipped to smelters for further processing.

To the east across the highway from these mines are the Santa Rita Mountains, with Mt. Wrightson (9453 feet) as their high point. Here as in many southern Arizona mountains the geologic pattern includes enigmatic thrust faults with slices of Paleozoic sedimentary rocks sitting astride or leaning up against a Precambrian core. The overthrust school subscribes to broad movement of a thin sheet of rocks from as much as 100 miles to the southwest. Thrust faulting in the Santa Ritas occurred 75 to 80 million years ago. Because their sedimentary sequence is relatively complete and only slightly deformed, these mountains contain more clues than most to the geologic history of the region. Both Paleozoic sedimentary strata and Precambrian core are intruded by Tertiary porphyry associated with scattered copper deposits.

An especially large alluvial fan spreads out below the Santa Rita Mountains. It is grooved by stream-cut channels, many of them the work of post 1880 arroyo cutting. Near the west end of the Santa Ritas, Elephant Head is part of an intrusion that occurred 68 million years ago. There were at least four episodes of intrusion in the Santa Ritas, and many of the dikes that were formed are visible from the rest stop near kilometer 31, in tilting ridges at the base of the mountains.

South of Green Valley the highway cuts through another alluvial fan and exposes the layered fine and coarse sandstone and gravel of which it is made. The layers are channeled and refilled in places. Watch for reddish soil zones covered over by younger gravel.

South of the Santa Ritas, again in sight from the kilometer 31 rest stop, is another high gravel surface. Deposits below it fill a narrow valley created by faulting; they are about 600 feet thick. The Patagonia Range farther south extends on beyond the Mexican border; this range is almost entirely Precambrian and Cretaceous granite.

The Tumacacori Mountains west of Tubac contain volcanic rocks similar in composition and appearance to those of the Tucson Mountains. Tubac lies near the site of a fortified Spanish presidio established in 1752. Between short-lived silver discoveries, farming, ranching, Apache raids, and nearby copper deposits, both the presidio and the town that succeeded it have led varied lives. Tumacacori Mission, a national monument, is farther south at kilometer 29.

The Atascosa and Pajarita Mountains just west of I-19 are volcanic southward extensions of the Tumacacori Mountains, with some intrusive and sedimentary rocks.

As the highway enters Nogales, note the large roadcut exposures of the Nogales Formation; in them, watch for faults and filled channels. A Spanish settlement originally, then a mining and railroad town, Nogales is now an important international port of entry and tourist center.

The log for Arizona 82 and 83 returns to Tucson around the other side of the Santa Rita Mountains, a cooler and geologically interesting route.

Ruby, west of Nogales, once prospered as a mining town. Ore lay along a contact between Paleozoic limestone and a Tertiary intrusion. Tad Nichols photo.

Interstate 40
Colorado River — Kingman
(43 miles)

Arizona's western boundary lies along the Colorado River, at midstream. This powerful river, stilled here by Parker Dam about 40 miles downstream, has its headwaters in the Rocky Mountains of Colorado and Wyoming. Its history and evolution are described in the introduction to Chapter IV.

For perhaps five million years — since Pliocene time — the river has carved its way through this western desert, gouging a channel through mountain blocks and gravel-filled troughs of the Basin and Range region. Because at the beginning of that time the mountain blocks were nearly drowned in a sea of gravel and sand, the river's course at first lay well above the present one. Easily carving a way through these still unconsolidated sediments, it later cut down also through the rocks of the hidden mountains, with its floodplain widening upstream from each dam-like range and narrowing again as the river encountered the resistant mountains themselves. You can see an example as you cross the bridge to Topock — the wide floodplain upstream, the narrow canyon downstream.

To the south, jutting above already rugged terrain, a cluster of volcanic spires represents magma that cooled and hardened in the conduits of now-eroded volcanoes. Mapped as Cretaceous, these rocks and the basalt that surrounds them may actually be early Tertiary, along with other newly dated volcanic rocks of western Arizona.

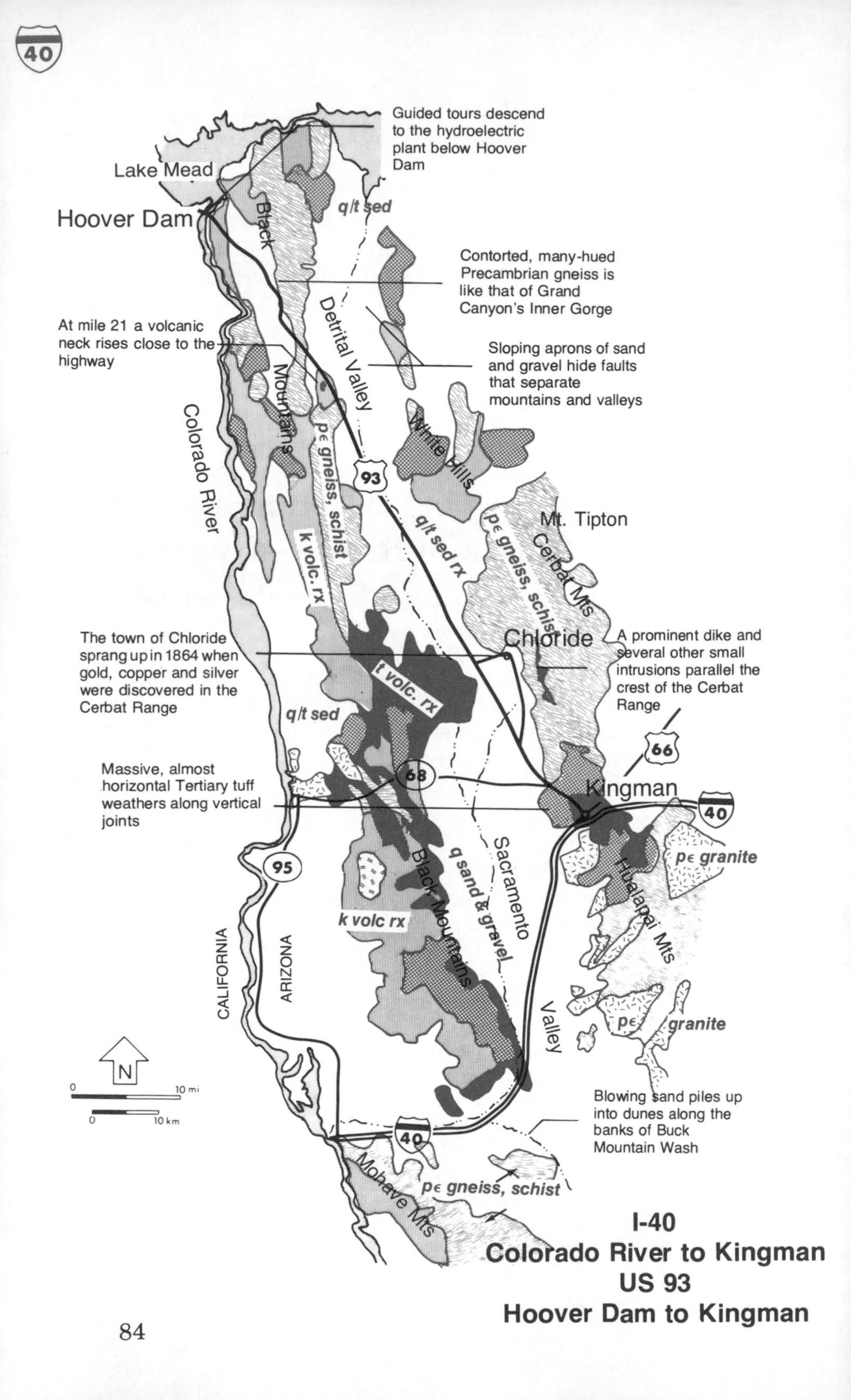
40
Lake Mead
Hoover Dam
Guided tours descend to the hydroelectric plant below Hoover Dam
Black Mountains
q/t sed
Detrital Valley
Contorted, many-hued Precambrian gneiss is like that of Grand Canyon's Inner Gorge
At mile 21 a volcanic neck rises close to the highway
Sloping aprons of sand and gravel hide faults that separate mountains and valleys
Colorado River
pє gneiss, schist
White Hills
93
k volc. rx
q/t sed rx
Mt. Tipton
pє gneiss, schist
Cerbat Mts
Chloride
The town of Chloride sprang up in 1864 when gold, copper and silver were discovered in the Cerbat Range
A prominent dike and several other small intrusions parallel the crest of the Cerbat Range
t volc. rx
q/t sed
66
Massive, almost horizontal Tertiary tuff weathers along vertical joints
68
Kingman
40
pє granite
95
Black Mountains
q sand & gravel
Sacramento Valley
Hualapai Mts
k volc rx
CALIFORNIA
ARIZONA
pє granite
N
0 10 mi
0 10 km
Blowing sand piles up into dunes along the banks of Buck Mountain Wash
40
Mohave Mts
pє gneiss, schist
I-40
Colorado River to Kingman
US 93
Hoover Dam to Kingman

The road rises immediately through the gravel of an alluvial apron that surrounds the north end of the Mohave Mountains. You will see the coarse gravel in roadcuts. At the surface, close-packed pebbles of desert pavement form a protective veneer. Wind, the desert's most active agent of erosion, has blown away the fine sand and silt, leaving only the pebbles that were once part of gravel layers.

North of this stretch of highway, at the south end of the Black Mountains, is a chaotic jumble of Tertiary and Cretaceous volcanic rocks — products of explosive volcanic eruptions. Above them are almost flat-lying lava flows less than 2 million years old. The prominent tooth-shaped peak is another volcanic neck.

Between mileposts 13 and 16 the interstate crosses Franconia Wash and Buck Mountain Wash, both usually dry. A few hardy trees and shrubs — mesquite, catclaw, and the green-barked palo verde — do seem to find enough moisture to grow in the seemingly dry sand and gravel. The desert is a precarious environment, and only superbly adapted vegetation can eke out an existence on it. Cactus, creosote bush, tall-branched ocotillo, and palo verde seem able to cope with the scarcity of water, each by a different method: cacti by storing water in their fleshy, thick-skinned stems; creosote bush by growing only tiny, waxy, water-conserving evergreen leaves; ocotillo by sprouting abundant but small and short-lived leaves after soaking rains; and palo verde by its slender, needle-like leaves and green, waxy bark.

Only lack of moisture makes the desert a desert. Note the more luxuriant vegetation right along the roadway, where extra rain water drains from the pavement. Newcomers are often surprised to see the many flowering plants that make the desert their home.

Rains when they come may be deluges so heavy that desert soil, surfaced with desert pavement, cannot absorb them. Washes like Franconia and Buck Mountain Washes may be swept by swollen, muddy floods that begin as the "wall of water" described in western literature. Such flash floods redistribute the sand of the watercourse or change the route of the stream, carry fresh loads of mud and sand down to a main stream (here the Colorado River), and then dry so rapidly that only the damp, smooth sand in the washes reveals that a storm has passed. A few days later new leaves sprout from desert plants, and the desert wears a soft new carpet of green.

Efficient downcutting by the Colorado River has lowered the level to which these and other tributary streams can erode, thereby strengthening their erosive power so that they in turn cut down through earlier surfaces. Three terrace levels along Buck Mountain Wash show that during alternating cycles of deposition and erosion it has three times cut down through its older deposits. Such cyclic behavior may relate to climate cycles, or it may reflect changes in Colorado River downcutting as natural barriers — lava dams or mountains of hard, resistant rock — were worn through.

Ahead at milepost 15 are the Hualapai Mountains, a range quite different in geologic style from the southern part of the Black Mountains. It contains far older rocks: Precambrian gneiss, schist, and granite. The irregular contact between dark gneiss and light-colored granite that intruded it shows clearly.

Here again sloping, stream-deposited aprons surround the desert range. Upper parts of the aprons are cut back into the upfaulted mountain block, then covered with a sparse mantle of sand and gravel. In contrast to the eroded upper pediments, lower parts of the aprons consist of hundreds, commonly thousands of feet of sand and gravel deposited by streams coming off the mountains. Some also contain fine lake deposits of sand, silt, limestone, and salt, showing that some valleys when first formed did not drain into through-flowing streams.

Most major changes on the gravel surfaces occur during and after sudden rainstorms, results of the "cloudburst climate" of the desert. But subtle changes are ongoing: Desert winds remove sand and silt, frost and rapid temperature changes crack rocks, desert varnish gradually darkens rocks and pebbles, and plant roots and burrowing animals loosen both rock and soil.

More clustered volcanic peaks mark the northwest skyline as I-40 turns north up Sacramento Valley. The Black Mountains extend north to Lake Mead, changing in that direction to Precambrian igneous and metamorphic rocks like those of the Hualapai Range. The Cerbat Range north of the Hualapais extends northward too, but not as far. The east dip of its Precambrian rocks is especially apparent from mile 42. Although this range is composed almost entirely of Precambrian gneiss, some Tertiary and Quaternary volcanic rocks partly fill a gap between it and the Hualapais. The highway swings through this gap into Kingman. There are many copper-silver-gold mines in the Cerbat Mountains, though most of them are inactive now.

The boundary between the Basin and Range Province and the Colorado Plateau is about 25 miles east of Kingman. It is really a three-province boundary, in that this is also the western tip of Arizona's Central Highlands.

In southern Arizona, broad valleys between the ranges are dotted with creosote bushes, admirably adapted to the desert climate.

U.S. 60/89
Wickenburg — Phoenix
(53 miles)

Flowing through Wickenburg, the Hassayampa River rises in Arizona's high, forested Central Highlands and flows south to join the Gila River west of Phoenix. The highway follows the river around the east end of the Vulture Mountains, a complex, much-faulted range composed of Precambrian and late Cretaceous granite (evidence of Precambrian and Laramide mountain building) surrounded by tilted Tertiary volcanic rocks (evidence of the Mid-Tertiary Orogeny). At the summit of the range are cliff-walled peaks of basalt, surviving remnants of a once vast lava plateau. All the rocks except the basalt are cut by faults and visible ridges of dark dikes.

The range as a whole trends northeast, parallel to its granite core. The faults and dike swarms trend roughly north-south, parallel to most nearby Basin and Range faulting.

Volcanic rocks at the east end of the range extend across the highway, showing up as pinkish volcanic tuff, massive volcanic breccia full of broken lava and tuff fragments, and thick, tilted beds of rhyolite lava. The same rocks appear in railroad cuts across the river.

US 60/89
Wickenburg to Phoenix

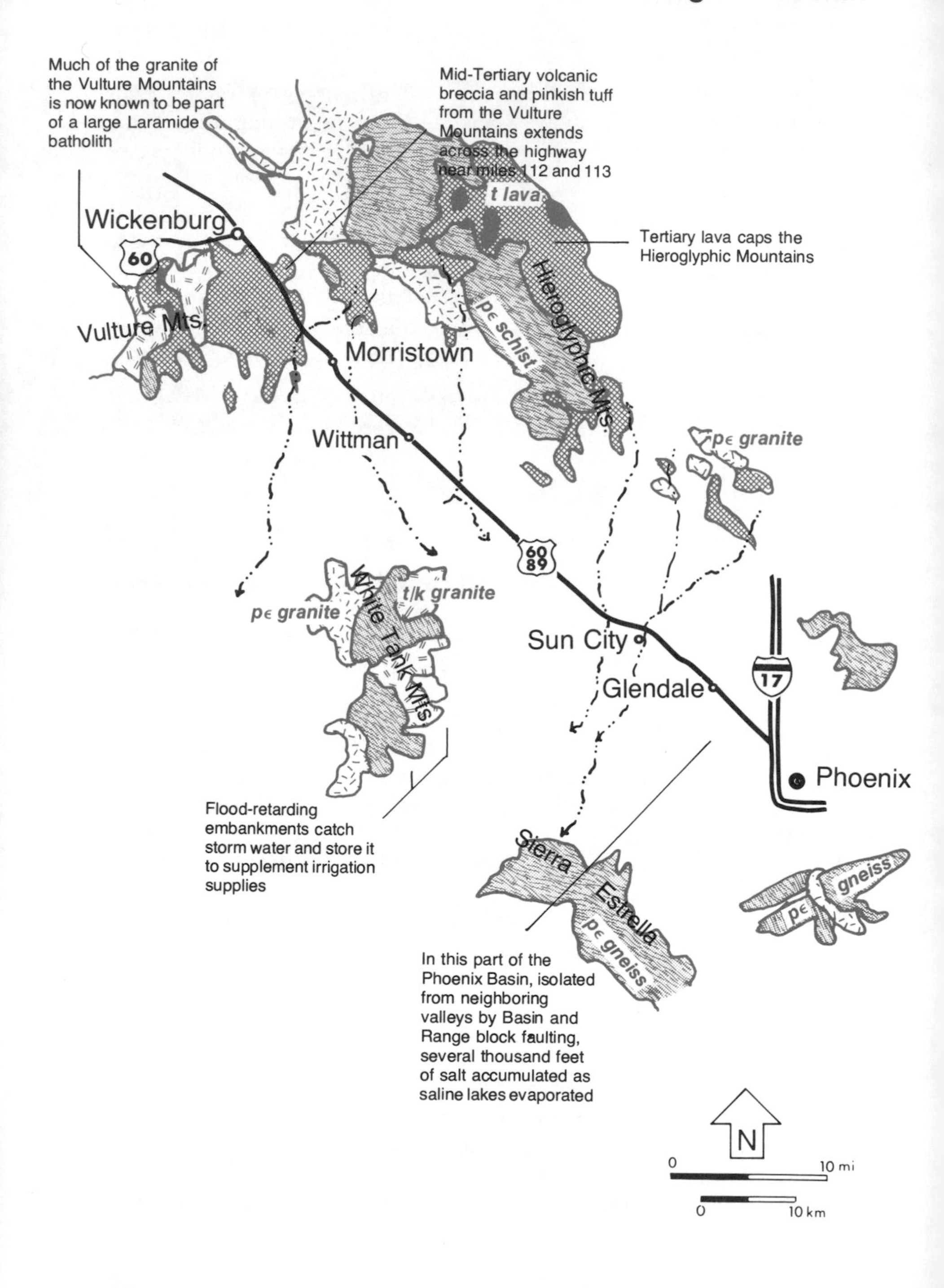

Where they are undisturbed by man, the rocks are brown with desert varnish. Many are blotched with light green **lichens** (as near milepost 117), each circular blotch a community of algae and fungi. Acid by-products of these primitive plants initiate the slow processes that turn rock into soil.

South of the mountains, on the flat, almost level floor of the Phoenix Basin, the highway straightens out for a long, direct run toward Phoenix. This basin is now irrigated with water from the Agua Fria and Salt Rivers. It serves as good evidence that only the lack of water makes the desert dry and barren. Add a little water and the fine, sandy soil becomes fertile and farmable. Apparently desert climates also prevailed here several million years ago, before the Ice Ages' cooler rainy cycles, for under parts of the basin are thick deposits of salt and gypsum-like anhydrite deposited in saline lakes similar to Great Salt Lake in Utah and the Salton Sea in California.

The highway more or less parallels the southwest edge of Arizona's Central Highlands (see Chapter III), represented between Morristown and Beardsley by the Hieroglyphic Mountains, a range composed largely of ancient Precambrian schist. From mile 132 to 133, far away ahead and to the left, rise Four Peaks in the Mazatzal Mountains, the largest range of the Central Highlands.

Southwest of miles 130-139 are the White Tank Mountains, a metamorphic core complex range with a core of Precambrian gneiss intruded by both Precambrian and Tertiary granite. Some of the Precambrian rocks were subjected to high enough temperatures to reset their radiometric dates to about 30 million years ago.

Beyond the White Tank Mountains desert vegetation gives way more and more to irrigated farmland, date and citrus orchards, and the burgeoning metropolis of Phoenix. For many years Phoenix and communities north of it drew on water from the Salt River for irrigation, industrial, and domestic use. In southern parts of the basin the Gila River furnished a seemingly unlimited water supply. As the population grew and farms, orchards, and fields spread farther through the basin, wells were drilled to supplement the surface supply. Now a critical stage has been reached: Need has outgrown available surface and well water, and wells must be drilled ever deeper to reach a declining groundwater table. More water will come from the Central Arizona Project, which taps the already much-tapped Colorado River. But Arizona is forced to look more and more toward conservation measures. As water becomes increasingly expensive, domestic and farm use are curtailed by the rising cost. Farmers and industrialists, with government advice and support, seek ways to improve efficiency of their water use. Croplands and crops are selected carefully to bring the largest return for the least water outlay. Some fields now lie idle, reverting to desert. The Soil Conservation Service is building experimental flood-retarding embankments that will catch and store water from heavy rains; one of these embankments partly surrounds the White Tank Mountains.

Phoenix and Vicinity

Near the confluence of the Gila and Salt Rivers, Phoenix is surrounded by the largest irrigated agricultural area in Arizona. Sunny days and warm winters insure all-year farming, and irrigation, largely with water from the Salt River and its tributary the Verde, has converted the desert into an emerald valley of lush fields and orchards. Valley sediments include deposits of freshwater and saline lakes.

Small mountains and rocky hills within the metro area reveal very little of the geology within the basin. The weathered granite, gneiss, and schist of South Mountain, dark brown with desert varnish, follow on a small scale the metamorphic core complex pattern. Good samples of its rocks can be seen in South Mountain Park. On Black Mountain similar Precambrian rocks are exposed. In Papago Park, coarse Tertiary conglomerate with large, angular granite boulders, cobbles, and pebbles weathers with many wind-carved alcoves and recesses. The head of the camel in Camelback Mountain is composed of Tertiary sediments — coarse gravel washed from the Central Highlands in Eocene and Oligocene time. The camel's hump, though, is Precambrian granite and schist. The whole sequence of Paleozoic and Mesozoic rocks that should come between the Tertiary head and the Precambrian hump is missing!

Camelback Mountain, composed of Precambrian and Tertiary rocks, gives us an idea of the immense amount of erosion that took place here prior to Tertiary deposition. This photo shows Camelback in 1902.
W.T. Lee photo, courtesy of USGS.

With Phoenix as base, there are a number of interesting geologic excursions:

• A loop trip to Tucson via Florence Junction and Oracle Junction (see US 60/89 Mesa to Florence Junction and US 89 Florence Junction to Tucson), returning via I-10.

• A loop to Globe via US 60, Payson via AZ 188, and return via AZ 87, a route that will introduce Arizona's Central Highlands.

• A visit to the mining towns of Superior, Miami, and Globe via US 60/89 and US 60, or a shorter trip just to the Superstition Mountains, where trails lead into the region of the legendary Lost Dutchman Mine.

• A scenic route on I-17 to Cordes Junction, to Prescott via AZ 69, and return via US 89 to Phoenix.

• The Apache Trail, AZ 88, a scenic adventure through volcanic rocks of the Superstition Mountains and granite of the Mazatzal Mountains.

Not much rain falls on the desert — less than 10 inches annually in the Phoenix area. About 95 per cent of the rain that does fall evaporates or is transpired, returned to the atmosphere, by plants. Only 5 per cent — less than half an inch — remains in the soil or recharges the groundwater supply.

Phoenix originally drew all of its water from the Salt and Verde Rivers. More than 1000 years ago the Salt River provided irrigation water for Hohokam Indian people, who dug large canals and worked out an intricate irrigation system for watering their beans, corn, squash, tobacco, and cotton. The Hohokam developed no water storage reservoirs, however.

More recently, the first white settlers refurbished and reused their predecessors' "ditches." In 1890 the first reservoir was constructed south of Phoenix near Picacho. Others followed in the 1930s and 1940s, on both the Salt and Gila Rivers. Water from the Colorado River along Arizona's western boundary, a source already heavily committed to serve cities and agricultural areas in California and Mexico, will soon be added to that used in the Phoenix region. More will come from storm catchment reservoirs now being constructed at strategic sites on the surrounding desert.

Water for the Phoenix area also comes from deep wells. However, since 1940, groundwater, the part of subsurface water that completely saturates all pore spaces of rock or rock material, has been used faster than it is naturally replaced. Each year wells must be drilled deeper to reach the dwindling supplies. Eventually recovery of groundwater will become so expensive that it will no longer be used for agricultural purposes.

Both river and well water contain, in this area, large amounts of dissolved mineral matter — chemical salts like calcium carbonate and common salt (sodium chloride). Both plants and evaporation selectively consume the water but not the mineral matter. More and more salts therefore become concentrated in the soil, until it loses its fertility and is no longer useful as farmland. In the desert, with or without irrigation, calcium carbonate and some other soluble salts accumulate as caliche. Any farmer or home gardener in the Phoenix area will describe the difficulties of digging through the crusty caliche layers.

Despite the seeming bounty of the Salt River Valley, despite the warm climate and sunshine, man has had to face several other geologic problems in living here. Termed **geologic hazards**, these problems relate to perfectly normal geologic processes that become geologic hazards only when man intrudes and does the wrong thing at the wrong place at the wrong time. By settling in the Phoenix area, stripping the broad basin of its normal vegetation, plowing the desert surface, man has aided and abetted the dust storms that several times each year sweep northward from Mexico. By building on the river floodplain he has invited flood damage, and has had to spend millions of dollars in efforts to control the seasonal fluctuations of

Clay soils form poor footage for pavement. Here swelling clays have pushed both asphalt and concrete out of line. Troy L. Péwé photo 4253, Jan. 19, 1979.

desert rivers. By withdrawing underground water to supplement his irrigation needs, lowering the water table, he causes the valley fill to compact and the land surface to subside and in places to open up in deep, wide earth cracks. In paving streets and sidewalks, in building homes and swimming pools and tennis courts on soft deposits of the valley floor, he risks property damage as near-surface clays, which swell when they become wet and shrink as they dry, disrupt what he has constructed. By cutting into and building on steep slopes — such few as there are in the Phoenix area — he toys with the danger of rock-slides and land slips as nature seeks to establish a more stable angle of slope.

With rockfalls and landslides nature seeks to restore slopes over-steepened by man. This rockslide occurred after several weeks of above-normal rainfall. Troy L. Péwé photo 4075, March 2, 1978.

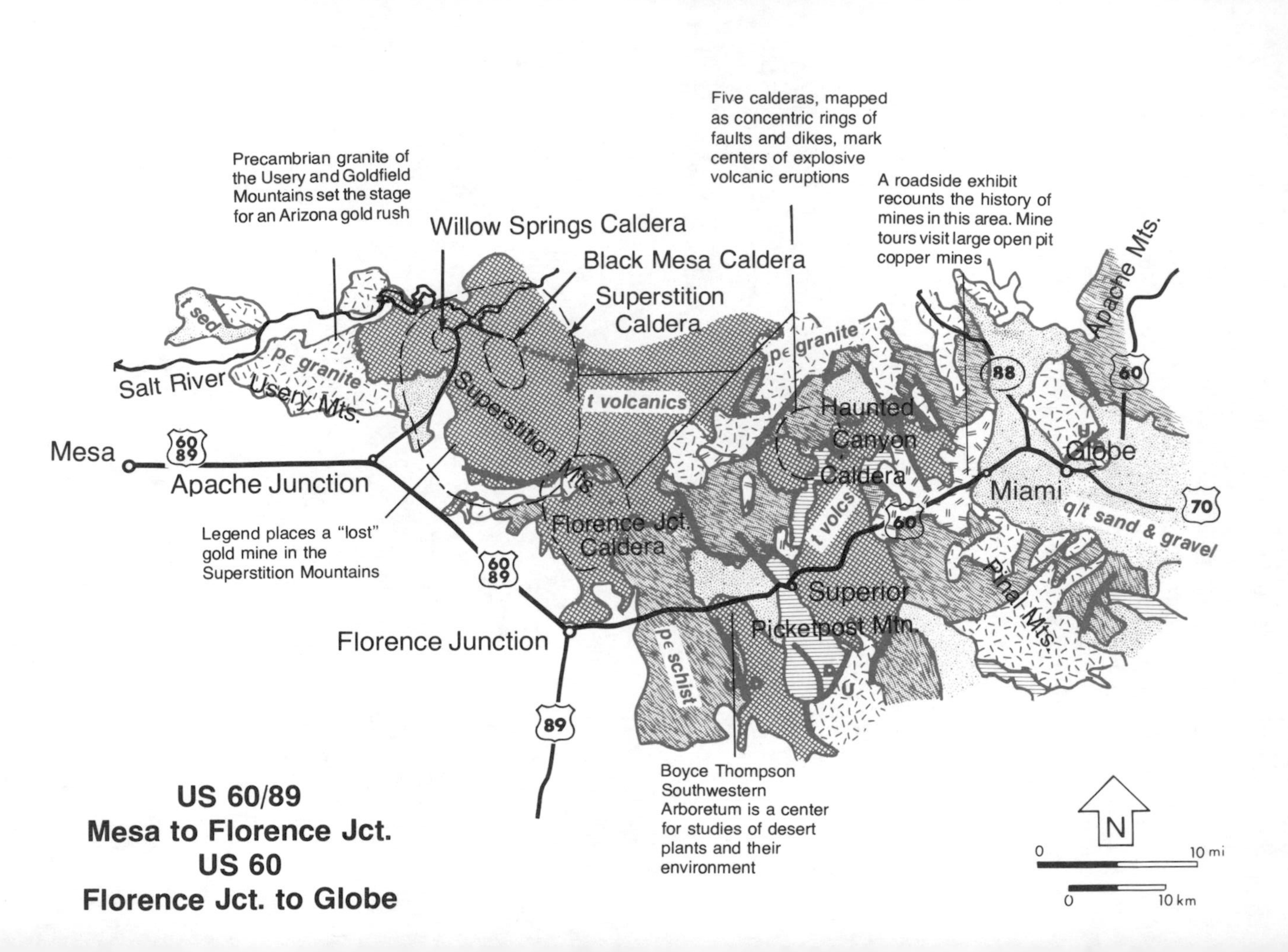
Precambrian granite of the Usery and Goldfield Mountains set the stage for an Arizona gold rush
Willow Springs Caldera
Black Mesa Caldera
Superstition Caldera
Five calderas, mapped as concentric rings of faults and dikes, mark centers of explosive volcanic eruptions
A roadside exhibit recounts the history of mines in this area. Mine tours visit large open pit copper mines
Apache Mts.
t sed
pє granite
Salt River
Usery Mts.
Superstition Mts.
t volcanics
pє granite
Haunted Canyon Caldera
88
60
Globe
Miami
70
q/t sand & gravel
Mesa
60 89
Apache Junction
Legend places a "lost" gold mine in the Superstition Mountains
Florence Jct. Caldera
t volcs
60
60 89
Superior
Pinal Mts.
Picketpost Mtn.
Florence Junction
pє schist
89
Boyce Thompson Southwestern Arboretum is a center for studies of desert plants and their environment
N
0
10 mi
0
10 km
US 60/89
Mesa to Florence Jct.
US 60
Florence Jct. to Globe

U.S. 60/89
Mesa — Florence Junction
(33 miles)

The most interesting geologic feature along the short distance between Mesa and Florence Junction is the Superstition Mountain Range. Composed almost exclusively of volcanic rocks erupted in mid-Tertiary time, 35 to 15 million years ago, these mountains were born in a volcanic frenzy that repeatedly swept thick palls of volcanic ash over this and other parts of southern Arizona, some of it so hot that it welded itself together where it fell. In a long-drawn-out series of such violent explosions, volcanoes in the Superstition Mountains emitted about 2500 cubic miles of ash and lava. Then the roofs of the partly emptied **magma chambers** collapsed, forming circular or oval **calderas**.

In the Superstition volcanic field, five partly overlapping calderas have been identified. Many more may be hidden beneath the thick layers of ash — now tuff and welded tuff — that these volcanoes emitted. Though the calderas are no longer high-rimmed, deep-floored, circular basins that can be identified at a glance, they can be recognized by careful mapping as circular or oval areas outlined by rings of faults and containing broken volcanic rock (**breccia**), with other features normally associated with calderas.

The largest is the Superstition caldera. In part it is overprinted by three others that seem to lie along the line of a NW-SE fracture cutting the earlier, larger caldera, which may have originated along a parent fracture. Its precursors, destroyed by the eruption, were a ring of volcanoes, mountains several thousand feet high. After the eruption, after collapse of the big caldera and before formation of Black Mesa and Willow Springs calderas, a central upthrust of thick, doughlike lava created a **resurgent dome**. This dome, an exceptionally large one as such domes go, now makes up most of the Superstition Mountains. The craggy northwestern prow of the range consists of two earlier domes, parts of the initial volcanic ring.

An alternate route to Globe, AZ 88 (the Apache Trail) branches northeast from Apache Junction. Though scenically and geologically interesting it is not described in detail here.

Other parts of the mountains are visible from US 60 southeast of Apache Junction. Thick layers of tuff stretching south from the resurgent dome now lie in a large syncline higher at its north end because of tilting during Basin and Range block faulting. Large alluvial fans below narrow canyons indicate the youthfulness of the range.

Many legends are told about the Superstitions. The most appealing is that of the Lost Dutchman Gold Mine. It's the mine that is lost, not the Dutchman. One Jacob Walz, a German (not Dutch) miner, apparently entered these mountains from time to time, returning after each trip with his burro laden with rich gold ore. As far as we know, no such ore occurs in the Superstitions. Though Walz was real, some think the story is pure fiction; others give Walz credit for "high-grading" the ore piece by piece from mines where he had been employed, then hiding his booty in one of the many caves in the Superstitions. Other tales range from Spaniards seeking gold and Indians seeking revenge to vanishing acts of those who sought to follow Walz, and to cryptic footprints and gunshots in the dark. Despite several mysterious and perhaps fanciful disappearances associated with these mountains, geologists who have studied the range have all come out alive.

U.S. 80
Benson — Bisbee
(49 miles)

Between Benson and St. David this highway parallels the San Pedro River, bordered by a grand staircase of terraces that represent alternating stages in the development of the river: periods of lake and stream deposition followed by periods of downcutting, perhaps correlated with on-again, off-again Ice Age glaciation farther north. The oldest, uppermost terrace was established in Pliocene time; others are Pleistocene to Recent. The older ones contain an impressive array of fossil skeletons of Pliocene and Pleistocene vertebrates: salamanders, frogs, turtles, lizards, snakes, birds, bats, rabbits, chipmunks, ground squirrels, shrews, gophers, muskrats, rats and mice, porcupines, foxes, wolves, dogs, martens, weasels, badgers, wolverines, sloths, skunks, bears, lynx, jaguars, peccaries, beavers, mastodons and mammoths, llamas, pronghorn antelopes, camels, and horses! The abundance and variety of wildlife, including water-loving beavers, muskrats, frogs, and turtles, implies an environment much wetter than that which prevails here today.

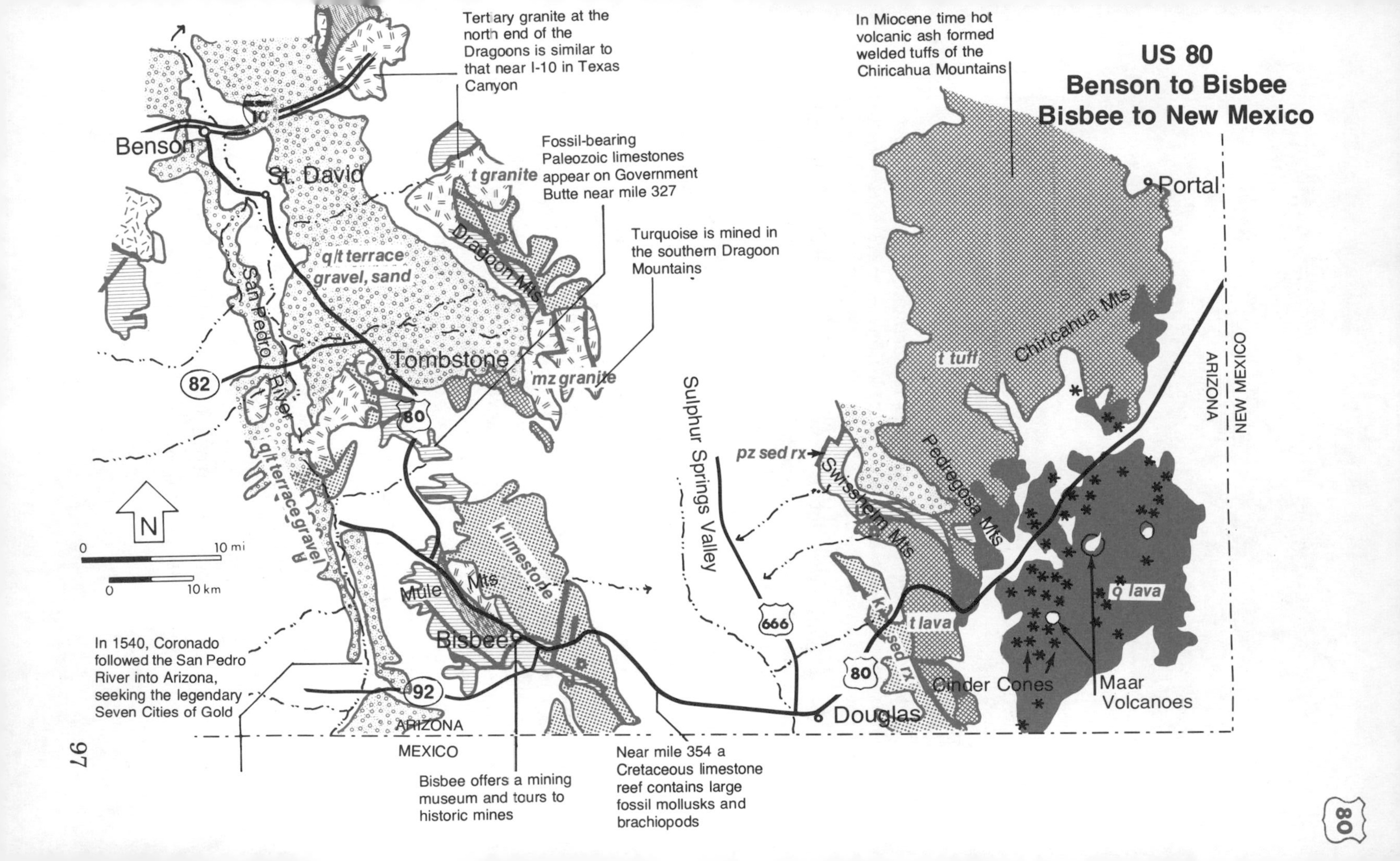
US 80
Benson to Bisbee
Bisbee to New Mexico
In Miocene time hot volcanic ash formed welded tuffs of the Chiricahua Mountains
Tertiary granite at the north end of the Dragoons is similar to that near I-10 in Texas Canyon
Fossil-bearing Paleozoic limestones appear on Government Butte near mile 327
Turquoise is mined in the southern Dragoon Mountains
Benson
St. David
10
t granite
q/t terrace gravel, sand
Dragoon Mts
San Pedro River
Tombstone
82
80
mz granite
q/t terrace gravel
N
0
10 mi
0
10 km
k limestone
Mule Mts
Bisbee
92
ARIZONA
MEXICO
Sulphur Springs Valley
pz sed rx
Swisshelm Mts
666
Pedregosa Mts
Chiricahua Mts
t tuff
Portal
ARIZONA
NEW MEXICO
q lava
t lava
k sed rx
Douglas
Cinder Cones
Maar Volcanoes
In 1540, Coronado followed the San Pedro River into Arizona, seeking the legendary Seven Cities of Gold
Bisbee offers a mining museum and tours to historic mines
Near mile 354 a Cretaceous limestone reef contains large fossil mollusks and brachiopods

Vertebrate paleontologists uncover a mastodon skeleton buried in Pleistocene deposits of the San Pedro Valley. Kirk Bryan photo, courtesy of USGS.

Spear points were found embedded in a mammoth skeleton from one of the lower terraces formed only 11,000 years ago, evidence that man the hunter was already prowling these parts. Human artifacts also occur in a 5500- to 1550-year-old terrace. And the youngest floodplain, near the river, yielded metallic objects emblazoned with the insignia of the U.S. Army! Several famous old fossil sites, quarried in the 1920s and 1930s, were quite near this highway.

The town of St. David is fortunate among southern Arizona communities: It has an abundant supply of **artesian** water — water that rises in wells without pumping. Valley sediments here are arranged in permeable and impermeable layers that slope toward the valley center. Water flowing down the permeable layers, prevented from rising by impermeable layers, develops enough hydrostatic head to resurface through wells. Irrigated fields occupy the lower terraces; upper terraces are used for pasture. Deep gullying in the terraces results from a widespread but unexplained post-1880 erosional episode that has affected much of southern Arizona.

A plentiful supply of artesian water *has made St. David an oasis in the desert.*

Deep gullying of terraces near St. David results from widespread but unexplained post-1880 erosion. J. Gilluly photo, courtesy of USGS.

Apache Powder Plant, across the valley from St. David, manufactures fertilizer and explosives from nitrates and potash.

A few miles south of St. David the highway rises onto the terrace deposits, with roadcuts displaying them well. From higher levels, there are good views of the mountains that border the great San Pedro Valley graben: the Whetstone Mountains to the west, the Dragoons to the east, and the Huachucas farther away to the southwest. The Whetstones contain the most complete sequence of Paleozoic and Cretaceous sedimentary rock in southern Arizona, all in proper order, all tilted southwest.

The gray hills around Tombstone make up a low dome of well stratified, light gray Paleozoic and Cretaceous sedimentary rocks, folded, faulted, and intruded with granite during Laramide mountain-building. Silver- and magnanese-rich solutions came in with the intrusions. From 1879, when silver was discovered here, to 1889, Tombstone was one of the boomingest towns in the West, boasting some 15,000 residents. But after 1890, when the mines flooded, the town became pretty much of a ghost town. The community itself is at the top of a pediment cut on tilted Cretaceous sedimentary rocks of the Bisbee Group, in which the silver ore occurs; rock beneath the town is riddled with tunnels. There are tours of abandoned mines, and a small mining museum; the town seeks to preserve the flavor of the Old West. Some small mines on adjacent hills are being worked again now.

To the northeast about 10 miles away are the Dragoon Mountains, famous as the last stronghold of the Apache warrior Cochise. The 73 million-year-old Stronghold Granite, a rock that weathers into rough, craggy topography, can be distinguished even from this dis-

The Pennsylvanian Nac Limestone north of Bisb may be part of a large overthrust.

J. Gilluly photo, courtesy of USGS

tance. Paleozoic rocks form smoother slopes north and south of the granite. The Paleozoic rocks of the Dragoons, as well as those of the Mule Mountains and the Whetstones, may be part of the postulated Cochise overthrust. If such large-scale overthrusting occurred, these rocks came from many miles southwest of here.

Pennsylvanian and Permian limestones in the hills south of Tombstone are bent, tilted, and cut by innumerable faults, seemingly the result of at least two episodes of faulting. Some of the faults are exposed in roadcuts, as at mile 322-323; bending and tilting of the limestone layers can be seen nearly everywhere. Many of the limestones are fossil-bearing, and **brachiopods**, **bryozoa,** tall-spired snails, and corals are fairly easy to find.

Beyond mile 327 the Mule Mountains near Bisbee can be seen ahead. More Paleozoic limestone layers make up an easily distinguished anticline arching across less visible Precambrian schist and granite. Because some of the Paleozoic rocks were eroded away before the Cretaceous Bisbee Group was deposited, the anticline is thought to have developed in Triassic or Jurassic time. The **unconformity** between the Paleozoic rocks and the Bisbee Group can be seen from mile 330-331, looking back along the highway. Renewed folding in Laramide time, plus possibly later overthrusting, complicated the structure.

Paleozoic strata here are quite similar to those exposed near Superior, some 140 miles northwest of here; Paleozoic seas were stable for long periods of time, and sedimentary rocks deposited in them can be correlated over large areas.

Roadcuts near mile 338 reveal coarse, cobble-filled gravel overlying Precambrian schist that forms the core of the Mule Mountains.

Between miles 335 and 339 the highway runs along the contact between this schist and some granite that intruded it. Both rocks are exposed at the entrance to the Mule Pass tunnel. Beyond the tunnel the same rocks are fractured and mineralized.

The Bolsa Quartzite, a Cambrian rock unit, appears in roadcuts on both sides of the highway at mile 341. This brittle rock unconformably overlies the Precambrian mountain core. The rock is a conglomerate near its base, with fragments of older rock imbedded in it.

Rocks in the Bisbee area are not only highly folded and broken, they are also altered **hydrothermally** — by heat and hot water. Herein lies the clue to the rich mineralization of this district, because hot solutions from intruding magma brought in copper and other minerals. The ores occur at contacts between Paleozoic limestone and Jurassic intrusive rocks. Rich copper-lead-zinc ores accumulated where the chemistry of the limestone caused the minerals to come out of solution in natural caves and **vugs** or cavities dissolved by groundwater in the limestone.

Built in a narrow, picturesque canyon, the old mining town of Bisbee now thrives as a tourist center. Mines there produced ores that accumulated in Tertiary time as cavity fillings in Paleozoic limestone.

Brilliant green malachite flowstone and dripstone — the same types of formations we find in limestone caves — are spangled with bright blue crystals of azurite — some of the most beautiful ore minerals in the world. Minerals are on display in the Bisbee mining museum and in the University of Arizona geological museum in Tucson.

Near the south end of Bisbee is Lavender Pit, one of the smaller open pit porphyry copper mines in Arizona. Developed around the shafts of earlier underground workings, it closed in 1975 when remaining ore could no longer be mined economically. Some of the tailings are now being reprocessed using new methods that remove more of the valuable minerals. The ores replaced Paleozoic limestone along the contact with a small Jurassic intrusion, or stock. The intrusion at the center of the pit was also mineralized, and mineral veins radiated from it like the spokes of a wheel.

As you leave town, watch the roadcuts for kinky folds in the Abrigo Formation, a Cambrian limestone. Other large roadcuts show green copper mineralization in Jurassic granite porphyry. A bright red oxidized zone occurs here, of a type known to be present above many porphyry copper deposits. Prospectors and geologists learned to use this "red thumbprint" as an indicator of copper mineralization.

U.S. 80
Bisbee — New Mexico
(47 miles)

Like other southern Arizona mountains, the Mule Mountains surrounding Bisbee are a fault block range created late in Tertiary time by Basin and Range faulting. On the southwest side of the mountains lie many small fault slices of Paleozoic sedimentary rocks. It is where these rocks are in contact with Mesozoic granite intrusions that the mines of Bisbee are located.

South of Bisbee the highway passes among good exposures of Paleozoic and Cretaceous sedimentary rock, most prominently the gray Cretaceous limestones of the Bisbee Group turned up sharply along a fault. The limestones are well exposed between miles 346 and 352. The cement plant and quarry east of the road near mile 356 provides lime from the calcium carbonate of Cretaceous rocks to neutralize acid copper ores in smelters at Douglas.

Near milepost 354, Cretaceous limestone thickens into a resistant, ridge-forming unit full of fossils, thought to have accumulated as a reef. The ridge extends south into Mexico.

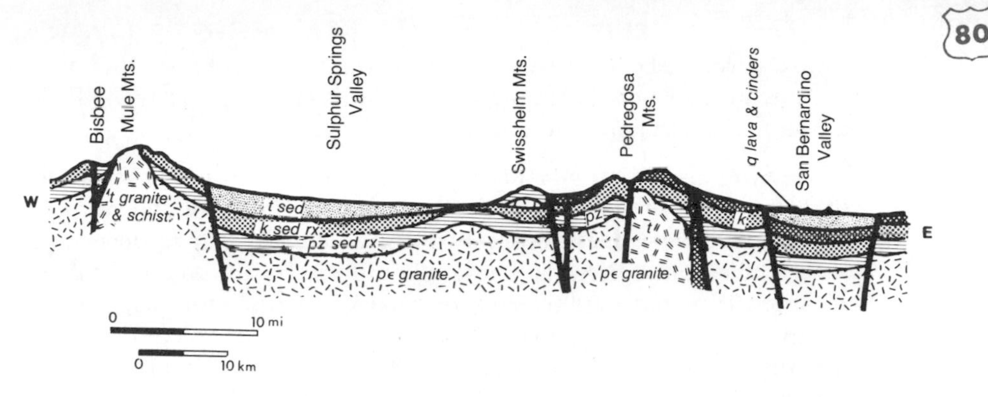

Section north of US 80 Bisbee to New Mexico, much simplified.

A broad pediment surrounds the Mule Mountains, and is continuous with the surface of the valley fill in this southern part of Sulphur Springs Valley. The thin layer of gravel that covers both pediment and valley fill contains pebbles and sand derived, as we might expect, from either Precambrian metamorphic rocks or Paleozoic limestones, the rocks exposed in the mountains. In places the pediment and valley gravel are tightly cemented with caliche.

East of Douglas in the Peralta Mountains are more Cretaceous sedimentary rocks; their westward dip can be seen clearly. Beyond mile 278 they are covered with volcanic ash — now tuff — that is part of the large volcanic outpourings that characterize the Chiricahua Mountains, several thousand feet of welded tuff from explosive eruptions of Miocene Time (see Chiricahua National Monument in Chapter V).

Stop and look around at milepost 380. There is a light-colored dike to the northwest; farther west a cliff of Escabrosa Limestone marks the front of the Swisshelm Mountains. This Mississippian limestone is part of a sheet of Paleozoic sedimentary rocks — mostly flat-lying here — that overlies younger Cretaceous volcanic rocks. Such clear-cut superposition of Paleozoic rocks over Cretaceous ones of course favors the overthrust interpretation of southern Arizona geology. In this area, detailed studies suggest not one but two overthrust sheets, the Hidalgo overthrust shoving Pennsylvanian rocks over Cretaceous, and the Cochise overthrust bringing Precambrian and lower Paleozoic rocks over Cretaceous. Whether or not such broad-scale overthrusting occurred here, Paleozoic and Cretaceous rocks of this region must have been involved to some degree in thrust faulting.

Smelters at Douglas were developed to process ore from Bisbee and Tombstone mines.

Castle Dome, between the highway and the Escabrosa cliff, is a volcanic plug or neck intruded along a fault that separates the Pedregosa and Swisshelm Mountains.

A few miles farther northeast, some of the Cretaceous volcanic rocks involved in the thrust faulting appear in roadcuts: purple, maroon, and green dacite scarcely distinguishable as lava flows and masses of volcanic breccia. These are distinctly the kinds of silicic volcanic rocks that tend to erupt explosively, causing immense outpourings of ash similar to but much greater in volume than the 1980 outpourings of Mt. St. Helens in Washington. These rocks are 72 to 68 million years old.

Near milepost 384 we come on the first of the much younger lava flows of the San Bernardino volcanic field. Volcanoes in this little field are mostly grass-covered cinder cones, though there are also a few **maar craters** — broad, shallow, low-profile craters resulting from steam explosions. By and large, eruptions here were much quieter, involving basalt magma. Small flows emerged from the bases of some cinder cones or from vents now obscured by erosion and infilling of the valley. Drill holes show that lava flows make up a large portion of the valley fill, and radiometric dating shows that the field was active from 3 million to 275,000 years ago. Some of the basalt contains white or honey-colored crystals of **feldspar**, large black rods of **pyroxene**, and green crystals of **olivine**. Such lavas derive from sources below the earth's crust, in the upper mantle, and rise through the crust with little or no contamination by crustal material. Laboratory tests show that rocks of this composition develop similar mineral crystals at over 2000° F in pressure conditions comparable to those 40 miles or more below the surface.

The steepness of the Chiricahua Mountain front, bordering the San Simon graben, is particularly well displayed near Portal.

San Bernardino Valley, bordered on the west by the Pedregosa and Chiricahua Mountains and on the east by the Peloncillo Range, is a downdropped block bounded on both sides by faults — a **graben**. It drains south into Mexico. The faults along its margin may still be active, as triangular facets at the lower end of the Chiricahua ridges have not had time to wear away, and an 1887 earthquake centered south of the border created a new 12-foot scarp along the eastern side of the valley in Mexico. The Rio San Bernardino cuts headward rapidly, and may eventually cut through the pass between the Chiricahua and Pedregosa Mountains to drain the northern, now undrained part of Sulphur Springs Valley.

North of the volcanic field the valley and its bordering mountains are offset eastward several miles. The road crosses at the offset into the San Simon Valley, another graben. Abrupt mountain fronts with truncated ridges again testify to ongoing faulting.

U.S. 89
Florence Junction — Tucson
(83 miles)

US 89 south of Florence Junction descends gradually toward the Gila (HEE-la) River, one of the few large streams in southern Arizona. Between Florence Junction and Florence, mountainous country visible 6 to 10 miles east of the highway is the northwest end of the Tortilla Mountains and the southwest edge of Arizona's Central Highlands. For much of this route the highlands can be seen behind low granite hills that more or less parallel the highway.

The Gila River flows from headwaters in the mountains of southeastern New Mexico through long, narrow valleys and tortuous canyons in the mountain highlands. Where it emerges onto the desert east of Florence, its gradient suddenly decreases, and the river drops its load of rocks and sand. At Florence its channel, choked with some of the debris, is broad, sandy, and irregular — a **braided channel** that divides and comes together around sandy temporary islands. Even though upstream dams and diversion canals partly control its waters, the river's banks and channels often bear the signs of recent floods — cutaway banks, piled branches, and debris-tangled vegetation. Much of the Gila's water is diverted for irrigation of areas north and west of the Picacho Mountains. Plans are now developing to dam the upper drainage of the Gila and its tributary the San Francisco

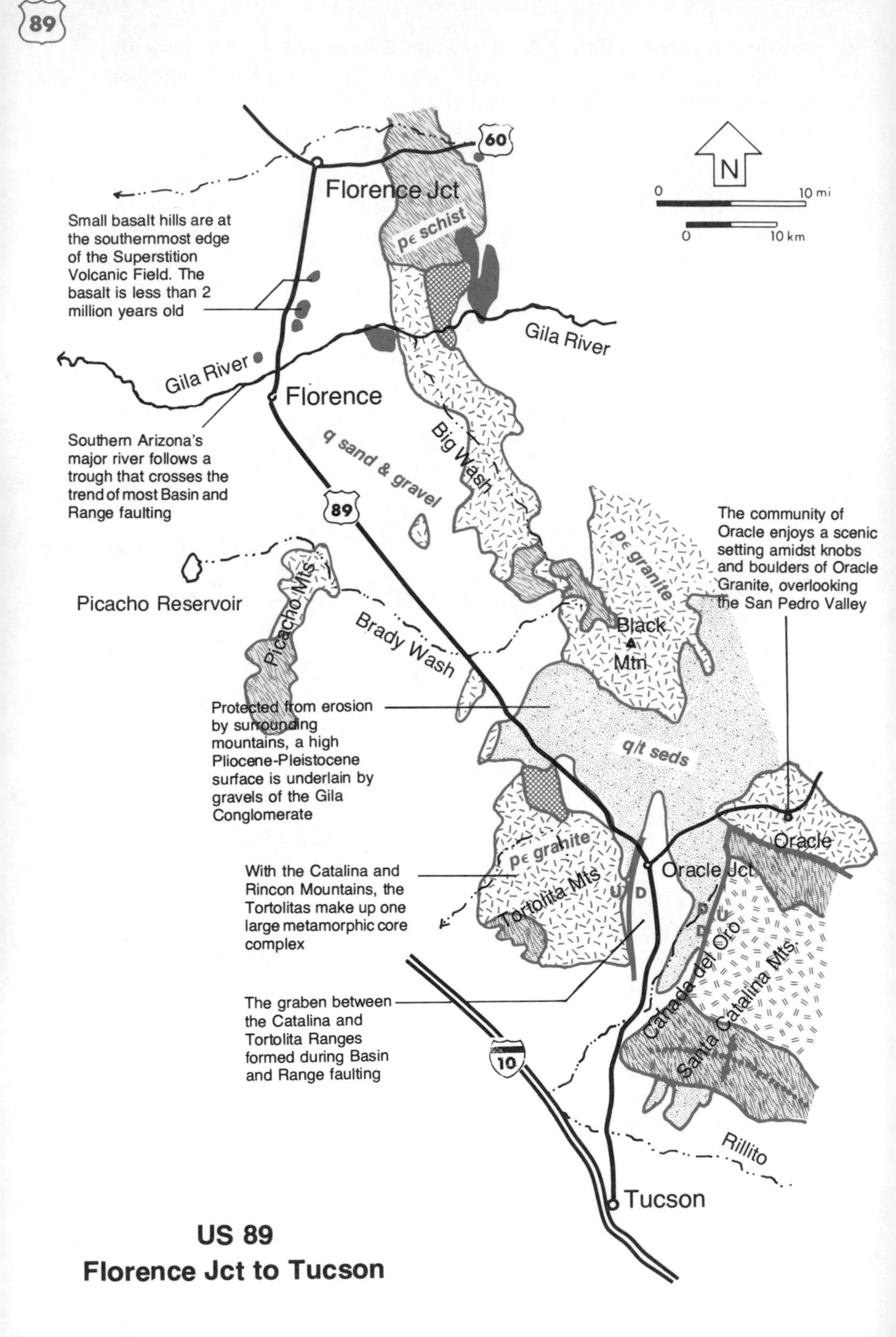

US 89
Florence Jct to Tucson

River in order to supply water to New Mexico, to maintain a sufficient water supply for mines, smelters, and an Indian Reservation in its drainage basin, and to decrease flood damage along the entire river route.

The granite mass of the Tortilla Mountains extends as a pediment far out onto the plain southeast of Florence, where it is only thinly covered with valley gravels. Where Big Wash and other desert washes sweep away the surface gravel, the granite is exposed; it also makes up the low hills between the highway and the mountains of the Central Highlands. The Precambrian granite resembles the Oracle Granite of the northern Santa Catalina Mountains, now coming into view to the south. In Precambrian time it may have connected with them as a single giant batholith — the roots of a mountain system extending far beyond Arizona's boundaries.

At mile 107 the highway passes through a little patch of glistening Precambrian schist and emerges onto a plain of sand and gravel protected from erosion by the Tortolita and Santa Catalina Mountains. Coated with pebbles of desert pavement, this surface supports such a variety and abundance of plants that it seems at odds with the definition of a desert. Rising 2000 feet higher than the Gila River floodplain at Florence, and more than 1000 feet higher than Tucson, the area attracts more rainfall than either. The pebble-covered surface is Pleistocene; it merges with known Pleistocene terraces of the San Pedro Valley. Its coarse sand and gravel layers are well exposed farther south, where the highway drops down through them to Oracle Junction. Pliocene sediments almost certainly underlie the Pleistocene ones.

The Tortolita Range to the south does not seem particularly high from this side, where it is surrounded by a broad pediment and half buried by Pliocene-Pleistocene gravels. From the south and southwest, though, where the Santa Cruz River has swept away all the thick, poorly consolidated Pliocene and Pleistocene layers, there is a greater elevation difference. The Tortolitas are a metamorphic core complex range cored with Tertiary granite and including just a few patches of Paleozoic sedimentary rocks.

In the Santa Catalinas, directly southeast along the line of the highway, Oracle and Catalina Granites make up the main part of the range, the broad, rounded, pine-darkened dome of Mt. Lemmon (9157 feet). South of this dome, on the side of the mountains that faces Tucson, there is a forerange anticline of metamorphic rock, the Catalina Gneiss. A fault-edged graben separates both the granite core and gneiss forerange from the Tortolitas.

South of Oracle Junction, US 89 passes through the graben between the ranges. Although the Tortolita Mountain front above its pediment is quite irregular, the west face of the Catalinas rises straight and steep above the great fault at the edge of the down-dropped block. It is thought that large alluvial fans once buried an eroded pediment on this side of the graben, but due probably to climate changes and to downcutting by the Santa Cruz River south of the mountains, the fans were washed away. There is no evidence of movement on the graben faults since Pleistocene time.

Farther along the graben the highway crosses the Canada del Oro — Canyon of Gold. The name is now given to the stream rather than to the valley down which it flows. Gold *can* be panned from its sandy streambed, but historically there has rarely been enough water, and there is probably not enough gold, to make processing the gravels worthwhile.

As you pass the end of the forerange you can easily distinguish the anticline that forms it — an anticline not in layers of sedimentary rock but in *foliation* of the Catalina Gneiss, the alignment of mineral grains and tiny fractures, and the tendency of the rock to split or weather along parallel mineral faces lined up by subterranean pressures at the time the metamorphic core complex formed. Cut by many joints, the gneiss of the forerange is less massive than the granite of the mountain core, and erodes into crags and turrets above a sort of amphitheatre formed by the west end of the forerange. The dark gneiss alternates with lighter granitelike rock, their color differences depending on the relative amounts of black mica (biotite) and white mica (muscovite). Both seem to have originated in Laramide time.

Rounding the corner of the forerange the road climbs once more onto the Pliocene-Pleistocene terrace. Then it drops down into Tucson Valley, crossing the Rillito (ree-YEE-toe: Little River), a usually dry desert river that drains the east and north parts of the nearly flat-floored basin. The Rillito may flow during the winter rainy season, and is quite capable of flooding under the right conditions — usually summer "cloudbursts." It has more than once destroyed the US 89 bridge.

Tucson Basin and the surrounding mountains are discussed under Tucson and Vicinity, I-10 Willcox to Tucson and Tucson to Casa Grande, and Saguaro National Monument (Chapter V).

Hoover Dam, sited on the Colorado River between Nevada and Arizona, stores water, controls floods by equalizing the river's flow, and generates hydroelectric power. Tunnels on either side carried river water while the dam was being constructed.
Courtesy of U.S. Bureau of Reclamation.

U.S. 93
Hoover Dam to Kingman
(73 miles)

For a geologic map of this part of US 93, see I-40 California to Kingman.

Hoover Dam, built between 1931 and 1936, impounds the waters of the Colorado River in Lake Mead, the largest man-made reservoir in the United States. Lake Mead holds nearly two years' worth of the river's normal flow. The dam is sited in a narrow gorge where the river has cut down through faulted Miocene volcanic rocks of the Black Mountains. Formerly the river foamed through the deep, narrow canyon in wild white-water rapids. At times of high water it brought disastrous floods to low-lying lands farther downstream. Now, with its floodwaters controlled and with a chain of smaller dams between here and the Mexican border, the river furnishes municipal and industrial water to Arizona and California cities, irrigates a broad strip of Arizona and California desert, and generates hydroelectric energy for cities in Arizona, Nevada, and southern California. In addition of course it provides recreation facilities along its course.

The river did not always flow through this narrow gorge. Through most of Tertiary time the Ancestral Colorado River, rising in the new-formed Colorado and Wyoming Rocky Mountains, probably did not even come through Grand Canyon, but held a southward course from Utah into eastern Arizona. Upriver from Hoover Dam, Miocene gravels tell part of the river's story. They lie in Arizona, east of the river, but include gravelly alluvial fans derived from areas west of the river, in California and Nevada. Because the gravels must have washed across the area where the river now flows, we know that the river could not have been here in Miocene time. (For further discussion of the history of the Colorado River, see the introduction to Chapter IV.) There are strong reasons for believing that the Colorado River was not diverted to its present route earlier than in Pliocene time, 5 to 2 million years ago.

Careful surveys before and after Lake Mead filled show that the weight of its 28 million acre feet of water has depressed the earth's crust here about 7 inches, setting off mild earthquakes in the Lake Mead area.

Leaving Hoover Dam, US 93 cuts through dark Tertiary volcanic rocks, topped with younger volcanic rocks, that makes up the northern part of the Black Mountains — banded, contorted, many-toned rocks that register repeated episodes of bending, breaking, and partial melting through Precambrian time. In this part of the Basin and Range Province, intermountain valleys filled with Tertiary and Quaternary sand, gravel, clay, and in some places salt are exaggeratedly long, lying as they do between equally long, slender fault block ranges. The highway enters Detrital Valley, a good example, at mile 15. In this as in other desert valleys, alluvial aprons — called **bajadas** in the Southwest — slope out from the bordering ranges. Valley fill is commonly several thousand feet deep, and somewhere below its surface lie the great faults along which the mountain blocks rose.

At mile 21 the lava-filled and later eroded conduit of a volcano, a volcanic neck, rises quite close to the highway, an interesting little peak with several radiating dikes.

The White Hills stand out east of mile 28 — not white, but much lighter than the Black Mountains to the west. Their massive, tan layers of Tertiary tuff front some more dark Precambrian gneiss. At their south end are reddish and buff-colored layers of older, probably Cretaceous volcanic ash, steeply tilted during both Laramide and Basin and Range mountain-building.

South of the White Hills another long, narrow range of Precambrian metamorphic rock, certainly the dominant rock type along this route, borders the valley. This is the Cerbat Range. Its high point, pine-spangled Mt. Tipton (7148 feet), marks this north end of the range. Notice how angular and ragged the rocks that make up these mountains are, and the absence of smoothly convex summits or bulbous knobs and boulders. The rough, angular weathering pattern is particularly characteristic of gneiss and schist — metamorphic rocks. You'll see farther south that granite — an intrusive rock — tends to weather with rounded or knobby contours. Bajadas along the Cerbat Mountains make up a low pass that separates Detrital Valley's northward drainage from the southward drainage of Sacramento Valley.

Just north of this divide a side road leads east to the little town of Chloride, pretty much of a ghost town now. It once had a population of 2000 or more, with 50 mines operating nearby. There is still one mine that produces copper, gold, silver, lead, and zinc, but only when prices for such metals are high! Other ghost towns along these mountains are not so fortunate. Mineral Park, for instance, was razed and sold to pay back taxes.

Cretaceous, Tertiary, and Quaternary volcanic rocks make up most of the south end of the Black Mountains, one of the longest ranges in Arizona, 100 miles from northern tip to southern toe. Some of the youngest of the volcanic rocks come close to the highway near mile 67; horizontal Quaternary basalt lava flows that came into being after Basin and Range tilting had taken place. As a rule of thumb, applicable only very generally, Quaternary lava flows lie horizontally, while Tertiary volcanic layers tend to be tilted or warped by Basin and Range faulting. Cretaceous ones are even more steeply tilted or quite obviously folded and faulted because they have come through not one but three mountain-building episodes: the Laramide orogeny of late Cretaceous to early Tertiary time and the Mid-Tertiary and Basin and Range orogenies of Tertiary time. In much of southern and western Arizona, moreover, most of the Quaternary volcanic rocks are black or dark gray basalt, whereas most of the Tertiary ones are thick buff-colored or pinkish beds of volcanic ash welded into tuff.

At mile 67 the highway turns eastward through some of this tuff, material that burst forth in an explosive eruption similar to that of Mt. St. Helens in 1980, still so hot when it touched down that it welded together at once. Vertical cooling cracks then broke it into columns and formed avenues for erosion. If you could keep going southeastward across Arizona, in about 360 miles as the crow flies you would reach Chiricahua National Monument, at the other end of the Basin and Range deserts. There, similar vertically jointed welded tuff has been carved by erosion into a fantastic forest of columns and pinnacles.

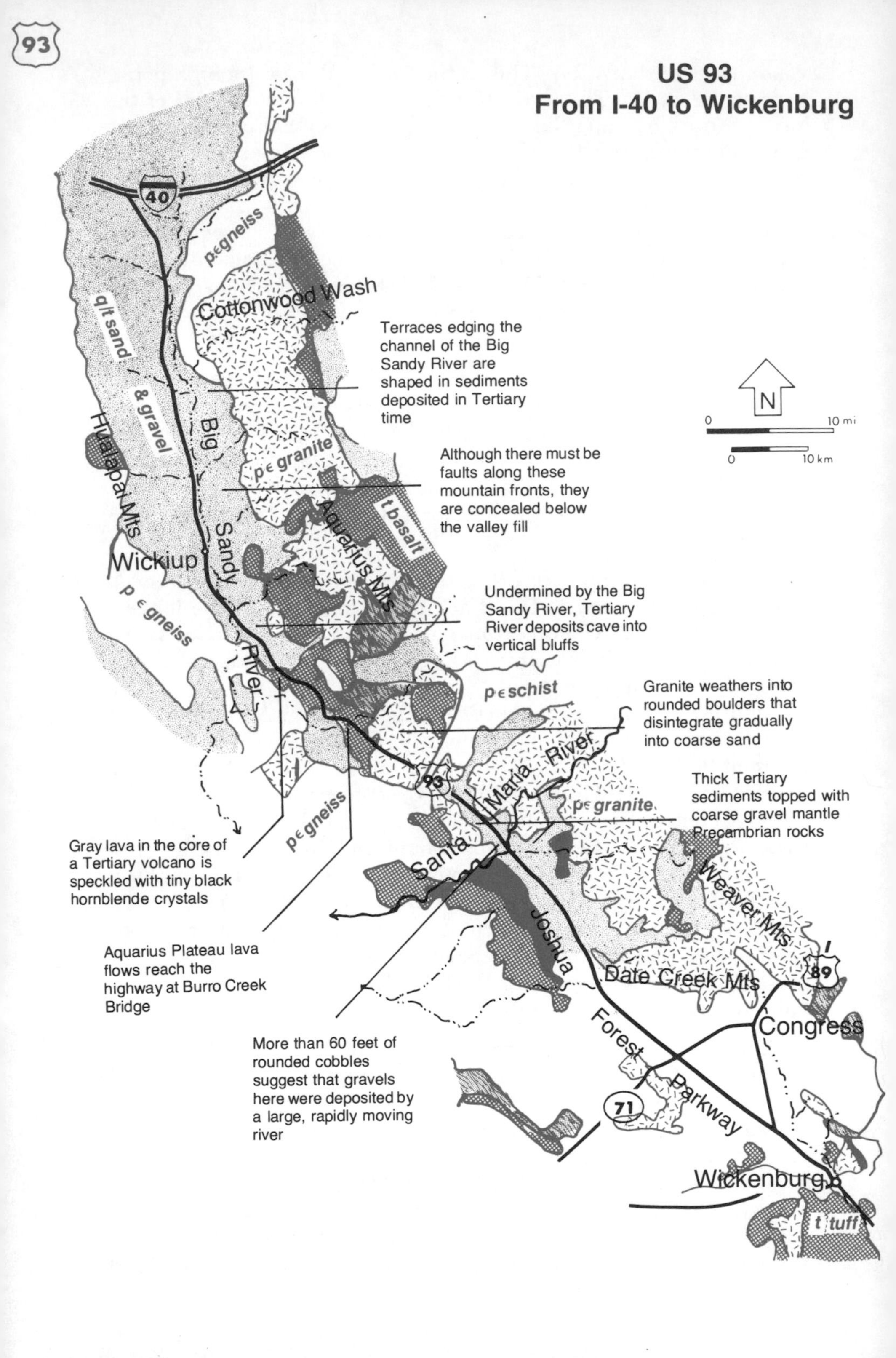
93
US 93
From I-40 to Wickenburg
40
pєgneiss
Cottonwood Wash
qlt sand
& gravel
Terraces edging the channel of the Big Sandy River are shaped in sediments deposited in Tertiary time
N
0
10 mi
0
10 km
Hualapai Mts
Big
Sandy
River
pє granite
Although there must be faults along these mountain fronts, they are concealed below the valley fill
t basalt
Aquarius Mts
Wickiup
p є gneiss
Undermined by the Big Sandy River, Tertiary River deposits cave into vertical bluffs
pєschist
Granite weathers into rounded boulders that disintegrate gradually into coarse sand
Santa Maria River
93
pєgneiss
pє granite
Thick Tertiary sediments topped with coarse gravel mantle Precambrian rocks
Gray lava in the core of a Tertiary volcano is speckled with tiny black hornblende crystals
Weaver Mts
Joshua
Aquarius Plateau lava flows reach the highway at Burro Creek Bridge
Date Creek Mts
89
Congress
Forest Parkway
More than 60 feet of rounded cobbles suggest that gravels here were deposited by a large, rapidly moving river
71
Wickenburg
t tuff

U.S. 93
From I-40 to Wickenburg
(109 miles)

Leaving I-40 at mile 71, US 93 curves southward down the valley of the Big Sandy River, between the gneiss and granite heights of the Hualapai and Aquarius Mountains. Although the Hualapai Mountains to the west are composed primarily of very old Precambrian gneiss, a younger Precambrian granite intrudes their north end, weathering in typical granite fashion into bare, rounded knobs. Farther south, some knobs of Tertiary granite jut through the gneiss. In the Aquarius Mountains to the east, the reverse is true: a similar Precambrian granite makes up most of the range, with some older gneiss here at the north end. Both granite and gneiss are similar to the rocks of the Inner Gorge of Grand Canyon.

Sediments that fill the valley of the Big Sandy River are both Tertiary and Quaternary. Flat-topped hills on either side of the highway are remnants of former floodplain and alluvial apron sediments, the highest ones probably dating back to Pliocene time, when streams in this region had not yet begun the great downcutting that followed regional uplift and diversion of the Colorado River to this side of Arizona. As the Colorado River cut down, so did its tributaries. Gravels on these Tertiary surfaces reflect the make-up of the adja-

luffs along the Big Sandy iver, as well as road cuts along US 93, present lessons in Tertiary and Quaternary geologic history. They expose sediments that record times of upstream uplift and of greatly increased stream flow.

cent mountains, with varying proportions of granite and gneiss pebbles corresponding with the proportions of granite and gneiss in the Hualapai and Aquarius Mountains. If you have a chance to look closely at some of these rocks, you will see that their color varies from light to dark gray with different quantities of glassy quartz, white feldspar, and bright flakes of white mica (muscovite), as opposed to black mica (biotite) and black rodlike crystals of hornblende.

Farms along the valley of the Big Sandy River are maintained with groundwater that has flowed from surrounding mountains through the Tertiary and Quaternary gravel aprons that surround them. Some of this water emerges as springs at the edges of the river; other water is obtained from wells.

Sandy, silty Tertiary sediments in bluffs along the river are still in their original position. The soft rock collapses easily when it is undermined by the river or its tributaries. In places their cut-away surface shows in cross-section that channels in the fine Tertiary siltstone are filled with much coarser, even cobbly Pleistocene gravel, reflecting both the uplift of Pliocene time and the vastly increased rainfall and heavy runoff of Pleistocene time.

When it is flowing the Big Sandy wanders back and forth across its bed, creating a braided channel, with shifting islands of sand and gravel temporarily blocking or diverting its flow. In floods, the whole channel is filled. The river has a fairly low gradient here, which encourages sideways cutting rather than downward cutting. Farther downstream, near milepost 136, it is partly blocked by hard ridges of gneiss that make up the southern end of the Hualapai Mountains. It does break through these ridges, but downcutting is slowed by the hardness of the rock.

South of Wickiup the Aquarius and Hualapai Mountains converge, and the vast expanse of Tertiary basalt that covers the Aquarius Plateau east of the Aquarius Mountains seems to spread westward, reaching the highway at mile 135. There is a sudden change in scenery, with steep-walled, narrow gorges incised into nearly flat-lying basalt layers. Gravel visible in roadcuts is composed almost exclusively of well-rounded basalt fragments.

The highway crosses a tongue of the Aquarius lava flows near Burro Creek Bridge, a worthwhile stop. The view down-canyon shows layers of gray lava, darkened by desert varnish, alternating with lighter layers and random patches of tuff. Some of the lava flows show some **columnar jointing** caused by shrinkage of the rock as it cooled. Cracks and crevices in the basalt are coated with white caliche deposited by groundwater, a coating also evident on many of the loose lava blocks. In the basalt itself can be seen tiny rodlike crystals of

Burro Creek has carved a canyon through basalt lava flows at the edge of the Aquarius Plateau.

black hornblende. Much of the basalt also contains small round **vesicles** formed as bubbles of gas were trapped in cooling lava.

To the west, also visible from the viewpoint, are the gneiss ridges of the south end of the Hualapai Mountains.

A few miles south of Burro Creek the route enters some of the granite of Arizona's Central Highlands. Here this rock reveals all stages of disintegration from original blocky masses separated by joints, to bulbous, flat-faced but round-cornered rock clusters, to partly fallen piles of well rounded boulders. Little by little the boulders flake away, leaving only sandy mounds. Later the mounds acquire a cover of soil able to support vegetation. This easily recognized type of disintegration, known as **spheroidal weathering**, is typical of even-grained intrusive rock and occurs throughout Arizona wherever granite is found.

At mile 150, higher parts of the Central Highlands are visible to the east, with their knobby outcrops of granite. A narrow ridge of gneiss — recognizable by its banding and color variety as well as by its close-together, near vertical fractures and vertical grain — crosses the highway at mile 151. It, too, is part of the Central Highlands.

Having no more than touched the northwest end of these mountains, the highway descends to the Santa Maria River across Quaternary/Tertiary stream gravels of one of its small tributaries. Roadcuts along this stretch of highway show layers of poorly consolidated sand and cobbles, totalling 100 feet thick or more, in cross section. The Santa Maria River, at mile 161, is a permanent stream with headwaters in mountains near Prescott.

South of this river the highway becomes Arizona's scenic Joshua Forest Parkway, and you may want to pay more attention to the unusual desert flora than to the geology. Suffice it to say that the Joshua trees grow most thickly on another high terrace, this one lying between the Central Highlands, now to the northeast, and a small range of sloping volcanic rocks to the southwest. An earlier base level of the Santa Maria River, before rerouting of the Colorado River, was probably close to the level of this terrace.

The sharp double peak visible to the south during the last part of this route is a slim block of basalt that is thought to be a remnant of a much larger overthrust sheet.

Sloping lava flows west o the Joshua Tree Parkwa are Tertiary in age. Joshuas seem to thrive o loose Quaternary and Tertiary terrace gravels.

Arizona 82-83
Nogales to I-10
(55 miles)

For a geologic map of this route see I-19 Tucson to Nogales.

Northeast of Nogales, Arizona 82 passes roadcuts exposing Miocene sandstone — the Nogales Formation. These rocks mark one limb of an anticline, with strata here dipping steeply east while those west of Nogales dip west.

The Santa Cruz River, at mile 6, flows north from Mexico, but only after flowing south *into* Mexico about 20 miles east of here. Northwest of the AZ 82 bridge the river flows for several miles along a fault that separates the Nogales Formation from younger Pliocene and Pleistocene gravels that together make up a high, flat-topped terrace, a remnant of a Pliocene-Pleistocene plain that once surrounded the Santa Rita Mountains and extended across the Tucson Basin. Though streams from the mountains cut downward through it, its highest surface retains long-weathered, oxidized, reddish soil that developed during Pleistocene time. One can imagine how it must have looked at that time: a lush plain with large herds of animals roaming upon it, among them the ancestors of today's horses, elephants, and camels. Fossil bones have been found in the terrace sands and gravels, which are about 1000 feet thick near Patagonia Lake State Park.

Like many ranges in this area, the Patagonia Mountains are pocked with small mines that produced copper, lead, zinc, silver, and gold. Most are inactive now. Ores occur in slices of Paleozoic limestone altered by the intrusion that forms the long crest of the Patagonia Range.

In stream valleys and roadcuts near mile 13 there is a change in roadside rock type, to volcanic tuff, breccia, conglomerate, and sandstone layers, all tilted by faulting. Some of these rocks, particularly those near mile 14 and on Red Mountain, are quite colorful. The tuff came from some nearby volcano that, having blown out all this ash, probably collapsed into its own partly emptied magma chamber. However no caldera, not even a severely eroded one, has been recognized in this area.

The highway follows Sonoita Creek, a small watercourse that is cutting headward into the old Pliocene-Pleistocene plain. Erosion has gradually supplanted deposition here as changes in climate, possibly the alternation of rainy and dry cycles in Ice Age time, caused alternating times of downcutting and refilling, bringing about the succession of terraces now visible along Sonoita Creek.

The Canelo Hills east of Patagonia consist of faulted Paleozoic rocks, with long, narrow slivers of rock that may be gravity-glide blocks that skidded down the sides of the rising Santa Rita Mountain core.

Grassy uplands north of Sonoita represent a Pleistocene surface underlain by Pliocene and Pleistocene gravels.
Kirk Bryan photo, courtesy of USGS.

Sonoita lies on the surface of the old Pleistocene plain, in a basin bordered on the east by the Whetstone Mountains and on the north by the smaller Empire Mountains. Deep, fertile Pleistocene soils support grass-rich, non-desert types of forage, making for ideal ranch country. Along gullies, well developed turf holds soil together, overhanging the gullies in places. Rainfall here is 17 inches; elevation is over 4000 feet. The fertile plain at one time was continuous northward all the way to the Rincon Mountains, but Cienega Creek to the north and the Babocomari River to the east are, like Sonoita Creek, gradually whittling away its edges.

North of Sonoita, Arizona 83 crosses part of this plain, and Pleistocene gravel is visible in roadcuts. Together with the Pliocene gravel that underlies it, it is 5000 feet thick here — testimony to the great quantities of rock debris that washed from surrounding mountains.

As erosion attacked, mountain fronts wore back, so that parts of the plain are pediments covered only thinly with Pleistocene gravel. Although the Sonoita Basin is downfaulted, the faults are well hidden, lying some distance from the mountain bases.

The Whetstone Mountains present a thick, southwest-tilted array of Paleozoic and Cretaceous strata, the most complete such sequence in southern Arizona. Cretaceous sedimentary and volcanic rocks occur at the south end of the range; the north end is Precambrian granite. Between the two are Cambrian, Ordovician, Devonian, Mississippian, Pennsylvanian, and Permian rocks — layers of fossil-bearing marine limestone, sandstone, and shale. A small Tertiary intrusion breaks through these rocks at Granite Peak.

Farther northward you'll see more gullies cutting through the Pleistocene surface, as well as through former stream floodplains. Early records show that the gullies came into existence after 1880, reflecting increased erosive power due perhaps to a lowering of groundwater levels and downcutting by the Gila River and its tributary the Santa Cruz.

A low divide west of milepost 38 marks the Sawmill Canyon fault zone, which separates the central part of the Santa Rita Mountains from their northern arm. The zone is complex, with many more intersecting faults than can be shown on the map. Nearby in this direction are Triassic volcanic rocks interbedded with layers of sandstone closely resembling wind-deposited Triassic sandstone of northeastern Arizona. Some of the volcanic rocks have been dated at 220 million years.

Greaterville, a few miles west of this route, marks a placer mining area where gold, discovered in 1874, was mined in a brief 12-year boom. Placer mining was hampered by lack of adequate water, and dry separators had to be used instead of more common and more efficient sluice boxes.

At the gap between the Santa Ritas and the Empire Range at mile 44, the highway enters Gardiner Canyon, where another headward-cutting stream is attacking the edge of the Pleistocene plain. Reddish soils of the old plain can be seen atop nearby hills. In its course toward

Gullies that crease the desert today developed since 1880. Whether they reflect a real climate change or some man-made alteration such as destruction of vegetation by overgrazing is unknown.

the Santa Cruz River this creek loses about 1200 feet of elevation in less than 12 miles — a gradient that gives it plenty of strength for downward cutting and headward erosion. Terrace levels along Gardiner Canyon show that periods of floodplain deposition have alternated with downcutting periods. Watch for more post-1880 gullying as the highway winds through rolling hills between Gardiner Canyon and I-10. Pliocene-Pleistocene gravels edge the road, some of them quite well consolidated and with stringers or veinlets of calcite running through them. Near milepost 46 this valley lies on steeply tilted, reddish, bouldery conglomerate that may have been a landslide or mudflow deposit.

To the west, out of sight, is the Helvetia-Rosemont Mining District, with a porphyry copper deposit estimated to contain 340million tons of copper ore. This district also contains placer gold and is a weekend favorite with amateur gold panners.

Like the Whetstones, the Empire Mountains display Paleozoic and Cretaceous sedimentary rocks, but they are not as neatly laid out here. Paleozoic and Cretaceous limestones between miles 47 and 51 are cut by light-colored porphyry dikes, part of dike swarms that sweep across the north base of the Santa Rita Mountains and that are thought in some way to be related to nearby copper deposits. Some of the dikes appear in roadcuts near mileposts 49 and 51. Larger masses of porphyry form stocks that in the Empire Mining District are enriched with copper.

The geology edging the Empire Mountains is quite complex: Precambrian granite, Paleozoic sedimentary rocks, Mesozoic and Tertiary sedimentary, volcanic, and intrusive rocks, all chopped in an immense chopping bowl and altered by intrusion of dikes and stocks. Then erosion, followed by the piling on of thick layers of Pliocene and Pleistocene gravel. And finally erosion of the gravel plain. Little wonder that geologists have come up with several different interpretations of the geologic wheres and hows of this area. Some would have these broken blocks shoved in from far to the southwest. Others think that the fault slivers slid down flanks of rising mountains or occupy small, wedge-shaped fault blocks that shifted primarily up or down rather than horizontally.

As this route nears I-10, the Rincon Mountains come into view ahead. They are a typical metamorphic core complex, with a large granite dome — round-topped Mica Mountain — flanked by gneiss-schist ridges and slide blocks of Paleozoic rocks. The range is discussed more completely under Tucson and Vicinity, as is the Santa Catalina Range, a continuation of the same metamorphic core complex north of Tucson.

AZ 85
Gila Bend — Lukeville
(77 miles)

Leaving Gila Bend, Arizona 85 crosses the southeast part of the Gila Bend Plain. Water for irrigation here comes from the Gila River and from wells tapping groundwater in valley sediments. The **water table**, the groundwater surface, becomes deeper southward, away from the Gila. More seriously, it becomes deeper with time because more water is being pumped from these wells than is naturally replaced.

Near mile 10, hills on both sides of the highway show nearly horizontal lava flows probably once continuous over much of this area. Still farther south, light-colored hills rise abruptly from the desert floor, remnant high points in an eroded fault block of Precambrian granite and gneiss at the north end of the Sauceda Mountains. Most of the range, including all that part east of the highway, consists of jagged, jumbled, irregular lava flows, layers of tuff, and volcanic breccia produced during an eventful, explosive volcanic history. Flat-topped Hat Mountain (2716 feet) to some extent indicates how much material has been eroded from this range. Its flat lava cap was originally far more extensive.

The Crater Mountains west of the highway are composed of similar rocks. However, lava and ash flows in these mountains are more nearly in their original positions, and the range as a whole is still flat-topped. The mountains get their name from a crater-like circular valley crossed by the highway near milepost 29. Dramatic spires within the "crater" are, like its walls, erosional features, and this is not a real crater at all. Some of the spires here mark ancient volcanic conduits, as magma cooled in conduits is more resistant; the same can be said of ridge-forming dikes within the "crater."

Irregular rock spires in the "crater" of the Crater Mountains are remnants of volcanic conduits and dikes. Their dark color is due to desert varnish.

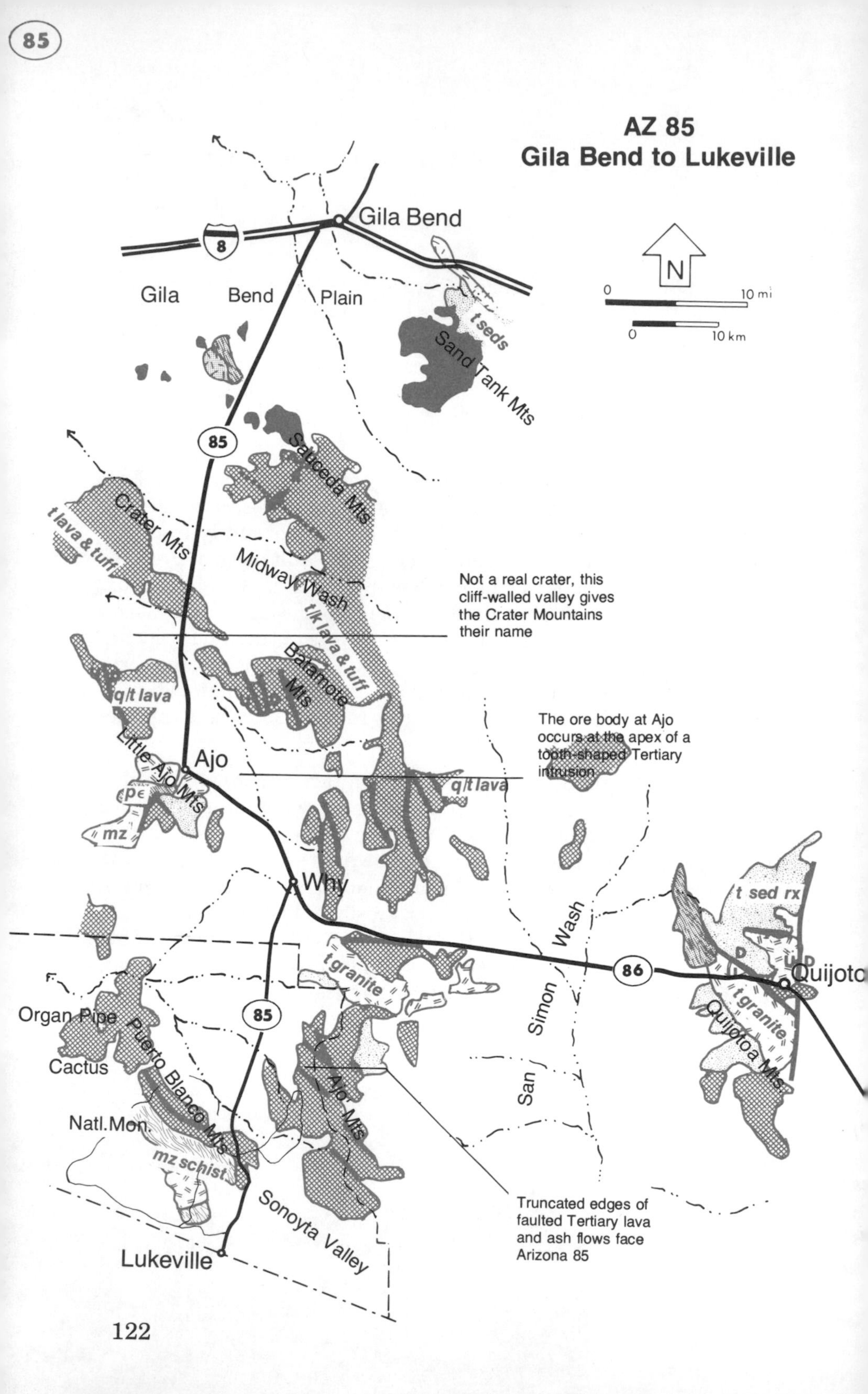
AZ 85
Gila Bend to Lukeville
Gila Bend
8
Gila
Bend
Plain
N
0
10 mi
0
10 km
t seds
Sand Tank Mts
85
Sauceda Mts
Crater Mts
t lava & tuff
Midway Wash
t/k lava & tuff
Not a real crater, this cliff-walled valley gives the Crater Mountains their name
Batamote Mts
q/t lava
The ore body at Ajo occurs at the apex of a tooth-shaped Tertiary intrusion
Little Ajo Mts
Ajo
pє
mz
q/t lava
Why
t sed rx
Wash
86
Quijotoa
t granite
Simon
t granite
Quijotoa Mts
Organ Pipe
Puerto Blanco Mts
85
Cactus
Ajo Mts
San
Natl.Mon.
mz schist
Sonoyta Valley
Truncated edges of faulted Tertiary lava and ash flows face Arizona 85
Lukeville

Broad Batamote Mountain, visible to the east from mile 35, is a shield volcano. *Its once fluid lava flows reach nearly to the highway.*

A broad cone-shaped mountain visible to the east from mile 35 is Batamote Mountain. This **shield volcano** was a center of Tertiary and Quaternary volcanic activity, the source of basalt lava flows that reach nearly to the highway. Its bulging central summit appears to be a volcanic dome, a mass of thick, pasty lava which effectively plugged the volcano's conduit.

Ajo (pronounced AH-ho) is a copper-mining town. Pure or native copper and colorful copper carbonate ores such as malachite were discovered here by Spanish settlers, perhaps as early as 1750. Their underground workings tunneled into rock directly above today's big open pit mine, where mining began in 1917. Turn right at the plaza stoplight to view the colorful New Cornelia pit, 850 feet deep and half a mile across.

The New Cornelia ore body is a typical porphyry copper deposit, closely related to a shallow Tertiary intrusion, of which it seems to form the apex. The first ores mined were chalcocite, malachite, and chrysocolla. Deeper ores mined since 1930 are copper-iron-sulfur compounds such as chalcopyrite and bornite. Ore is concentrated in the mill north of the pit, where it is crushed, finely ground, and mixed with water and frothing agents. The froth brings copper-rich particles to the surface, where they are skimmed off. The concentrate thus

'he New Cornelia Mine at Ajo yields copper ores ıssociated with a shallow Tertiary intrusion. A isitor viewpoint overlooks this pit.

obtained is smelted (melted) in large retorts. Almost pure copper sinks to the bottom of the melt and can be drawn off and cast into flat slabs for shipment. In addition to the copper, small quantities of gold and silver are recovered.

Tailings stretching northward and eastward from the mill contain both the unwanted overburden and finely milled material from

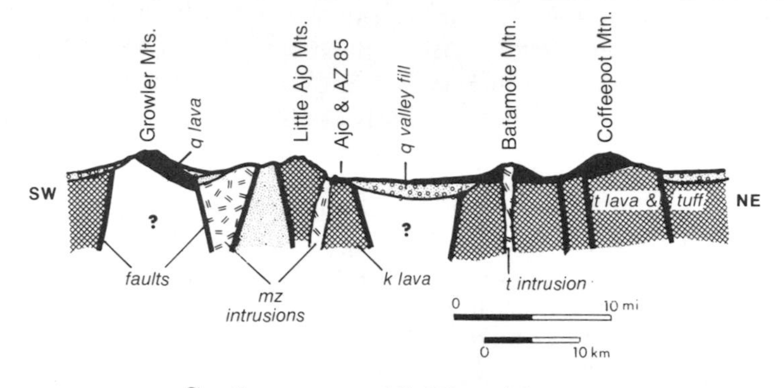

Section across AZ 85 at Ajo.

which the copper has been removed.

The southern part of the Little Ajo Range, visible from a few miles east of Ajo, contains faulted, tilted blocks of Precambrian gneiss, Cretaceous-Tertiary (Laramide) intrusive rocks, and some Cretaceous or Tertiary volcanic rocks, tipped every which way by Basin and Range faulting. Like other ranges in this part of Arizona, these mountains were once completely covered with valley gravel and volcanic flows that have now been eroded off.

East of the highway as it approaches the Y-shaped intersection at Why are the Pozo Redondo Mountains, piles of basalt and andesite lava and tuff faulted into north-south slivers, tilted westward, and then eroded.

At Why, Arizona 85 turns south toward Organ Pipe Cactus National Monument (discussed in Chapter V), Lukeville, and Gulf of

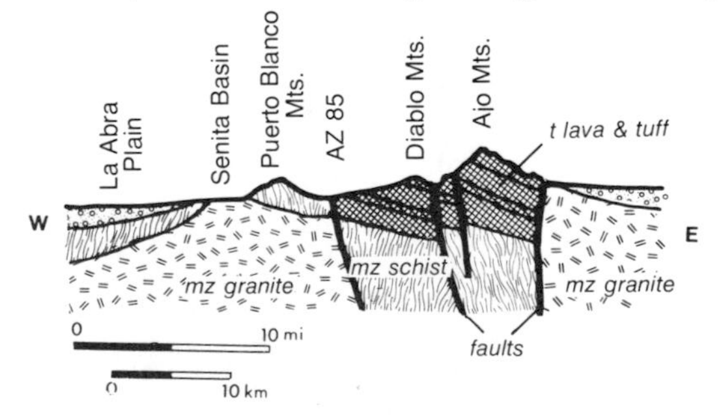

Section across AZ 85 at Organ Pipe Cactus National Monument.

California ports. The highway travels along the west-sloping bajada of the Ajo Mountains, crossing a number of large washes draining this range. All of the drainage from the west side of the Ajo Range funnels toward Growler Pass, visible to the west between the Growler and Puerto Blanco Mountains.

Rugged and forbidding in appearance, the Ajo Mountains are another of the many faulted volcanic ranges which characterize this part of Arizona. Most of the range is composed of rhyolite and andesite lava flows and layers of tuff faulted and tilted to the northeast. From the highway one can clearly see the sloping, vertically jointed, cliff-forming lava flows and softer, orange-tinted layers of volcanic ash. Dikes cut across some of these layers, forming sinuous ridges. Montezuma's Head, the prominent peak at the north end of the range, is capped with resistant rhyolite lava, as are other prominent peaks along the crest of the range.

The Puerto Blanco Mountains west of the highway at first glance appear to be another range of faulted, east-tilting volcanic rock. But in their southern and western parts are exposed Mesozoic granite and schist on which lie the younger volcanic rocks. East-dipping volcanic layers show up well in Twin Peaks near the entrance to Organ Pipe Cactus National Monument.

Near Twin Peaks, the highway crosses a low divide into Sonoyta Valley, which extends south to the Sonoita River in Mexico. Bajadas sloping toward this broad valley from mountains of two nations wear a stony armor of desert pavement, for dust devils, common here, sweep away silt and sand, leaving only a tight-packed pebbly surface. The area receives light rains in winter and heavier ones from summer thunderstorms, and with the mulch afforded by desert pavement a wide variety of desert plants grow here, including many species unknown in other parts of the country. Denser and different vegetation along washes indicates that moisture is more abundant there.

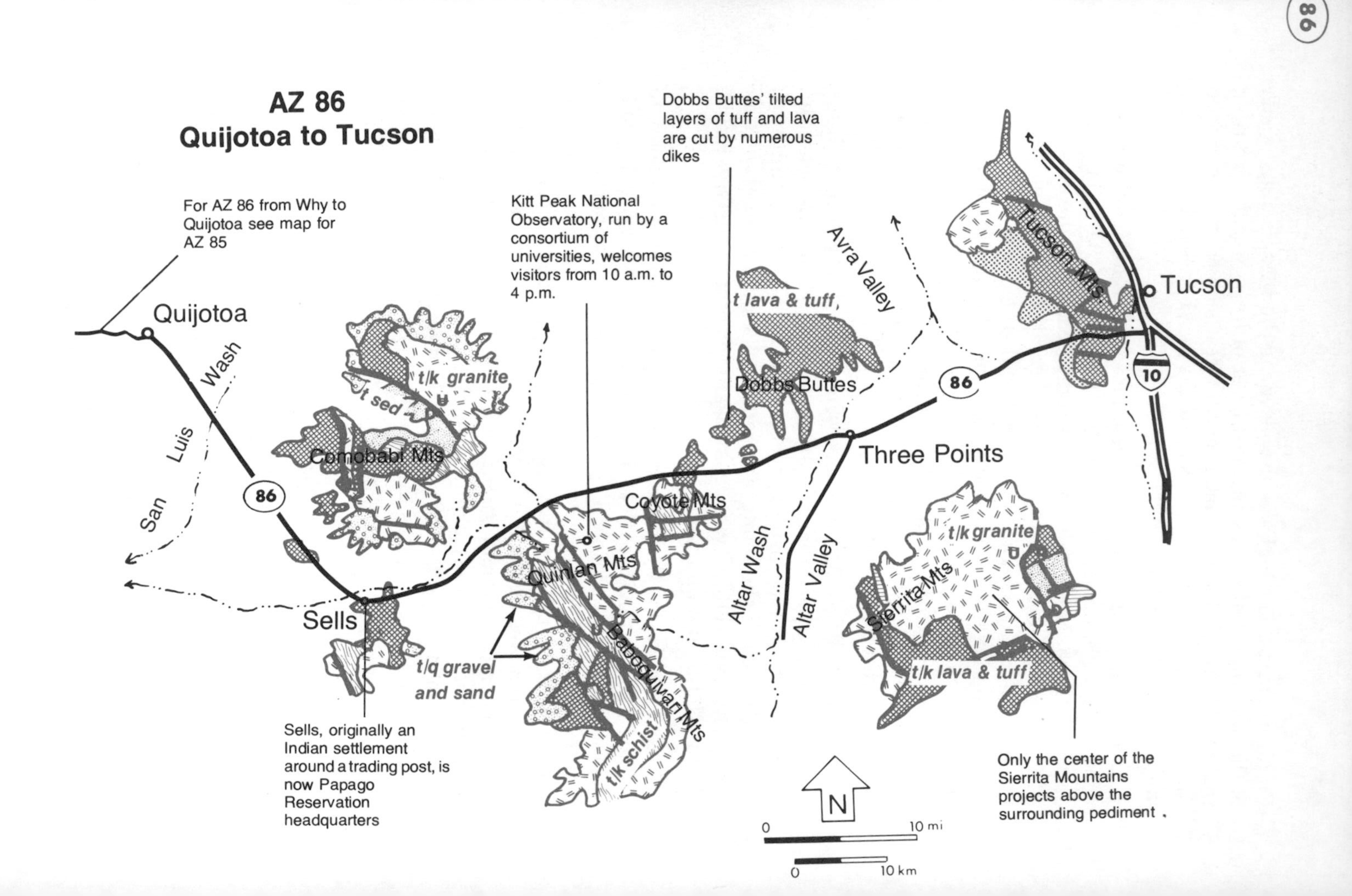
AZ 86
Quijotoa to Tucson
For AZ 86 from Why to Quijotoa see map for AZ 85
Kitt Peak National Observatory, run by a consortium of universities, welcomes visitors from 10 a.m. to 4 p.m.
Dobbs Buttes' tilted layers of tuff and lava are cut by numerous dikes
Quijotoa
San Luis Wash
86
Comobabi Mts
t/k granite
t sed
Sells
t/q gravel and sand
Quinlan Mts
Baboquivari Mts
t/k schist
Coyote Mts
t lava & tuff,
Dobbs Buttes
Avra Valley
Altar Wash
Altar Valley
Three Points
Sierrita Mts
t/k granite
t/k lava & tuff
Tucson Mts
Tucson
10
N
0
10 mi
0
10 km
Sells, originally an Indian settlement around a trading post, is now Papago Reservation headquarters
Only the center of the Sierrita Mountains projects above the surrounding pediment .

With deep-searching roots mesquite trees are able to survive the desert climate, growing along washes where underground moisture accumulates.
L.C. Huff photo, courtesy of USGS.

AZ 86
Why — Tucson
(118 miles)

As Arizona 86 leaves Why it circles the south end of the Pozo Redondo Mountains. Broken by faults, this range appears to be hardly more than a jumbled mass of volcanic blocks. To the south are the Ajo Mountains, faulted and deeply eroded lava flows, tuff, and thick, irregular masses of breccia that originated in violent explosive eruptions early in Tertiary time. Between the highway and the north end of the Ajo Mountains an arcuate band of Laramide granite forms low, light-colored hills.

East of Why the route descends the Ajo Mountain bajada into San Simon Valley. This area is part of the Papago Indian Reservation and contains small settlements, usually near springs or wells, which serve as summer homes for Papagos farming and ranching here during the summer rainy season. Portions of the area draw enough drainage to pond during the rainy months. Near mile 75 the highway acts as a dam, holding stream drainage and sheetwash to the north side of the road, where mesquite trees and other vegetation thrive on the added moisture. The predominant plant on most of the valley floor is creosote bush or greasewood, dark green, widely spaced shrubs with tiny resin-covered leaves. Before the white man's arrival, the valley was a sea of grass. It is thought that overgrazing in Spanish hacienda days destroyed the original grass-dominated flora in this and many other southern Arizona basins, allowing creosote bush, with bitter, unpalatable foliage, to become established. Each creosote's wide-spreading root system secretes poisons preventing or retarding growth of competing plants.

A marker at mile 79 indicates the Gila and Salt River Meridian, a precisely surveyed north-south line used as a base for land surveys in most of Arizona.

The Sierra Blanca and Quijotoa Mountains north and south of Quijotoa together make up a single fault block range with a central core of granite edged with slivers of Paleozoic limestone and quartzite. Both ranges are much lighter in color than the dark volcanic mountains farther west. Quartzite has been quarried here for use as a flux in the Ajo smelter, and small, rich pockets of copper, gold, and silver ores here and farther south along the range were mined at about the turn of the century. Notice the nearly straight eastern front of these ranges, the approximate line of the fault at the edge of the uplift block.

East of Quijotoa the highway descends another broad bajada to San Luis Wash, and then ascends still another to Sells. Numerous small, usually dry washes drain the Comobabi Mountains northeast of the highway.

In hills east of milepost 106, silver was mined as early as the 18th Century by Spanish and Indian miners. The hills between here and Sells are more erosional remnants of explosive-type volcanic rocks.

The Comobabi Mountains north of Sells are cored with Mesozoic granite and gneiss and bordered with volcanic rocks — a common picture, with variations, in many ranges along this route. The pattern seems to involve variations on the metamorphic core complex theme, a theme more fully developed in the Baboquivari and Quinlan Mountains farther east, in the Santa Catalina and Rincon Mountains near Tucson, and much farther north in other ranges.

The Baboquivari Mountains are in view now to the southeast, with

Baboquivari Peak's massive granite dome is part of a metamorphic core complex that extends north to Kitt Peak (background).
Peter Kresan photo.

the sharp turret of Baboquivari Peak (7730 feet) rising above them. Geologically, though not topographically, the range is continuous with the Comobabis, though there may be a down-faulted block between them. They are a long, narrow, sinuous range, and rise particularly sharply from the surrounding desert, with only a narrow band of foothills. Baboquivari Peak itself is Jurassic granite, whereas parts of the craggy, deeply dissected west slope are schist. The range as a whole consists of Jurassic, Cretaceous, and Tertiary granite with patches of schist and sedimentary rocks, in a variant of the metamorphic core complex pattern.

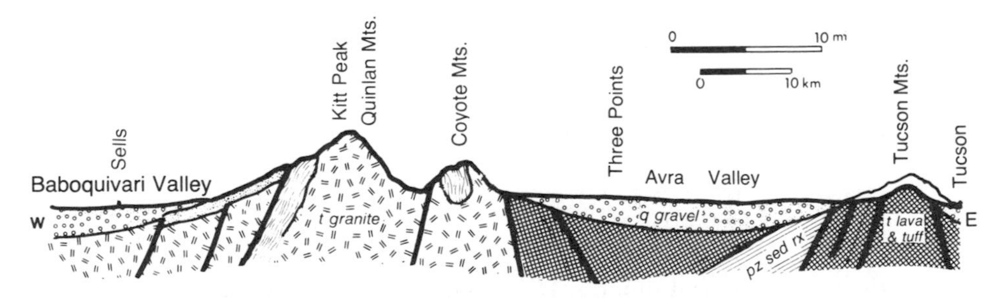

Section parallel to AZ 86 from Sells to Tucson. The highway runs north of the Quinlan and Coyote Mountains.

The Quinlan Mountains, with white observatory towers of Kitt Peak National Observatory, make up the northern end of the Baboquivari Range. The road to the observatory leaves this highway near milepost 134, and provides a good opportunity to see the dark diorite, white pegmatite, and salt-and-pepper granite of the range, as well as to look out over the surrounding Basin and Range country. Coyote Mountains south of mile 135 are part of the same geologic unit as the Quinlans, and are cut by dike swarms and edged with prominent faults. Farther east are some unaltered sedimentary and volcanic rocks; these are also exposed in a deep wash at mile 137.

Unlike the Baboquivari-Quinlan-Coyote metamorphic complex, Dobbs Buttes north of the highway are composed almost exclusively of volcanic tuff and thin lava flows. A little farther north these rocks rest on older sedimentary and volcanic rocks.

Originally, water used by the Papago Indians came from springs, some of them quite small. Early miners and traders dug wells that penetrated valley fill near the mountain fronts and encountered water at the bedrock surface; Indians settled near and became dependent on many of the wells. Although rainfall is sparse and seasonal, such water until recently has been adequate, even for supplementary agricultural use. As you leave the Papago Reservation, however, you will see that the the white man uses water much more extravagantly than do the Indians. In the Altar Valley between Three Points and Tucson, and in the Santa Cruz Valley near Tucson, groundwater is used freely for irrigation, and the once relatively shallow groundwater surface, the **water table**, has suffered for it. As depth to water increases, water for irrigation can be expected to become increasingly expensive. In addition, off-reservation pumping has brought dwindling groundwater supplies in wells on the Indian Reservation, wells essential to Indian livelihood. As a result, the Indians are asking for a share of the water from the Central Arizona Project, a massive water diversion network that will bring Colorado River water to Phoenix and Tucson.

The Sierrita Range southeast of Three Points is another granite-cored range, but in contrast to the long, narrow Baboquivaris the massive intrusion at its center is roughly circular in plan. On its east and west flanks are large fault slices of Paleozoic and Mesozoic sedimentary rocks and Cretaceous volcanic rocks, many of them altered by the intrusion. Some of these rocks are highly mineralized, and porphyry copper deposits are open-pit mined on the east side of the range. Broad slopes surrounding the Sierrita Range are particularly noteworthy. Their upper portions are a mountain pediment excavated in the fault block mountain mass, while their lower portions, flush with the upper, are valley fill, debris from the mountains.

A few islands of bedrock project through the thinly graveled pediment surface.

Kinney Road, going north from mile 166, leads to the Arizona Sonora Desert Museum and zoo, and to the west section of Saguaro National Monument in the northern Tucson Mountains.

The south end of the Tucsons, as well as smaller peaks south of the road, are little fault blocks of Cretaceous volcanic rock, a unit known as Cat Mountain Rhyolite, dated at 68 million years. The ridge of the Tucsons for some distance north of the highway displays two bands of resistant, cliff-forming volcanic breccia separated by softer, lighter-colored tuff. Dikes and volcanic necks are prominent.

Tucson's "central" business area, far west of the population center, lies in the valley of the Santa Cruz River. Early descriptions of this valley describe the Santa Cruz as a permanent stream bordered by cottonwood and mesquite trees. In about 1880 most streams in this part of Arizona began to entrench themselves in the valley sediments, bringing about a general lowering of the water table. Pumping from wells since then has further lowered the water table by several hundred feet, and the river as a result flows only in times of very heavy rains, and then only for a few hours or days.

AZ 95
Quartzsite — I-40
(90 miles)

North of Quartzsite this highway is almost immediately on a Tertiary sedimentary surface that forms the banks of Tyson Wash. Although it looks flat, this surface is trenched by desert washes; its pebble veneer is a good example of desert pavement, with the pebbles dark with desert varnish. Tyson Wash and its few tributaries are bordered with palo verde, ironwood, and other drought-tolerant trees, shrubs, and cacti that depend for their sustenance on occasional heavy rains. Drainage is to the west around the north end of the Dome Rock Mountains, but stream flow rarely reaches the Colorado River west of the range, though it may have in Pleistocene time. Instead it sinks into the desert, replenishing soil moisture along the watercourses, flowing very gradually through pore spaces in the coarse sand, gravel, and silt of the valley fill.

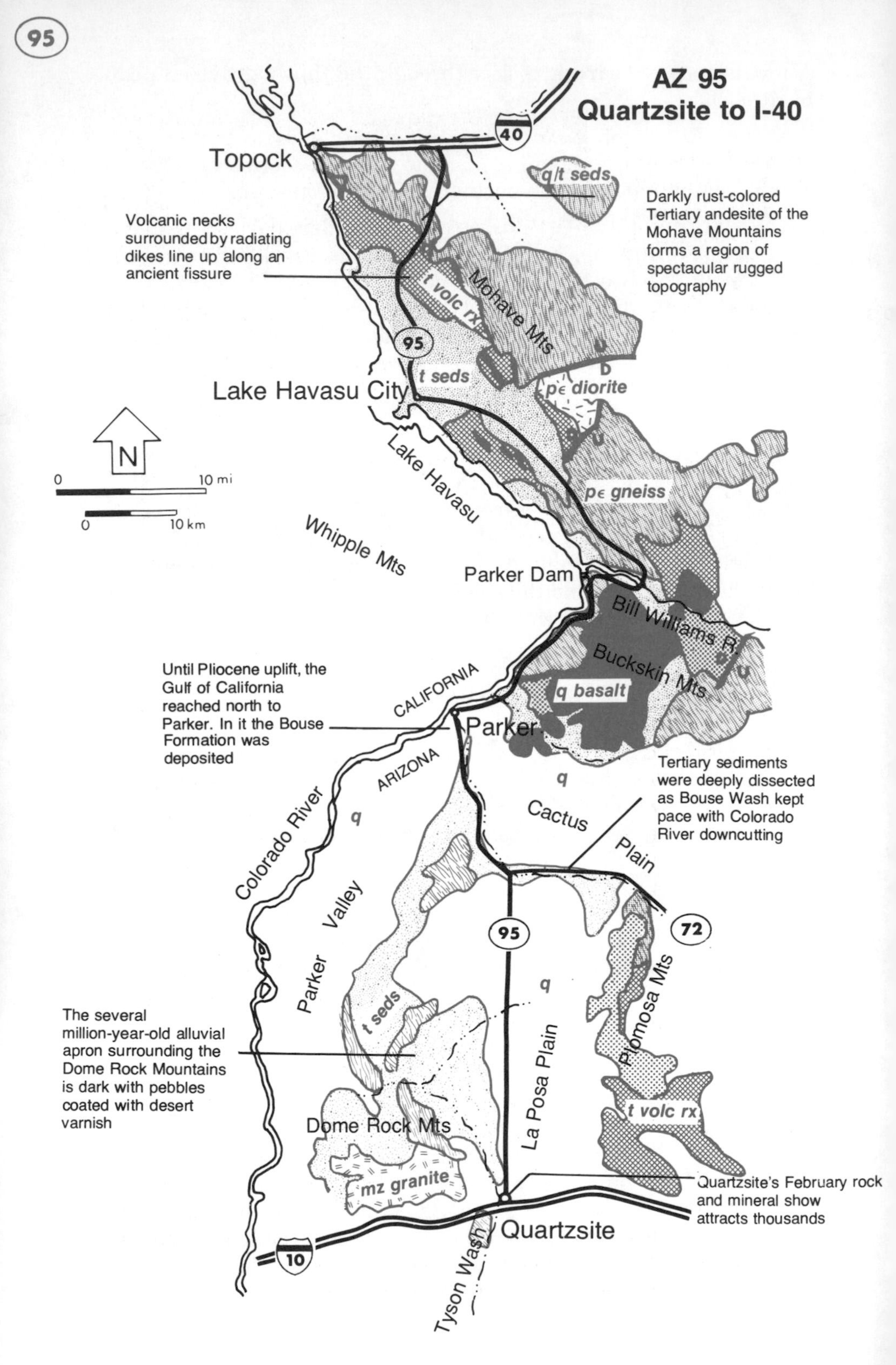
AZ 95
Quartzsite to I-40
40
Topock
q/t seds
Darkly rust-colored Tertiary andesite of the Mohave Mountains forms a region of spectacular rugged topography
Volcanic necks surrounded by radiating dikes line up along an ancient fissure
t volc rx
Mohave Mts
95
t seds
Lake Havasu City
pє diorite
N
0
10 mi
0
10 km
Lake Havasu
pє gneiss
Whipple Mts
Parker Dam
Bill Williams R.
Buckskin Mts
Until Pliocene uplift, the Gulf of California reached north to Parker. In it the Bouse Formation was deposited
CALIFORNIA
q basalt
Parker
ARIZONA
q
Tertiary sediments were deeply dissected as Bouse Wash kept pace with Colorado River downcutting
Colorado River
q
Cactus
Plain
Parker
Valley
95
72
q
Plomosa Mts
t seds
The several million-year-old alluvial apron surrounding the Dome Rock Mountains is dark with pebbles coated with desert varnish
La Posa Plain
t volc rx
Dome Rock Mts
mz granite
Quartzsite's February rock and mineral show attracts thousands
Quartzsite
10
Tyson Wash

The Bouse Formation consists of loosely consolidated Pliocene sediments deposited in and near a proto-Gulf of California.
D.G. Metzger photo, courtesy of USGS.

Mountains to the east are the Plomosa Range, a faulted, tilted block of Tertiary volcanic rocks and Mesozoic sedimentary rocks. Tilted sedimentary layers can be seen on northern parts of the range.

Hillocks of sand that border the highway near mile 125 look like sand dunes, and probably are, though they are grass covered now. A little less rain, a little more grazing, and the sand would again blow with the wind. Grayish valley sediments beneath them, visible north of mile 129, lie along the valley of Bouse Wash and are known as the Bouse Formation. They contain fossils that date them as Pliocene, 4 to 5 million years old. In part at least they are made up of fine-grained marine or brackish-water limestone, with siltstone and fine claystone in their upper part. As much as 2000 feet thick, and extending beyond the town of Parker, they were deposited when the Gulf of California extended much farther north than it does now. In the course of Basin and Range faulting and regional uplift they were raised a thousand feet above sea level. Originally they extended across this part of western Arizona and into California, lapping up onto the alluvial fans of mountains that were high enough at that time to have been islands or promontories. The Tertiary deposits slope right up to the north end of the Dome Rock Mountains, for instance, and interlayer with alluvial fans there.

Turning west at mile 131 the highway continues among these sediments, following Bouse Wash toward Parker. Here you can appreciate the thickness of the Bouse Formation and the way it merges with debris from the mountains.

Northwest of the highway are the Buckskin Mountains, a metamorphic core complex range superimposed with some rough and jumbled volcanic topography here near its western end. Structurally the range, as you will see north of Parker, crosses the Colorado River to become the Whipple Mountains in California.

Below Parker Dam the Colorado River threads through stark ranges of Precambrian gneiss and Tertiary volcanic rock.

In late Miocene and early Pliocene time just a few mountaintops jutted through the sea of limestone, sand, and clay that came to be deposited here. As the present mountains were lifted, the Colorado River was beginning to flow west to make its way through what would become Grand Canyon, and to turn south along the down-faulted valleys — grabens — seeking its way to the gulf. The river's course was at first high above most of the buried mountains, whose structural grain in some cases lay at an angle to its path. The plain on which it flowed formed a ramp, so to speak, to the Gulf of California, across which the river could pretty much pick and choose the easiest route. The Whipple-Buckskin Mountains, though, lay at an angle to the river, so that the river turned southeastward, found a way to pass between the ranges, and then turned southwestward again — all the while flowing on the Tertiary sedimentary blanket. As uplift continued, erosion intensified both here and in the faraway Rocky Mountains where the powerful river drew its nourishment, and the river's course was increasingly deepened. Thus entrenched, the Colorado continued to cut down, until it reached the buried ranges across its path. Erosion must then have slowed down, for the Precambrian rock of these ranges was far harder than the new-formed Bouse Formation sediments and their upstream equivalents. But the river carried powerful erosion tools — sand and gravel — and used them effectively to jackhammer and sandblast its path through the moun-

Coarse, gravelly terrace deposits, surfaced now with desert pavement, contrast with steep, sloping volcanic layers of the Mohave Mountains.

tains. Farther north it was also cutting into Precambrian rocks in the depths of Grand Canyon and in the Black Mountains near Hoover Dam. In such resistant rock, the river carved narrow, deep canyons; in regions of soft sediments it shaped wide, shallow valleys. From time to time, as downcutting slowed, newer gravel was deposited in the valleys, only to be dissected again to form terraces like those near Parker.

Between Parker and Parker Dam the highway rides on one of these lower terraces, following a trail obligingly pioneered by the river. Most of the rocks at the roadside are the Precambrian gneiss core of the Buckskin and Whipple Mountains, typical metamorphic core complex ranges. Dark purple volcanic rocks also match up across the river; they are Tertiary in age, and erupted before the river cut this canyon.

Precambrian gneiss, tough and hard, walls the narrow gorge at Parker Dam, forming at its narrowest point an ideal dam site. The gneiss is deep red here, heavily marked with yellowish or very dark brown veins. The dam shunts water to the California side, through the Colorado River Aqueduct to Los Angeles. Rising waters have made a shallow, marshy estuary of Bill Williams River just north of Parker Dam; 20 miles farther upstream, beyond the end of the narrow gneiss-walled canyon, they spread out as Lake Havasu.

North of Bill Williams River the highway skirts the southern part of the Mohave Mountains, more Precambrian gneiss bordered with Tertiary volcanic rocks, again matched on the California side of the river. (The view into California includes a prominent spine — a volcanic neck.) At miles 172-177 the sharp backbones of several dikes parallel the highway, and then the scenery opens up into Chemehuevi Valley and Lake Havasu. Tertiary sediments underlie the great terrace on which Lake Havasu City lies; the lowest, youngest terraces and the still lower river floodplain lie beneath the waters of the lake.

North of Lake Havasu City, volcanic necks punctuate the desert skyline.

In the Mohave Mountains east of Lake Havasu, a Precambrian gneiss core is edged with volcanic rocks that are Tertiary in age. Below them, alluvial gravels slope toward the river, with unusually well developed, light-colored flat-topped levees bordering the stream channels that cross them. Older portions of the alluvial apron have had time to darken with slow-forming desert varnish.

Arizona 95 continues north beyond Lake Havasu City to cut through volcanic rocks of the western Mohave Mountains. These ominous-appearing rocks almost certainly were deposited on irregular and steeply sloping terrain, in association with explosive volcanoes. In addition they have been tilted by faulting. As a result, they appear to follow no set pattern: here sôme lava, there some volcanic ash, and on beyond some conglomerate made up of lava boulders and ash deposited by mudflows and landslides. The lava is andesite, a thick, viscous type of lava that spreads into short, stubby flows.

Higher up the climb into the Mohave Mountains the road is edged with Precambrian gneiss well displayed in roadcuts. Beyond the highway summit at mile 196 the route continues through this gneiss, and then comes out of the mountains onto a gravel-covered pediment and alluvial apron that slopes away from the range into the valley of Sacramento Wash. The geology there is described under I-40 Colorado River — Kingman.

Chapter III
Mazatzal Land — The Central Highlands

The Central Highlands lie in a diagonal band almost bisecting the state and make up a transition zone between the southern and western Basin and Range deserts and the Colorado Plateau. In fact they are often called the Transition Zone. The fault-bordered mountains are in general higher and more closely spaced, but they display the same northwest-southeast trend as most of the desert ranges. The fault-bound valleys between are shallower and less broad, less filled in with sediment, than the desert basins. On the other hand, in parts of the Transition Zone where they are exposed the sedimentary rocks — both Precambrian and Paleozoic — are flat-lying or nearly so, like those of the Colorado Plateau.

Abundant evidence, much of it quite well exposed, shows that the Highlands underwent several pulses of uplift involving both folding and faulting, and finally were lifted higher than either the desert to the south or the plateau to the north. They have lost to the forces of erosion their once thick robe of Paleozoic and possibly Mesozoic rocks, eroding down to their very cores — hard igneous and metamorphic rocks of Precambrian age, 1 to 2 billion years old.

In addition to a number of small basins, a string of larger valleys runs lengthwise through the Central Highlands, dividing them or separating them from the Colorado Plateau to the north. Large or small, many of these valleys display well developed terraces, marginal ones of coarse gravel and more

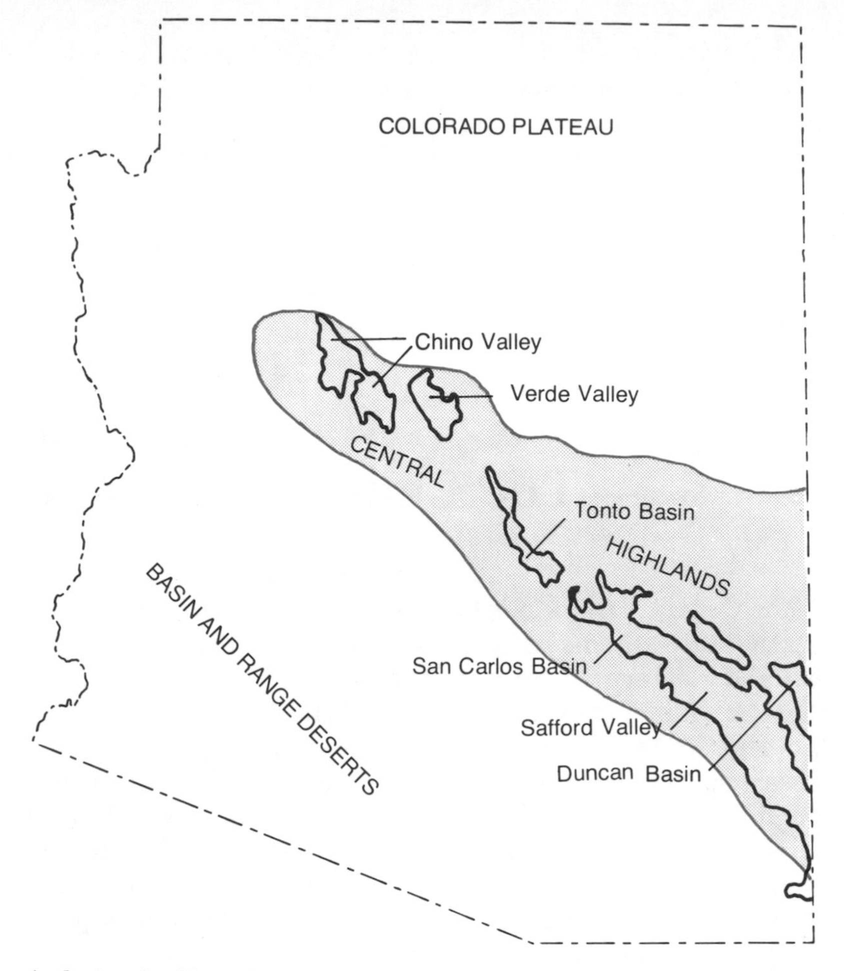

A chain of valleys divides the Central Highlands (shaded) or separates them from the Colorado Plateau to the north.

central ones of fine lake silt, lake limestone, water-laid volcanic ash, and in places salt or gypsum. Many of the deposits also contain fossil bones of Pliocene and Pleistocene mammals.

The terraces indicate a complex drainage history. When first formed the valleys seem to have been isolated from one another by faulting, by mountain uplift, and perhaps by volcanism. Only in comparatively recent time, well along during the Pleistocene Epoch, did the valleys interconnect into one vast network that drained ultimately across the desert to the Colorado River.

The high country receives more rain than the rest of the

state, and large parts of it are heavily timbered either with pygmy forests of piñon and juniper or with stately stands of ponderosa pine and Douglas fir. Streams and small lakes abound. Much of the stream flow, however, is trapped in man-made lakes and routed to agricultural areas in the desert.

Despite dams, floods are not uncommon in the mountain valleys. Streams that drain the larger valleys have had to erode their paths, with cobbles and pebbles and sand as abrasive tools, through resistant barriers of Precambrian rock. Most of this erosion has taken place in less than the last 2 million years.

As in the desert ranges, the ranges of the Central Highlands are in many places sharpened by the frost of winter nights and pried apart by the roots of trees. Particularly along the southwest edge of the Highlands, desert processes have crept up the mountainsides, and the mountains are skirted with alluvial fans and broad, gently sloping bajadas. Long, deep, in-place weathering has reduced near-surface granite and gneiss and **diabase** (a dark intrusive rock) to rounded boulders surrounded by coarse sand or **grus** — tan in the case of granite, dark brown in the case of gneiss and diabase. Where boulders of these rocks are exposed they are commonly rounded, as in the desert ranges, by exfoliation due to heating and cooling.

Because most of the intermontane basins are small in comparison with the ranges that surround them, more of the jigsaw puzzle of their history can be reconstructed than in the Basin and Range Province. There are still major gaps about which we know next to nothing: several billions of years in Precambrian time, 200 million years during the Ordovician and Silurian Periods, and more millions during parts of the Mesozoic Era. Nevertheless we can piece together this history:

Between 1820 and 1750 million years ago, more than halfway through Precambrian time, volcanic rocks were accumulating across Arizona: thousands of vertical feet of lava flows, welded tuff, and volcanic breccia. This long and apparently violent period of volcanism was followed by accumulation of sandstone and shale, such as the thick layers of sandstone that now, altered by metamorphism, make up the prominent Mazatzal Quartzite of central Arizona. Thereafter, between 1730 and 1650 million years ago, the Arizona region drifted southeastward against another crustal plate, as described earlier. As the colliding plates buckled and crumpled, in an episode of mountain-building known as the Mazatzal Revolution (named after a range in the Central Highlands), a great mountain range rose, mountains possibly on the scale of the Himalayas, with a SW-NE fabric extending far beyond the limits of today's Arizona. Intrusion of vast granite batholiths took on the same linear trend, still revealed in irregular bands of granite exposed across the Highland ranges. These rocks — sedimentary, metamorphic, and intrusive — make up the Older Precambrian sequence.

Gradually the rugged ranges were worn down by erosion. Then as the area subsided, layers of sedimentary rocks — the Younger Precambrian sequence — were deposited across the vast eroded plain. New sediments — sand, silt, clay, and limy mud — alternated with volcanic ash and floods of lava. Dark diabase magma then forced its way up into the granite and metamorphic rocks and squeezed between the sedimentary layers, hardening into sills now well exposed in the Salt River Canyon. And then the land was beveled once more by long erosion that lasted until the end of Precambrian time.

Soon after the Paleozoic Era began, Cambrian seas crept across the beveled surface and deposited layers of sandstone, now the Bolsa Quartzite. The sandstone, or quartzite, lies directly on beveled Precambrian rocks, though in places remnants of younger Precambrian rocks stuck up through the Cambrian seas as islands. The "Great Unconformity" between the Precambrian and Cambrian rocks represents close to a billion years of time.

The Cambrian quartzite and other sedimentary layers above it thicken to the west, thin eastward. By analyzing the distribution of rock types we learn that during Paleozoic time the sea transgressed from west to east, and that responding to slight fluctuations in the level of the land it came and went a number of times. Toward the end of the era, desert dunes and wide coastal floodplains left their mark on the sedimentary sequence. Here in Arizona the Paleozoic Era was remarkably free of volcanic activity.

We have little record of events in the Central Highlands during the Mesozoic Era, but the rudiments of what happened there can be reconstructed from rocks exposed to the north. In Triassic time the area seems to have risen enough to be called mountainous, and thereafter streams carried debris from the central Arizona mountains to a broad coastal plain farther north. We assume for lack of contrary evidence that the region was still well above sea level in Jurassic and most of Cretaceous time.

Late in the Mesozoic Era compressive or collision-type mountain-building began, mountain building that matured as the Laramide Orogeny early in Tertiary time. Between 75 and 50 million years ago the whole state, as well as adjacent states, became increasingly mountainous, with folding and faulting and volcanism each contributing to the change. By now the continent was drifting southwestward, bumping into several small Pacific microcontinents. Because the compression was in a northeast-southwest direction, the mountains took on a northwest-southeast trend, perpendicular to the compression, like folds in a piece of paper when opposite corners are pushed together. As in the plateau country to the north, some flat-lying blocks of crust were lifted or dropped without a lot of distortion. But the Central Highlands as a whole were lifted several thousand feet higher than either the desert or the plateau regions.

During the next 30 million years erosion again ruled. Mountains were worn down, valleys were partly filled. Fine rock debris from the Central Highlands was swept northward as far as broad interior lakes in the area where Wyoming, Colorado, and Utah meet today, and southward into lowland basins. Coarse gravel was deposited nearer to the mountains; some of it became today's Rim Gravels at the edge of the Colorado Plateau.

Recurrent uplift occurred in mid-Tertiary time 13 to 7 million years ago. In the Central Highlands several ranges rising separately disrupted drainage patterns. For a time the basins between ranges had no outlets, and as much as 3000 feet of sediment and volcanic rocks accumulated in them: stream gravel and sand, lake limestone and silt, gypsum and salt, volcanic ash and lava flows. Basin filling eventually brought integrated drainage — through-flowing streams — as streams connected the basins. And though lake and stream deposits continued to accumulate for a time, ultimately the streams, some of them by then sizeable rivers, began to cut down through them. The downcutting of the valleys lasts to this day, exposing to our view the thick, terrace-forming sedimentary layers and bringing us a picture, with the fossils they contain, of freshwater and saltwater lakes bordered by lush savannahs supporting herds of camels and antelope, as well as pigs, mastodons, and other animals, a scene more like that of central Africa than of present-day Arizona.

Parts of the Central Highlands, particularly areas around Jerome, Globe and Miami, and Clifton and Morenci, have a rich mining history. The Clifton-Morenci mines as well as the Globe-Miami district produce copper ores, generally chalcopyrite and chalcocite, that lie in the contact zone where early Tertiary intrusive rocks, most of which are granite porphyry intruded during the Laramide Orogeny, have penetrated sedimentary rocks — particularly limestones.In these regards the ores are similar to the porphyry copper deposits of southern Arizona. Such ore bodies are usually large, but their ores are low-grade and buried beneath considerable overburden. Hence large-scale open-pit mining is the most economical way to get the ore out.

Tours of a mine and smelter can be arranged through the Chamber of Commerce in Globe. At this writing, low copper prices have closed many of Arizona's mines.

Copper ores mined at Jerome, where the mines shut down in 1953, are not associated with either Tertiary intrusions or Paleozoic limestone, but are the result of Precambrian volcanism. The ores accumulated at the top of a massive pile of Precambrian undersea volcanic rocks.

Interstate 17
Phoenix — Camp Verde
(87 miles)

Once north of the Phoenix metropolitan area, this highway passes between hills and small mountains of Precambrian gneiss, schist, and granite, and several small mesas capped with lava flows, a foretaste of things to come in the Central Highlands. At mile 208 the route crosses the Arizona Canal, a man-made waterway that distributes Salt River water to this rich agricultural area. North of mile 220 another canal will eventually distribute water pumped uphill from the Colorado River. Volcanic rocks and red Tertiary sandstone are exposed near milepost 227.

Where the route appoaches the Central Highlands many small faulted blocks of Precambrian metamorphic rocks jut above the desert floor like miniature mountain ranges. Farther along, the same rocks are exposed near the road: both gneiss and schist. Other hills close to the highway are remnants of Tertiary volcanic rocks. Higher mesas behind them are capped with basalt lava flows both Tertiary and Quaternary in age.

Near New River the road emerges onto light-colored, well stratified Tertiary lake and stream sediments, the former fine-grained and limy, the latter with coarse fragments of volcanic rocks whose not yet rounded shapes suggest that they came from nearby lava flows. Stream-type cross-bedding is common in the uppermost parts of these sediments.

Crossing the Agua Fria River at mile 243 the highway enters the mountains proper. Basalt-covered tablelands ahead edge narrow gorges cut by mountain-fed streams through dark Precambrian metamorphic rocks. Natural outcrops and roadcuts north of this point

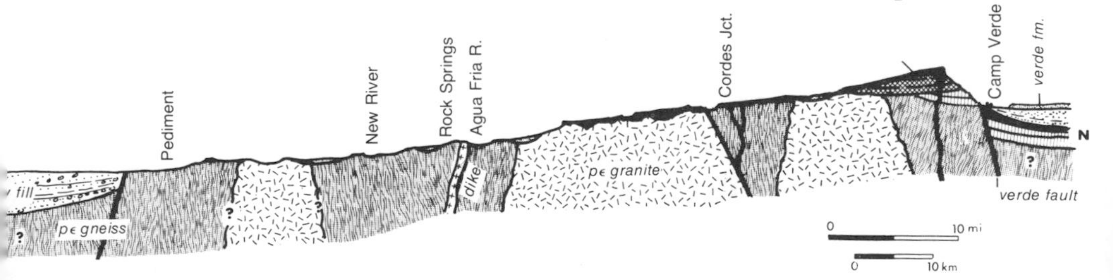

Section along I-17 from Phoenix to Camp Verde, greatly simplified. Paleozoic sedimentary rocks shown near the Verde Fault are exposed 6 miles southeast of I-17.

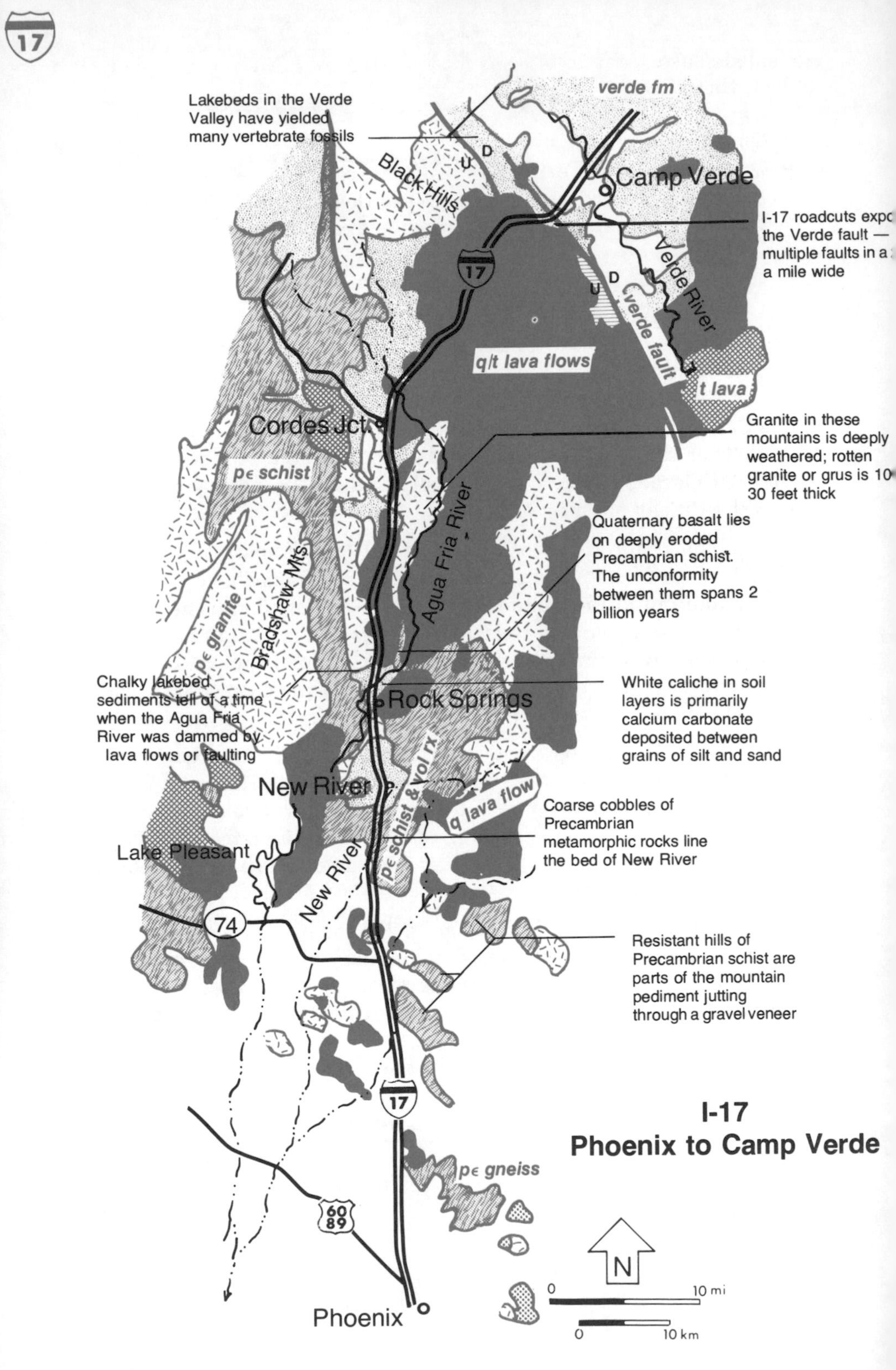

17
Lakebeds in the Verde Valley have yielded many vertebrate fossils
verde fm
Black Hills
U
D
Camp Verde
I-17 roadcuts expo the Verde fault — multiple faults in a a mile wide
17
Verde River
U
D
verde fault
q/t lava flows
t lava
Cordes Jct.
Granite in these mountains is deeply weathered; rotten granite or grus is 10 30 feet thick
pє schist
Agua Fria River
Quaternary basalt lies on deeply eroded Precambrian schist. The unconformity between them spans 2 billion years
Bradshaw Mts
pє granite
Chalky lakebed sediments tell of a time when the Agua Fria River was dammed by lava flows or faulting
Rock Springs
White caliche in soil layers is primarily calcium carbonate deposited between grains of silt and sand
New River
pє schist & vol rx
q lava flow
Coarse cobbles of Precambrian metamorphic rocks line the bed of New River
Lake Pleasant
New River
74
Resistant hills of Precambrian schist are parts of the mountain pediment jutting through a gravel veneer
17
I-17
Phoenix to Camp Verde
pє gneiss
60
89
N
0
10 mi
0
10 km
Phoenix

reveal six basic, easy-to-tell-apart rock categories. From youngest to oldest, these are:

- Quaternary gravel;
- Quaternary and Tertiary lava flows, commonly lying on soil zones baked brick red by the heat of the flowing lava;
- Tertiary stream deposits: sand, silt, and gravel with stream-rounded pebbles;
- Tertiary lake sediments: flat-lying layers of whitish limestone, siltstone, and water-laid volcanic ash;
- Precambrian granite: grainy tan or light gray rock that weathers to rounded boulders and knobs;
- Precambrian metamorphic rock, mostly flaky, silvery schist.

As you can see, Cenozoic rocks lie right on or against Precambrian ones here. The absence of Paleozoic and Mesozoic rocks was for a long time taken to mean that the Central Highlands had been a topographic high all though Paleozoic and Mesozoic time, an interpretation now disproven.

In places the irregular pre-lava surface of the granite and schist is quite well defined and shows up clearly in roadcuts. North of milepost 247, for instance, in the highway cut, two lava flows lie on an uneven purple or brick-colored baked soil zone. One of the flows appears to be **pillow lava**, with bulging protuberances that form when hot lava flows into water.

The highway climbs steadily and finally emerges on a grass-covered lava plateau. Sunset Point Rest Area looks down on Black Canyon and the dark metamorphic rocks about 1.7 million years old that give it its name. The Bradshaw Mountains, across Black Canyon, are walled with the same rock but have at their heart a large mass of granite that intruded the metamorphic rocks in Precambrian time.

The lava plateau north of the rest area forms the drainage divide between Turkey Creek (an Agua Fria tributary) and the Verde River. The basalt of the plateau shows up well here: dark gray, splotched and lined with tones of rust resulting from oxidation and deep weathering.

At milepost 256 the highway leaves the basalt and drops down onto a hilly surface of pale Precambrian granite. Below the surface this rock has weathered into rounded boulders surrounded with rotten granite or **grus**. Such deep weathering suggests long exposure not far below the surface. Watch for other knobs rounded by **exfoliation**, the ongoing surface phenomenon that takes up where subsurface weathering leaves off, peeling off thin layers of rock. In places here the granite is so finely broken by a mêlée of joints that it weathers and disintegrates directly into soil, rather than forming knobs and boulders.

Near Cordes Junction are more loosely consolidated Tertiary stream and lake deposits, some of them with basalt caps. Faulting in Tertiary time created many isolated lakes in these mountains, as these sediments now document. Similar deposits are being formed today in undrained depressions on top of the lava plateau, as well as in man-made lakes.

At mile 281 look carefully across at the large roadcut on the south-bound part of I-17. The irregular surface filled in by basalt flows shows up well, and several prominent near-vertical faults are beautifully displayed. Farther north, other highway cuts show faults and broken fragments of volcanic rock. This zone of cracked and faulted rock, nearly a mile wide, marks the Verde Fault, a major break that separates the Central Highlands from the Verde Valley north of it. Movement on this fault started in mid-Tertiary time and continued, off and on, until Pleistocene time. Actually there is evidence of much earlier Precambrian movement too.

The Verde Valley is demarcated almost as sharply on its northeast side by the Mogollon Rim, the south edge of the Colorado Plateau (see Chapter IV). The Mogollon (MUGgy-own) escarpment, though, is an erosional break rather than a fault scarp. It became established probably in Oligocene time, before the Grand Canyon or the Central Highlands existed as such, when an ancestor of the Verde River began to cut down through weak red rocks that underlie the harder layers of the plateau surface. Minor faults may have initiated the downcutting.

Flat-topped, grayish-white hills in the center of the Verde Valley are chalky lake limestone and siltstone deposited when the Verde River was dammed at its southeast end by uplift of the Central Highlands. Streams that drained into this valley were, and still are, laden with calcium carbonate dissolved from abundant Paleozoic limestones of the Plateau. In the still waters of a succession of lakes, algae and microscopic animals absorbed the calcium carbonate, using it for shells and skeletons. As they died, this material, mixed with silt and sand washed in by streams, accumulated on the lake floor as today's Verde Formation.

A national monument of archeologic and geologic interest, Montezuma Castle, lies in the Verde Valley. It is discussed in Chapter V. You may also want to visit Tuzigoot National Monument nearby.

Tilted Paleozoic strata east of Superior range in age from Cambrian to Pennsylvanian.

U.S. 60
Florence Junction — Globe
(39 miles)

For a map of this route see US 60/89 Mesa — Florence Junction, in Chapter II.

From Florence Junction, US 60 heads across a high Quaternary terrace toward the Central Highlands. North of the highway, rugged hills of the southern Superstition Mountains display thick beds of Tertiary tuff accented by sharply jutting dikes and volcanic necks, sources of this region's explosive volcanic outpourings. Northeast of the volcanic rocks are Precambrian metamorphic and intrusive rocks of the Iron Mountains, part of the Central Highlands, separated from the basins and ranges of the desert by a complex array of faults that run beneath the broad slope that surrounds the mountains.

Beyond mile 217 the route climbs into some of these Precambrian rocks, which appear in roadcuts. Particularly noticeable is a shiny blue gray schist.

The prominent squared-off peak south of the highway near mile 224, Picketpost Mountain, wears a cap of lava over thick layers of

Glistening coils of copper wire are products of Globe's copper mills.

volcanic tuff that resemble those of the Superstition Mountains. The lava at the top flowed from a vent on the east side of the mountain about 18 million years ago. A light-colored, beady type of volcanic glass, **perlite**, is mined in hills east of the mountain. Puffed up and expanded by heating, perlite forms a good lightweight aggregate and is used in plaster and insulating materials and as a soil lightener for greenhouses. Apache Tears, dark marble-sized globs of **obsidian** or volcanic glass, are found nearby.

The Boyce Thompson Southwestern Arboretum, named for the mining magnate who founded it, lies just below Picketpost Mountain. Run by the University of Arizona, it is a center for botanical and ecologic studies of the desert. Open to the public, the arboretum displays about 1500 species of desert plants, as well as a "geological garden."

Superior, at the north end of Dripping Spring Mountains, nestles below mountain slopes striped with tilted Paleozoic sedimentary rocks pushed up along the fault that edges the mountains. These rocks range in age from Cambrian to Pennsylvanian. The layers of quartzite, gray limestone, and red-brown shale are cut by many secondary faults, some of which served as channels for copper-rich mineral solutions. Superior's underground mines are about 3000 feet below the highway.

Just west of the tunnel the highway crosses the contact between Paleozoic sedimentary rocks and overlying Tertiary welded tuff, resistant but brittle and highly fractured rock. East of the tunnel, as you will see, the welded tuff has eroded into sharp ridges and towers that overlook Queen Creek and its tributaries. Upward the tuff is less and less firmly welded, and where the highway emerges on top of it near milepost 231 it is deeply weathered.

The highway crosses the north end of the Pinal Mountains on a granite intrusion known as the Schultze stock. Dated at 58 million

In the Central Highlands as elsewhere, rounded boulders are characteristic of weathered granite terrain. Curved slabs break off in a process called exfoliation.

F. L. Ransome photo, courtesy of USGS.

Man-made strata – mine and mill tailings near Miami – are measures of the immense amounts of rock that have been removed from mines in this area, mostly from open pit mines hidden behind their own dumps.

years, the granite shows up in typical granite fashion as knobs and piles of rounded rock — good examples of spheroidal weathering. More finely jointed granite is visible just west of the Pinto Creek bridge, where one of the roadcuts follows parallel joints in the rock. Toward the top of the cut, well weathered granite gradually takes on a "rotten granite" texture and falls apart readily into coarse sand.

Mines near Miami, including the large multi-pit mine near milepost 237, obtain copper ore from ore bodies at the contact of the Schultze Granite with surrounding rocks. Over 2 billion pounds of copper have been mined here. If you have never seen an open pit copper mine or a copper mill, go on a tour here.

Between Miami and Globe the highway crosses the upper (northwest) end of a long graben that extends southeastward as the Safford Valley. During Pliocene time this graben received more than 1500 vertical feet of rock debris from mountains rising around it, debris that is now part of the Gila Conglomerate. With pulse-like episodes of upwarping in Pleistocene and to some extent in Recent time, through drainage was established and the valley sediments were gradually carved by the Gila River into the terraces that now edge the graben. Globe is built on several terrace levels, as you can see from mile 251.

Mill tailings near Globe display the gullying erosion that attacks steep slopes unprotected by vegetation.

US 60
Globe to Show Low

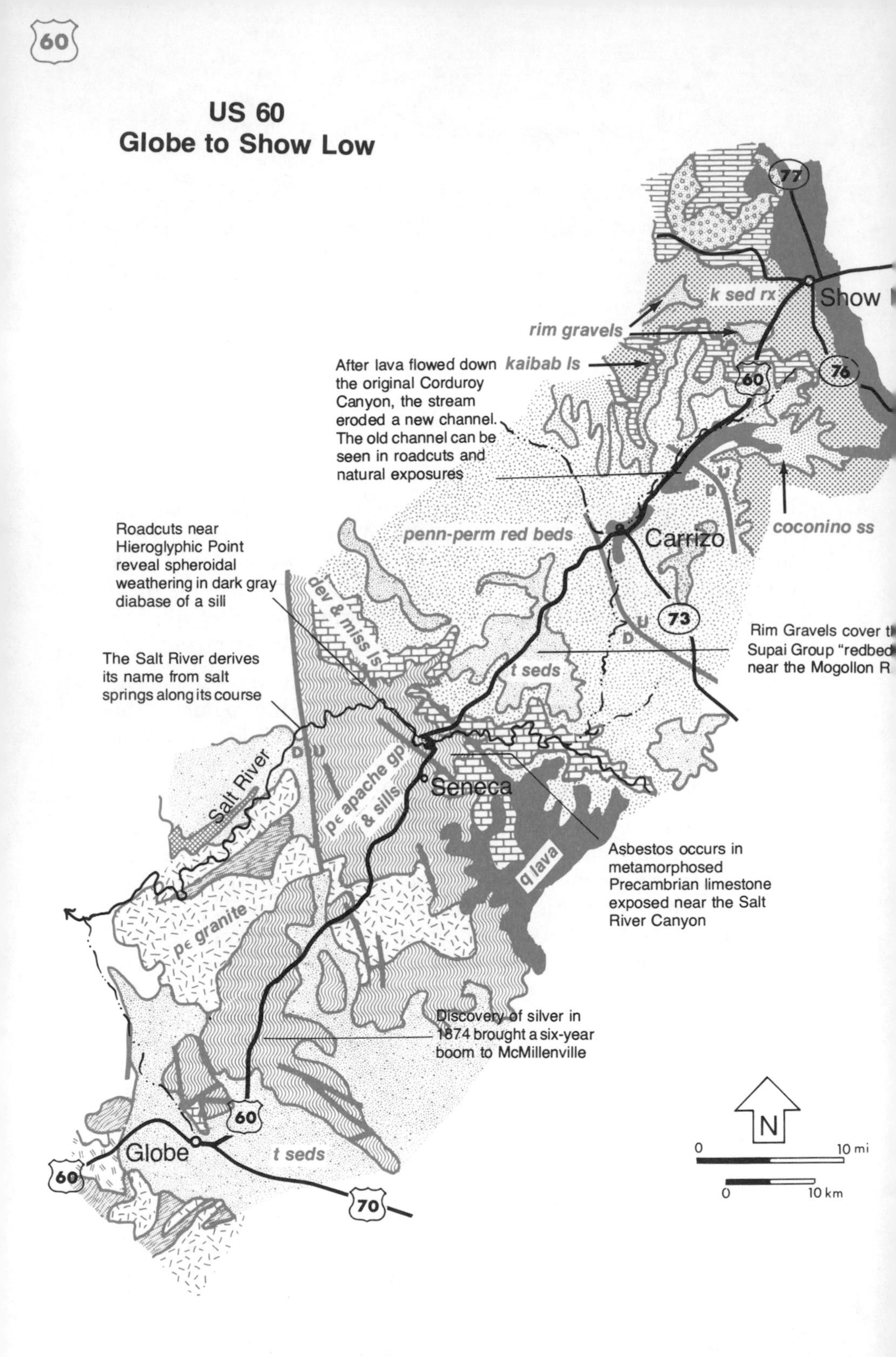

U.S. 60
Globe — Show Low
(87 miles)

North of Globe this highway climbs onto a terrace of Gila Conglomerate, a coarse Tertiary-Quaternary unit more fully discussed in the next two sections. Here you can see the coarse cobbles and pebbles of the conglomerate, and at mile 254 you can look east along the San Carlos graben and see the gravel terraces that edge it.

The Gila Conglomerate shortly gives way to hills of hard Precambrian sedimentary rocks. There are several Precambrian formations here, all belonging to the Apache Group. They range through quartzite, conglomerate, and shale to marbleized limestone. Very roughly a billion years old, they have had plenty of time to compact and recrystallize and harden. Where they first appear at mile 253, brittle quartzite and occasional layers of **round-pebble conglomerate** are cut by many joints so that they break into rectangular blocks. Close inspection shows that they are so well consolidated that they break *through* rather than *around* their sand grains and pebbles — a distinguishing feature between quartzite and sandstone.

Another type of rock makes its first appearance near milepost 259: dark gray diabase, an intrusive rock common in dikes and sills. Here the diabase forced a way between the layered rock units of the Apache Group. It is difficult to visualize the great pressures that forced molten magma between hard, deeply buried rock layers! The sills, like the sedimentary rocks, are Precambrian, but because they penetrate the Apache Group we know that they are the younger of the two. The diabase is quite hard, and caps and protects hills in which it occurs.

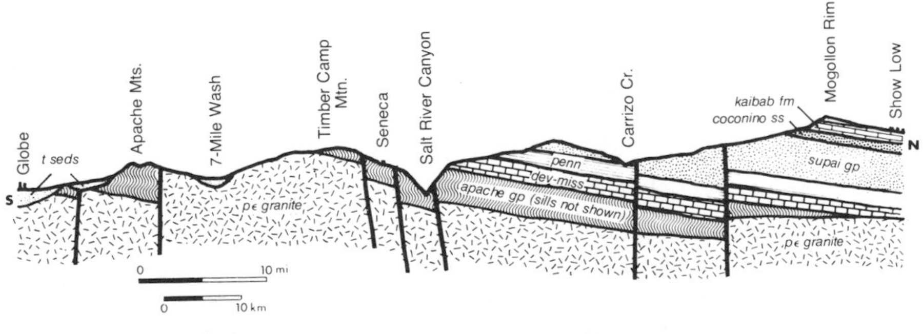

Section along US 60 Globe to Show Low.

The Apache Group rests on granite, which you'll see, deeply decomposed, farther along the highway. The sills cut the granite too, and shift from one sedimentary level to another, meeting and fusing with other sills in a complex pattern that has only been deciphered by detailed mapping. The granite and sedimentary rocks step up to the north along a series of faults that keep them at highway level for some distance.

Still farther north, sills east of the highway are topped with light-colored tuff, solidified volcanic ash. Lava flows cap some hills and appear in the valley bottom near Seven Mile Wash. However most rock for the next five miles is granite, deeply weathered, even to depths of 60 feet, and marked with veins of white quartz. Granite-derived soils are coarse, sandy, light brown. In places the granite is almost white, lacking the iron and magnesium that go into dark minerals. The very light granite is strikingly marked with dark diabase dikes in the roadcut just north of milepost 272. Near mile 274 the granite contains unusually large feldspar crystals, some more than 3 inches across.

Looking west from the big S-shaped highway fill at mile 278 you can see Four Peaks in the Mazatzal Mountains, also part of the Central Highlands. Geologists once thought that the Central Highlands existed as highlands throughout Paleozoic and Mesozoic time (an idea now disproved), and named their theoretical highland or island Mazatzal Land.

Near the top of the Timber Camp Mountains, at mile 278-279, sedimentary rocks of the Apache Group once more appear. These orange-brown rocks dip northward here, and in addition are faulted, so they again remain at roadside level as the highway gradually descends toward the rim of the Salt River Canyon.

In the Seneca area **asbestos** occurs where diabase sills penetrate and alter Precambrian limestone. Because its fine, fiberlike crystals will not conduct heat or electricity, asbestos can be spun and woven and made into fireproof clothing or theater curtains or electrical insulation. Mines are some distance from the highway.

Stop at Hieroglyphic Point to look at the Salt River Canyon and the rocks it exposes, as well as to see the prehistoric petroglyphs pecked through the desert varnish that coats blocks of diabase. Petroglyphs near the base of the steps, thought to have been chiseled about 1000 years ago, have barely begun to show new deposits of desert varnish.

The south wall of the canyon consists almost exclusively of Precambrian strata and their associated diabase sills. In the roadcut near the viewpoint one of the sills displays deep below-the-surface spheroidal weathering. Unweathered, the rock is dark gray to brown,

The view west from Hieroglyphic Point shows north-dipping Precambrian strata and sills overlain at the upper right by younger layers.

with very irregular columnar jointing. Farther down the highway some Apache Group limestone is altered to marble by the heat of the intruding sills.

Since the Apache Group maintains its northward dip, these rocks form only the lowest two thirds of the north canyon wall. The lighter-colored, more clearly stratified rocks in the upper slopes are Paleozoic: the Devonian Martin Limestone and above it the Mississippian Redwall Limestone.

Paleozoic limestones in the north canyon wall are best seen from Becker Butte Overlook at milepost 297. There are no Cambrian, Ordovician, or Silurian strata here. On Becker Butte and in roadcuts near the viewpoint the Devonian Martin Limestone, characterized by thin greenish mudstone bands between gray-brown limestone layers, lies above a thick sill which apparently was exposed on the sea floor in Devonian time. The Martin Limestone contains fossil seashells, but they are hard to get to here.

Above the Martin is a cliff of Mississippian Redwall Limestone, another fossil-bearing marine limestone. Here it displays **karst** features formed late in Mississippian time when the new-made limestone was lifted above the sea: breccia-filled sinkholes, solution caves, evidence of collapsed valleys, and residual red-orange clay soils. Blobby fist-to-football-sized lumps in the Redwall are nodules of **chert**, a form of silica.

Paleozoic sedimentary rocks, like the Precambrian ones below them, slope gently northward. This northward dip, plus a rising surface elevation, places progressively younger rocks at road level. The contact between the Redwall and Martin Limestones comes half a mile above the viewpoint, above a thick bed of blue-gray shale. That between the karst top of the Redwall and the overlying Pennsylva-

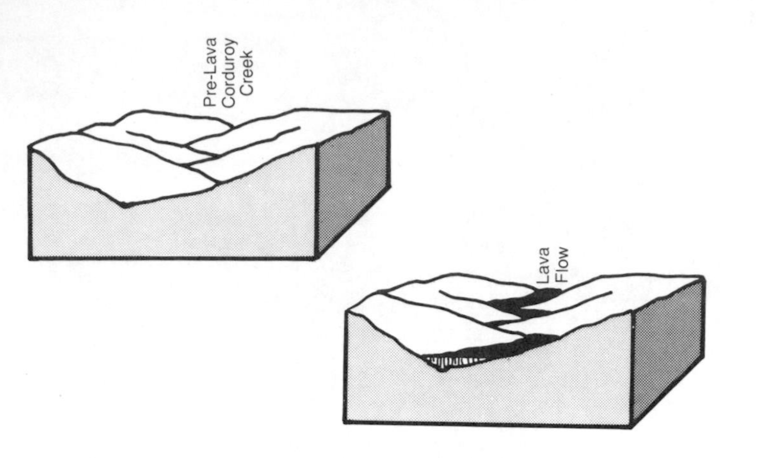

When lava flowed down Corduroy Creek it displaced the stream. Because rocks of the bordering ridges were weaker and eroded away, the lava now caps a long, narrow ridge adjacent to the stream's new course.

nian Naco Limestone is another half mile north. Buff-colored and many-layered, the Naco is flexed into several little anticlines and synclines between miles 299 and 301.

North of milepost 301 red mudstone and siltstone layers of the Supai Group soon appear. These Pennsylvanian-Permian marine and coastal deposits include a wide range of rock types — limestone, claystone, mudstone, sandstone, conglomerate — just what one might expect on a delta, river floodplain, or shore where environments change with every sea level or shoreline change.

After crossing Carrizo Creek the highway follows Corduroy Creek, with a long, narrow ridge of basalt capping Snake Ridge west of the highway. Forced from an older bed by the Snake Ridge lava flow, Corduroy Creek moved over and developed a new channel beside it. Watch roadcuts carefully in miles 320 and 321; as the road curves across the line of the flow and back again you can see the lava-filled channel in cross section.

Beyond milepost 325 are outcrops of yellow-white Coconino Sandstone and thin, buff-colored layers of sandy Kaibab Limestone, both Permian in age, both normally cliff-formers marking the edge of the Mogollon Rim, the escarpment at the southern edge of the Colorado Plateau. Both Kaibab and Coconino are atypical here, where they seem to represent a fluctuating near-shore environment. In this area, the exact position of the rim is concealed by lava flows that are part of the White Mountain volcanic field.

U.S. 70
New Mexico — Safford
(44 miles)

Where US 70 enters Arizona, thick river gravels fill the valley of the Gila River. The river swings north here, while the highway crosses the north end of the Peloncillo Mountains, a volcanic range nearly swallowed up in the sea of cobbly debris which surrounds it. The high-level, fairly well cemented gravels, with rounded pebbles and cobbles of river deposits, are mainly Pliocene and Pleistocene. The coarseness of the sediments reflects rapid stream flow and steep gradients when large-scale faulting lifted neighboring mountain blocks.

From Duncan an alternate route, scenic and historic, follows AZ 75 and US 666 to Clifton and Morenci, still active mining towns. Geology is shown on the map.

The town of Duncan is gradually being moved to higher ground because of floods that periodically sweep along the Gila. Plans for upstream dams, which could lessen flooding, are now being considered.

Volcanic rocks of the northern Peloncillo Mountains are both older and younger than the gravels. Except for some Quaternary basalt, the flat-lying lava flows north of the road near mile 371-370, for instance, the mountains are largely purple Tertiary **rhyolite**. Near milepost 361 several dikes and volcanic necks project above the general surface. Some old silver mine dumps are being reprocessed here now, using new and efficient techniques that extract more silver than before.

West of the Peloncillo Mountains we come to another sea of coarse gravel in the Safford Valley. These deposits, known (somewhat imprecisely, for they vary from place to place) as the Gila Conglomerate, were originally more than 1500 feet thick. As on the other side of the mountains they owe their existence to repeated uplift and faulting in surrounding ranges. The coarsest deposits, some with boulders up to 12 feet in diameter, are close to the mountains and can be considered

US 70
New Mexico to Safford

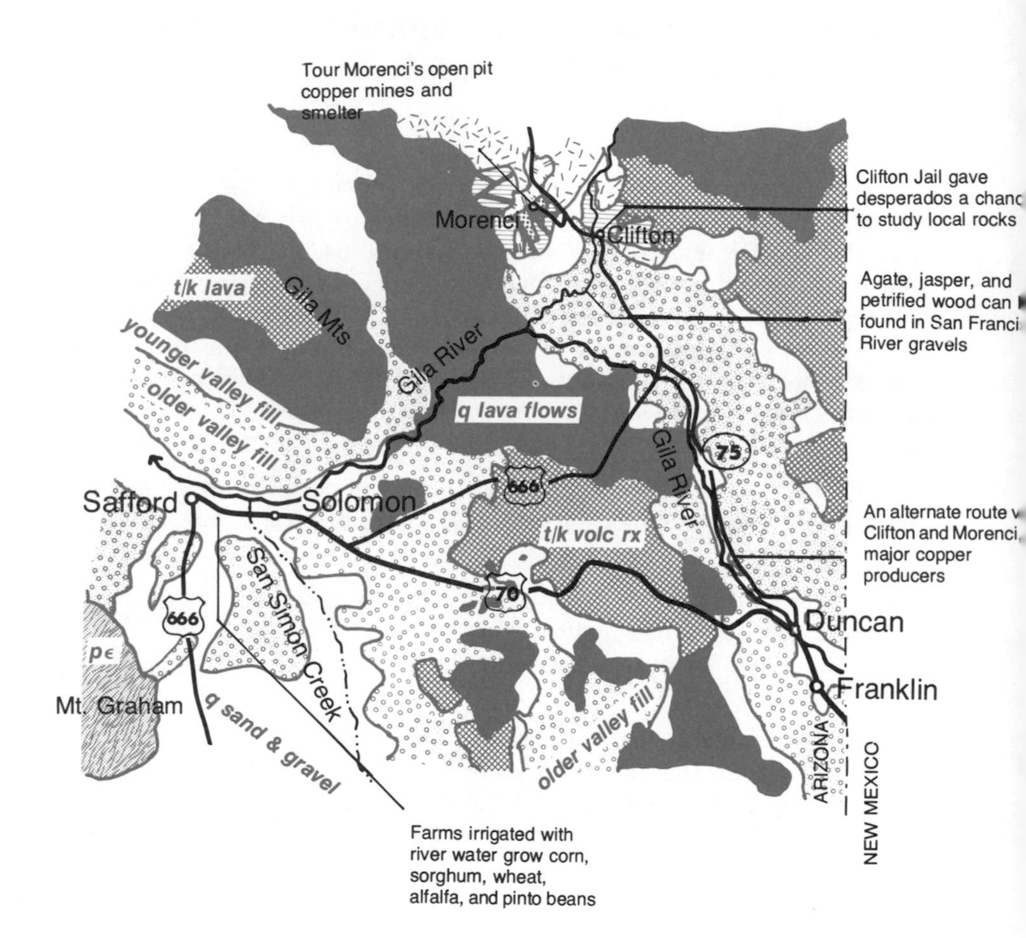

alluvial fan conglomerates, or **fanglomerates**. Farther from the mountains the deposits are finer, and in the center of the valley there are buried lake and playa deposits that may be thousands of feet thick.

The Gila Conglomerate here can be separated into older and younger layers, partly on the basis of whether or not the layers have been deformed by continuing uplift. Both older and younger deposits consist of fanglomerates, gravel, and lake deposits. Older and younger can be distinguished by degree of cementation, which is weaker in younger deposits; by color, which is reddish in younger deposits; and by vertebrate fossils, which are early to mid-Pliocene in the older deposits, late Pliocene and early Pleistocene in the younger. Older lake deposits include gypsum and salt, evidence that for a time the Safford Valley contained an undrained, salty lake. In places an irregular channeled surface separates the older and younger deposits.

The Gila River didn't exist as such when the older valley deposits formed. Streams tumbling from the mountains dropped their stony loads and ponded into lakes near Duncan and in the Safford Valley. Later, as the highlands were elevated again, the Gila and San Francisco Rivers deposited the younger gravels.

As in other parts of southeastern Arizona gradual drainage changes eventually brought through-flowing streams and rivers that cut down into the former deposits, a process that is still going on today. The Gila River has now cut through 900 feet of younger and older deposits, leaving remnants of younger deposits high and dry along the edges of the mountains. These remnants show up as unnatural-looking terraces high on the sides of the Pinaleno Mountains southwest of the Safford Valley and the Gila Mountains to the northeast.

What happened to all the material removed by the river? Most of it was put to use as tools for chiseling a deep downstream channel through the Mescal, Dripping Spring, and Tortilla Mountains. Pounded in the river's mill, it eventually wound up as sand and silt, spread out over the flat-floored desert around Florence, where the gradient and therefore the carrying power of the river suddenly decreased.

Mt. Graham (10,713 feet) and the Pinaleno Mountains are a metamorphic core complex type of range, with a core of Precambrian granite and metamorphic rocks. They are probably fringed with slivers of Paleozoic and Mesozoic rocks now hidden, if they exist, beneath thick valley gravels. (For a discussion of this type of range, see the Introduction to Chapter II.) On this side the range is mostly Precambrian gneiss; on the other side it is mostly granite.

The Safford Valley marks the boundary between Arizona's Central Highlands and the deserts of the Basin and Range Province. As with most natural boundaries this borderland between the two provinces has many characteristics of each. The mountains to the north, a mélange of Tertiary and Quaternary volcanic rocks, are classed as part of the Central Highlands. In fragments along their southern edge — in deep canyons like that of the San Francisco River — we find large masses of Precambrian igneous and metamorphic rock and smaller patches of much faulted Paleozoic rock similar to those that make up other parts of the Central Highlands. These little patches are especially tantalizing because where they are exposed, as at Clifton, Morenci, and Globe, they may show mineralization with silver, copper, and other valuable metals.

Both the Duncan Basin and the Safford Valley are links in the chain of large valleys that marches northwest through the Central Highlands or separates them from plateau country to the north. These valleys are not just products of erosion, like the Mississippi River Valley or the valley of the Colorado River. They are at least in part structural valleys, grabens, downfaulted along one or both sides.

Looking northeast from mile 352, you can see that the Gila River has cut a deep canyon through flat-lying basalt lava flows, as well as through some of the Pliocene-Pleistocene gravels which rise in terraces along either side of the valley. The dark lava flows extend as a long tongue, partly eroded away, from the volcanic region of the White Mountains of east-central Arizona. On this side of the river the highway rides a gravel surface of the highest terrace. Soon it drops to lower levels and to the modern valley floor and the town of Safford.

U.S. 70
Safford — Globe
(77 miles)

On either side of Safford and the Gila River, eroded bluffs expose gravelly, cobble-filled layers of the Gila Conglomerate. Southwest of town, high on the flank of Mt. Graham, what appears at first glance to be a big alluvial fan with sliced-off edges stands out from the mountain base. A fan it is — also part of the Gila Conglomerate. In Pleistocene time these layers of valley fill, more or less continuous with gravels in other southeastern Arizona valleys, filled in this valley to a depth of more than 1500 feet. They came into existence during Pliocene and early Pleistocene time, when repeated faulting and mountain uplift established the long valley or graben now occupied by the Gila River.

In mid-Pleistocene time, after the Gila Conglomerate was deposited, regional uplift of this part of Arizona brought on an erosional cycle that lasts, with minor fluctuations, to this day. The present integrated drainage system, with tributary streams flowing into the Gila River, is therefore quite young. Only within the last 2 million years has the Gila flowed northwest through the northern part of the valley, and then emptied southwestward through the mountains into the desert lowlands around Phoenix. Before that time, streams from the mountains disgorged into the long, undrained, often lake-filled valley, and flowed no farther. But as the Gila cut down through three

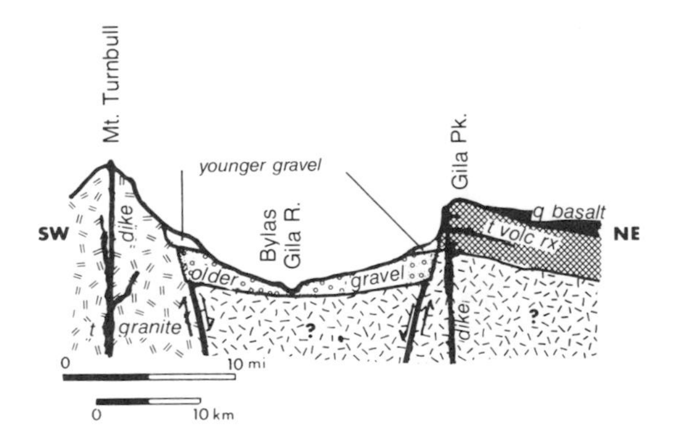

Section across US 70 Safford to Globe.

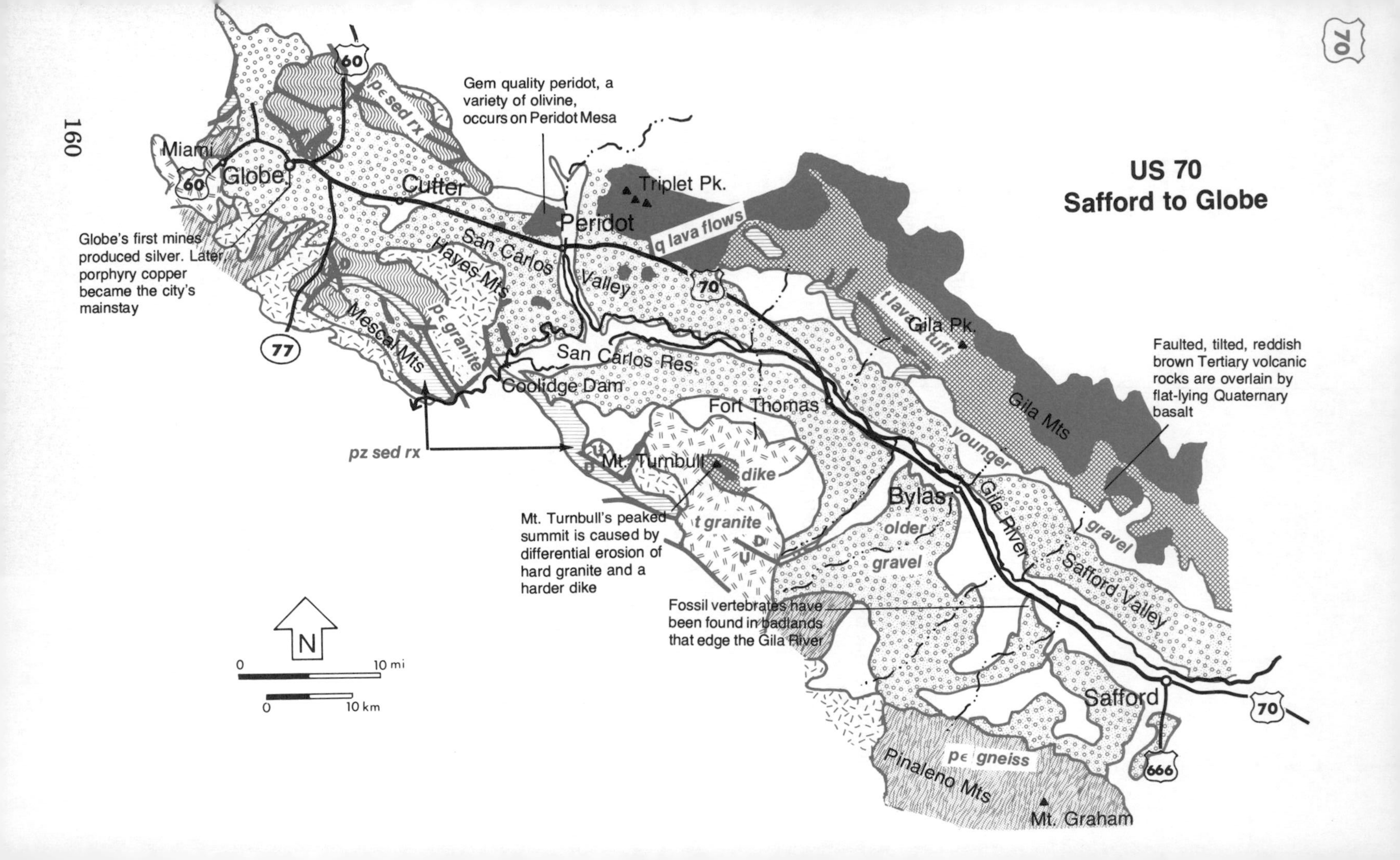
US 70
Safford to Globe
70
Gem quality peridot, a variety of olivine, occurs on Peridot Mesa
Globe's first mines produced silver. Later, porphyry copper became the city's mainstay
Faulted, tilted, reddish brown Tertiary volcanic rocks are overlain by flat-lying Quaternary basalt
Mt. Turnbull's peaked summit is caused by differential erosion of hard granite and a harder dike
Fossil vertebrates have been found in badlands that edge the Gila River
Miami
Globe
Cutter
Peridot
Triplet Pk.
q lava flows
San Carlos
Valley
Hayes Mts
Mescal Mts
pє sed rx
pє granite
pz sed rx
San Carlos Res.
Coolidge Dam
Fort Thomas
Mt. Turnbull
dike
t granite
t lava
tuff
Gila Pk.
Gila Mts
younger
gravel
Bylas
older
gravel
Gila River
Safford Valley
Safford
pє gneiss
Pinaleno Mts
Mt. Graham
60
77
666
N
0
10 mi
0
10 km

successive mountain ranges west of the present Coolidge Dam, step by step it also in its upper reaches cut through the Gila Conglomerate, washing away the boulders, cobbles, sand and silty lake deposits that had filled the valley.

Between Safford and Eden some of the finer valley deposits appear at roadside level: lake silt and fine volcanic ash. Erosion here is rapid. Clay minerals in volcanic ash, swelling and shrinking with moisture and dryness, prevent the growth of plants, and true soils do not develop. In the dry climate, infrequent but drenching rains cut steep, tortuous paths through the soft deposits, unprotected as they are by roots or leaves.

The terraces that border the Gila become more irregular northwestward. Those northeast of the river merge with modern alluvial fans along the edges of the Gila Mountains. Mt. Turnbull (7945 feet) southwest of Bylas is surrounded with similar fans, coalesced into an even apron of rock debris. Mt. Turnbull itself is a large intrusion of Tertiary granite tipped right at the summit with a dike of harder rock that gives the mountain a peaked-hat appearance.

The valley watered by the Gila River is a fertile one and benefits from a mild climate and a long growing season. Water for irrigation is diverted from the river by gravity-flow ditches.

North of mile 292 the floodplain has been cleared of salt cedars or tamarisks, dusty green, feathery, overly thirsty, non-native trees that have invaded many southwestern watercourses, in another effort to control the water supply here. Where the trees are gone the river's channel can be seen to advantage. The gradient is low, and as one winding channel fills with sand another is created, giving a braided design to the sand and mud on the floodplain.

In flood, the Gila River endangers farms on its low-lying floodplain. There is talk now of dams that would control the river's flow, providing a constant supply of irrigation water and at least to some degree alleviating periodic floods.

Near San Carlos Reservoir, Triplet Peak stands above flat-lying lava flows and terrace gravels. Here at their northwest end the Gila Mountains become subdued, with dark, nearly horizontal lava flows covering most of the surface. The three-tipped peak and a smaller peak east of it are made up of volcanic breccia, a common rock type in old volcanic conduits. Note that gravel layers below the volcanic rocks seem to sag under the great weight of the lava flows and breccia. The earth's crust is somewhat flexible, we know, and does sag when it is loaded. In volcanic regions there is an additional reason for such sagging: magma has been removed from magma chambers, and the crust may sink to fill the void.

Coolidge Dam, completed in 1928, is at the head of a narrow gorge where the Gila turns southwest through the Mescal Mountains. N the dam the gorge is edged with Precambr granite and steeply tilted layers of Precambrian and Paleoz sedimentary rocks.

The small town of Peridot gets its name from a clear, pale variety of an olive green mineral called, justly, **olivine**. Gem quality **peridot** crystals occur in a basalt flow on Peridot Mesa near the town. Basalt containing olivine or peridot comes from a particularly deep source, the semimolten upper mantle, below the crust. Its presence here indicates that breaks in the crust, pathways for the rising lava, must extend all the way down to the mantle.

San Carlos Valley's Pliocene and Pleistocene sediments, continuous with those of Safford Valley, include whitish lake limestone. You'll see more and more of this limestone as the highway nears Globe. In and beyond Globe the valley deposits are covered with monumental man-made terraces: mine dumps and mill tailings. Hidden from view behind the dumps, the open pit copper mines yield abundant low-grade porphyry copper ore on a scale impossible in underground operations. Tours visit some of the large copper pits.

Copper ore is processed by crushing and grinding it in nearby mills, mixing it with water and frothing agents. The froth brings copper-rich particles to the surface, where they can be skimmed off and relayed to the smelter. There the concentrate is melted. Almost pure copper sinks to the bottom of the melt and is drawn off and cast into slabs. After further refining, the pure, malleable metal is rolled into sheets or spun out into wire.

The mountains surrounding Globe expose complex arrays of Precambrian granite, schist, and quartzite, finely sliced Paleozoic limestone and quartzite, and several Tertiary granite intrusions, shallow, rapidly cooled masses of porphyry in which not all the minerals had time to grow into large crystals. Over and over again, here and in desert ranges farther south, we see a pattern of copper mineralization associated with such intrusions, especially where they have come in contact with Paleozoic limestone. Because of this association the ores have come to be known as **porphyry copper deposits**.

Near Globe the rocks are cut by multitudes of faults — many many more than can be shown on the map. Faults show up particularly well in mines, where there is incentive to map and understand them. Much to everyone's dismay, faults may sharply and abruptly cut off promising ore bodies. On the other hand, they may also serve as avenues for concentration of ore minerals. It's hard to know of course whether the same number of faults would be found elsewhere were one to dig big holes or honeycomb the mountains with tunnels, as in the mining areas, studying the faults in detail as one went along.

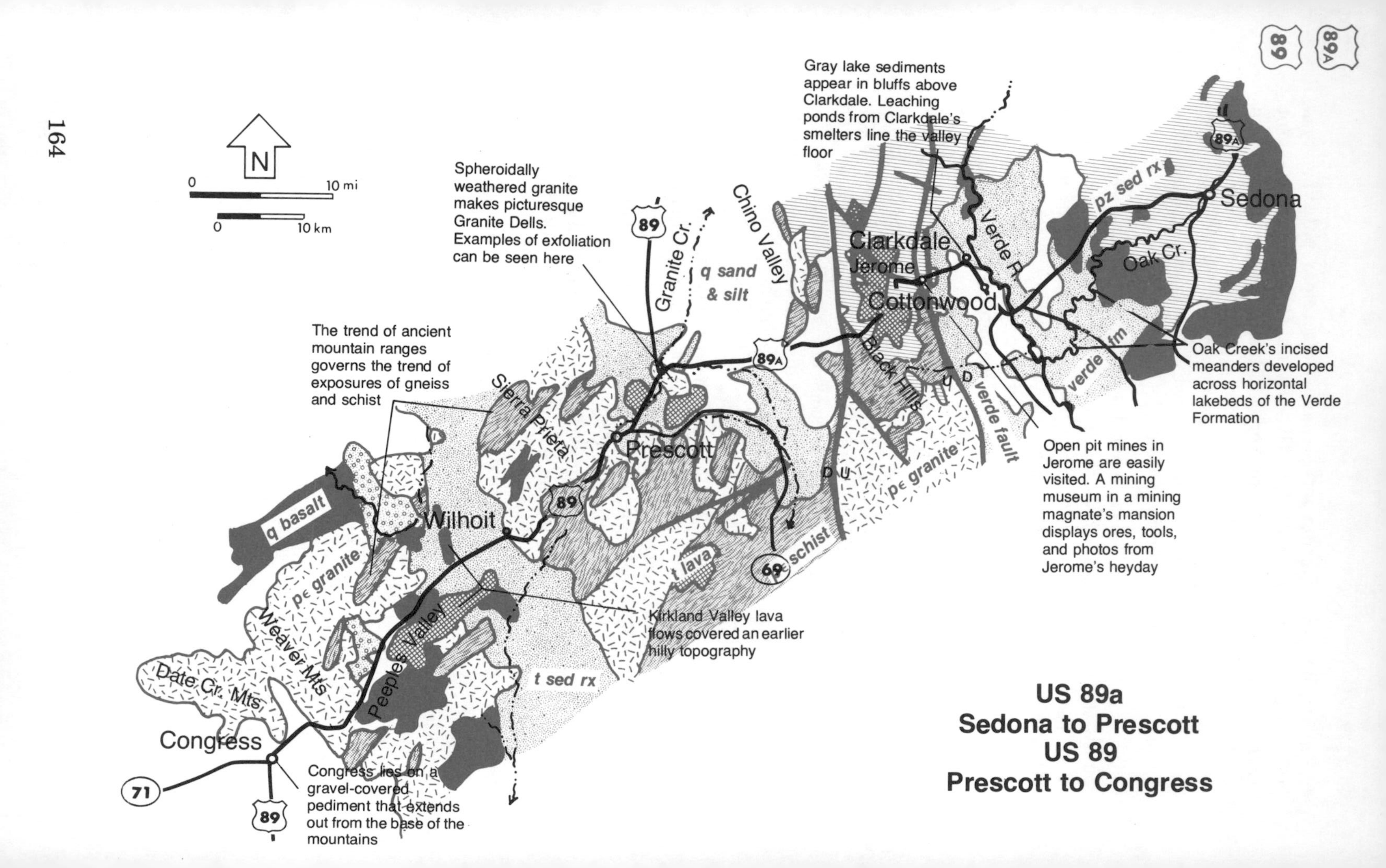

US 89a
Sedona to Prescott
US 89
Prescott to Congress

U.S. 89a
Sedona — Prescott
(60 miles)

As Oak Creek Canyon opens out south of Sedona into the Verde Valley, the brilliant panorama of red sandstone cliffs and towers is exchanged for gray lava-capped buttes and mesas and for chalky, gray-white lake deposits of the Verde Formation, containing fossil freshwater snail shells, plants, and mastodon bones and teeth that establish its age as Pliocene and early Pleistocene. The history of this interesting valley, bordered on the northeast by the Mogollon Rim (the south edge of the Colorado Plateau) and on the southwest by the Central Highlands, is given under I-17 Camp Verde to Flagstaff (Chapter IV). While here you may want to visit Montezuma Castle and Tuzigoot National Monuments; the former is discussed in Chapter V.

From a distance the Central Highlands appear dark and forbidding. Lava flows cap more or less horizontal Paleozoic and Precambrian strata. The town of Jerome clings to the steep slopes that gave it birth.

At Clarkdale, abandoned as a smelter town with the closing of Jerome's mines in 1953, the highway begins the climb to Jerome. Here it crosses one branch of the Verde Fault, which with 6000 feet of total displacement brings Precambrian and Paleozoic rocks to the surface in the Black Hills. Legend has it that a small party of Spanish explorers discovered gold here in 1583. However, not gold but the green copper mineral malachite is said to have given the Verde (green) Valley its Spanish name. And not malachite but chalcopyrite, another copper ore, was the real treasure here. Between 1876 and the closing of the mines in 1953 the area produced more than $375 million in copper, along with significant quantities of silver, gold and zinc.

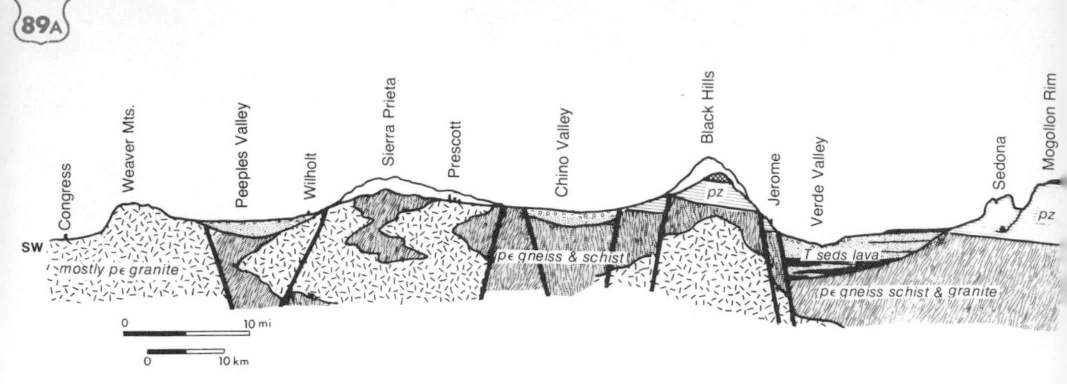

Section along US 89a Sedona to Prescott and US 89 Prescott to Congress

The ore bodies at Jerome differ from the porphyry copper deposits of southern Arizona and the Globe-Miami area. They occur at the top of a great pile of Precambrian submarine volcanic rocks, now so metamorphosed that they were at one time thought to be an intrusion. Some of the rocks in direct contact with the ore bodies have been dated as about 1800 million years old. The richest ores lie along folds in the ancient volcanic rock.

Both town and mines managed in their heyday to cling precariously to the steep mountain flank between two branches of the Verde Fault. Blasting in the mines at one time set off earthquake-like shocks that shifted the town and its unstable base downhill. Cracks in once-imposing buildings, now empty shells, tell of the geologic instability of the site. Always handicapped by lack of water, both town and mines were plagued by fire. The town burned to the ground three times.

Be sure to visit the mining museum in the former James Douglas mansion.

Above Jerome, outcrops of Precambrian gneiss mark the southwestern branch of the Verde Fault. Paleozoic rocks, from Cambrian to Permian, overlie the gneiss. The most obvious unit is a gray cliff-former, the Mississippian Redwall Limestone. Poorly exposed above it are reddish siltstones of the Supai Group. The brilliant red sandstone so magnificently exposed near Sedona doesn't extend to this side of the Verde Valley.

Jerome clings to steep slopes between two branches of the Verde Fault. Mines are close to the town.

Miocene lava flows form the dark crest of the Black Hills, both along the highway and at the summits of Woodchute and Mingus Mountains to the north and south. West of the crest, the Paleozoic sequence is repeated; the highway descends through it. The volcanic rocks rest directly on red shales of the Supai Group and the irregular, eroded surface of the Redwall Limestone. The Redwall is a marine limestone; in places it contains fossil corals, brachiopods, trilobites, bryozoans, and other marine shellfish. Its upper surface, exposed to the elements not long after it was deposited, is deeply marked with sinks and solution valleys formed by collapse of limestone caverns — a karst surface typical of limestone regions exposed to tropical or semitropical weathering, as this region was at the end of Mississippian time. Below the Redwall Limestone, in brownish layers of Devonian limestone, fossil fish plates and Devonian shellfish are found.

Between miles 335 and 334 the highway threads through a small canyon eroded along a fault that exposes Precambrian rocks on one side of the canyon and Paleozoic strata on the other. The little canyon ends at the prominent fault that defines the west edge of the Black Hills, and the highway enters the down-faulted graben that is the southern end of Chino Valley. Like the Verde Valley, this basin is edged with flat-lying Tertiary sediments.

Liesegang rings *are caused by precipitation of limonite in granite. Joint planes in the Dells Granite meet at nearly right angles*. M.H. Krieger photo, courtesy of USGS.

Southwest of Chino Valley is another fault-bound range, the Sierra Prieta. There are no sedimentary rocks at all in these mountains; as in most of the Central Highlands both Paleozoic and Mesozoic strata have long since been eroded away. At Granite Dells near the junction of US 89 and US 89a we glimpse the Precambrian granite that is the predominant rock type of the range. Here large outcrops of granite, cut into blocks and fingers by vertical joints, are rounded by spheroidal weathering. Just within Granite Dells one can see the successive stages of rounding and weathering: the solid granite, granite cut by joints, disintegration along joints, somewhat rounded blocks, and completely rounded, free-standing blocks surrounded by coarse granite-derived **grus**. Lichens add splotches of color to many rock surfaces. Granite Creek, flowing through the dells, contains placer gold — not very much, but enough to have brought Prescott into existence!

U.S. 89
Prescott — Congress
(43 miles)

For map, see the preceding section.

Leaving Prescott this highway climbs a little to cross a low pass in the Sierra Prieta, with good exposures of igneous and metamorphic rocks. The igneous rock — light gray, tan-weathering granite — appears as rounded knobs and boulders that are gradually but clearly disintegrating into coarse sandy soil. The metamorphic rocks — both gneiss and schist — are darker in color and commonly banded and quartz-veined. Normally they break down into fine, dark brown soil aglitter with flakes of mica. The granite, gneiss, and schist are similar to ancient rocks well displayed in the Grand Canyon's Inner Gorge, and to those known to underlie most continents. Roots of ancient mountain ranges, they tend to occur in SW-NE bands (see map), demonstrating that the trend of the ancient ranges was not the same as that of today's mountains, which trend NW-SE.

Between miles 310 and 300 both gneiss and granite are well exposed. The gneiss is marked with irregular, branching veins and with squiggly folds that show it must have been at least partly melted during metamorphism. There are good exposures of disintegrating "rotten" granite at mile 296. Gneiss in this area has suffered the same fate, turning into rounded cobbles surrounded by dark sand and soil.

Jointed granite turns into coarse sand or grus, *even before it is exposed at the surface – part of the process of spheroidal weathering. Note the dark rock fragment imbedded in the granite, and the fine-grained dike at lower right.*

At Wilhoit we come again into light-colored Tertiary sediments hardly well enough consolidated to be called rock. Kirkland Valley is probably edged with faults. Kirkland Creek, which drains the valley, is a tributary of the Santa Maria River. Lava flows cap extensive shelves along the south side of the valley, and several mid-valley ridges are lava-topped as well. Pink baked soil layers appear at the bases of some flows. Note that the basalt flowed over an already hilly surface. Other flows show the stockade-like appearance of columnar jointing created as the lava cooled and shrank.

Some of the flows are Tertiary, some possibly Quaternary. As in many other parts of Arizona the Tertiary flows are tilted, folded, and to some extent faulted, whereas Quaternary flows still lie in their original horizontal or near-horizontal position. In terms of rock types the flows differ, too. Some are silicic lavas such as rhyolite and dacite, lavas that are thick and sticky and that as a consequence tend to form thick, stubby lava flows. Others are much more fluid basalt lavas that erupted quietly and spread out into lava sheets only tens of feet thick. The south end of Peeples Valley is closed by such basalt flows.

The Weaver and Date Creek Mountains form the abrupt southwest edge of the Central Highlands, the boundary between mountain and desert. The gray granite of these ranges, with its distinctive salt-

and-pepper texture and color, is well exposed in roadcuts. The rounded granite knobs display many examples of exfoliation, a continuation at the surface of processes that contribute to spheroidal weathering. Because the arid desert climate to some extent climbs the mountain, heating and cooling contribute greatly to the gradual breakdown of rock, and wind plays a major part in sweeping away the finer debris from this breakdown.

The first good glimpse of the desert basins and ranges comes just south of Yarnell as the descent into the desert begins in earnest. A few highway cuts reveal patches of silver schist and banded gneiss, evidence of the irregularity of the granite-metamorphic rock contact here. When granite invades other rocks it commonly melts its way into them, leaving isolated blocks and lenses of the host rock more or less intact.

As the road descends the mountain front, or from the viewpoint near milepost 275, you can distinguish several types of desert ranges. Some are fault block mountains separated by down-faulted valleys. The dome-summited Harcuvar Mountains to the southwest are a metamorphic core complex range (see Chapter II). The slanting crest of the Vulture Mountains east of them consists of tilted Tertiary lava flows and volcanic ash, with the prominent tooth at their crest an erosional remnant of a once more widespread lava flow.

The major fault that edges the Central Highlands is some distance out from the mountains, beyond the alluvial fans and gravel-covered pediment that stretch out below the mountain front. Bedrock of the pediment — gneiss, granite, and shiny schist — can be seen in roadcuts and terrace edges near Congress, which lies on the pediment itself.

AZ 87
Pine — Mesa
(93 miles)

South of Pine one begins to see exposures of light gray, many-layered limestones of the Pennsylvanian Naco Formation and, below it, gray, massive Mississippian Redwall Limestone. Both these formations contain many fossils, mostly horn corals and brachiopods. Near Pine the original calcium carbonate shells are commonly replaced, molecule by molecule, with silica, and fossils in the Naco Formation may be coated with red jasper, which makes them easy to find. The red-tinted solution breccia at the top of the Redwall Limestone, which separates it from the Naco, shows well near milepost 259.

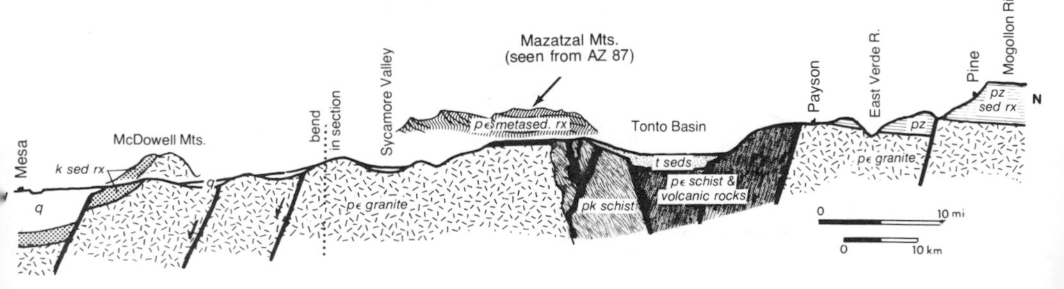

Section along AZ 87 Pine to Mesa.

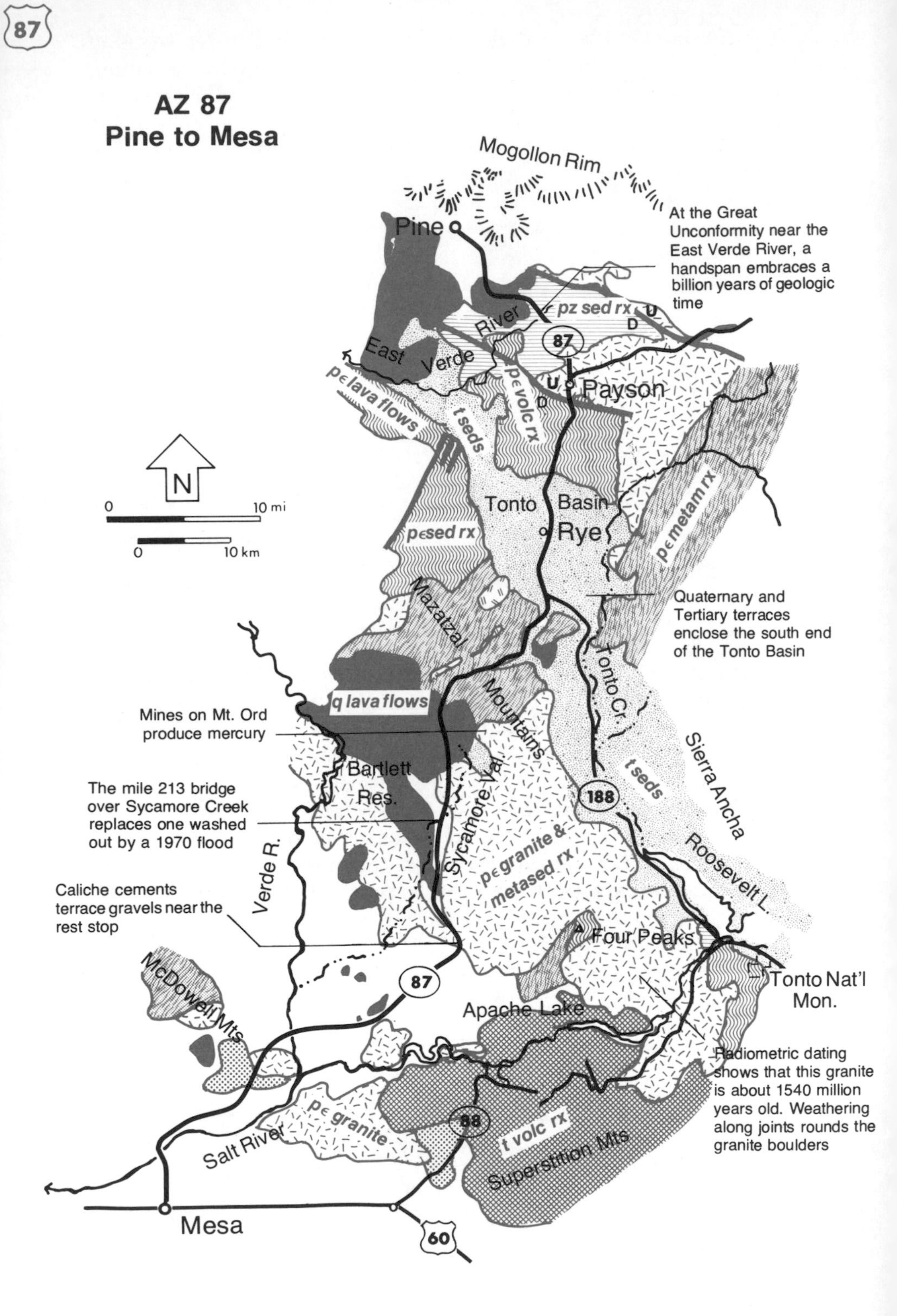
AZ 87
Pine to Mesa
Mogollon Rim
Pine
At the Great Unconformity near the East Verde River, a handspan embraces a billion years of geologic time
pz sed rx
U
D
East Verde River
87
Payson
pє lava flows
t seds
pє volc rx
N
0
10 mi
0
10 km
Tonto Basin
Rye
pєsed rx
pє metam rx
Quaternary and Tertiary terraces enclose the south end of the Tonto Basin
Mazatzal Mountains
q lava flows
Mines on Mt. Ord produce mercury
Bartlett Res.
Tonto Cr.
Sierra Ancha
t seds
188
The mile 213 bridge over Sycamore Creek replaces one washed out by a 1970 flood
Verde R.
Sycamore Val.
pє granite & metased rx
Roosevelt L.
Caliche cements terrace gravels near the rest stop
Four Peaks
Tonto Nat'l Mon.
McDowell Mts
87
Apache Lake
Radiometric dating shows that this granite is about 1540 million years old. Weathering along joints rounds the granite boulders
pє granite
Salt River
88
t volc rx
Superstition Mts
Mesa
60

Continuing south the highway drops lower and lower through Paleozoic rocks — brownish ledges of Devonian Martin Limestone and then the Cambrian Tapeats Sandstone. Other Cambrian strata are missing here. Near the East Verde River, at mile 258, watch carefully for the base of the Tapeats and the top of the unlayered or unstratified Precambrian granite. This contact, the Great Unconformity between Precambrian and Paleozoic rocks, represents a time interval of nearly a billion years. For most of that billion years erosion seems to have been king, beveling off the continents to an almost uniformly horizontal plain.

Between Pine and Payson and about 3 miles west of the highway is Tonto Natural Bridge, touted as "the world's largest travertine arch bridge." This interesting structure, 150 feet wide and 183 feet above Pine Creek, was shaped by natural solution and stream erosion in spring-deposited travertine, a hard, dense, finely crystalline form of limestone common in limestone caves and around hot springs in limestone areas. Beneath the bridge Pine Creek flows into a deep, clear pool, near which are several caves that display stalactites and stalagmites, also made of travertine. The site is surrounded by cliffs of hard Precambrian sedimentary rock.

South of Payson, faulted against the granite you saw at the East Verde River, there are Precambrian metavolcanic rocks: platy greenish rock often referred to as **greenstone**, and lighter but also platy quartzite that is thought to derive by long metamorphism from volcanic ash high in silica. These rocks are cut by numerous dikes, veins, and veinlets.

The Tonto Basin south of Payson is renowned for its many Tertiary and Quaternary terraces, some of which are quite apparent near the margins of the basin. They are composed of fine lake limestone and siltstone layers and coarse stream deposits washed in from surrounding areas. Many of the stream sediments contain basalt boulders and cobbles — conglomerate in the making. The community of Rye is surrounded by these terraces; the pinkish Tertiary sediments contain quite a bit of volcanic ash. Just south of Rye the highway rises onto the terraces — a chance for a good look at their gravels.

A short distance south of the junction with Arizona 188, more whitish lakebeds appear. Then the route climbs onto a platform of Precambrian schist, volcanic rocks, and granite. Here the road bisects the Mazatzal Mountains, largest range in the Central Highlands, a medley of very hard, erosion-resistant Precambrian metamorphic and igneous rocks. At one time the range was probably topped with Paleozoic and Mesozoic sedimentary rocks; these are all worn away now, though pebbles that may be derived from them sprinkle the Rim Gravels of the Colorado Plateau. West of the high-

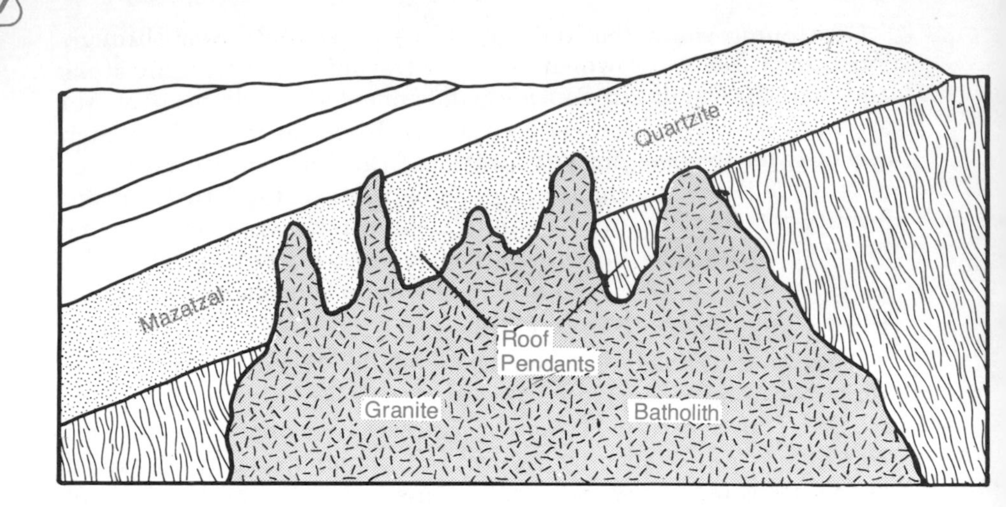

Hanging down into a molten batholith, quartzite roof pendants *were preserved as the granite crystallized. Today they form the Four Peaks.*

way, the Precambrian sedimentary and volcanic rocks are folded and in places faulted, though less than we would expect considering their long, long history.

The highway between miles 229 and 224 follows the NE-SW grain of the Precambrian rocks of Arizona, passing through the mercury mining district around Ord. In this area intermixed dark red conglomerate, slaty shale, thin-bedded quartzite, and shiny purple mica schist, all Precambrian, are cut by many dikes and fine criss-crossing veinlets.

Watch the highway cuts; they display several "textbook" faults: At mile 229 Tertiary gravels have been pushed up and over Precambrian rocks; at mile 225 Tertiary gravels are faulted against more Tertiary gravels.

After crossing the basalt-covered hills near Sunflower the highway is again in granite, red-tinted, coarse-grained, and weathering into giant rounded boulders. The route then descends into Sycamore Valley, a little intermontane basin downfaulted and partly filled with Quaternary gravel, sand, and lava flows. Continue to watch roadcuts for well displayed faults.

A particularly hard Precambrian unit, the Mazatzal Quartzite, is exposed in big roadcuts at mile 206. This rock also appears in the Four Peaks, now in view to the east, which were preserved as **roof pendants** hanging down into a magma chamber, source of the granite that now surrounds them.

The viewpoint south of mile 203 is on a gravel-veneered pediment sloping toward the Phoenix Basin. From it can be seen the Four Peaks, the granite hills below them, and to the south the Superstition Mountains, a rugged volcanic range, part of a large lava and welded tuff complex surrounding five overlapping calderas. The most pronounced of the calderas surrounds Weavers Needle, the prominent sharp peak east of the rounded prow of Superstition Mountain. The calderas, which originated as volcanoes collapsed into their own partly emptied magma chambers, no longer appear as circular basins ringed by steep walls, a la Crater Lake in Oregon, but have been identified by carefully mapping the ring of faults that surrounds them. The western prow of the range is a cluster of three volcanic domes, now much eroded.

South of the viewpoint, various rocky hills protrude through the gravel. The largest of these are the McDowell Mountains, partly composed of Miocene stream deposits. Much coarse gravel, with large rounded cobbles of Mazatzal Quartzite, basalt, and other hard rock types, are in the benches and terraces of the Verde and Salt Rivers, which come together near the McDowell Mountains and carve a channel between them and the granite of the Goldfield Mountains. Along the road are many good exposures of the gravelly sediments, some of them cemented with whitish caliche.

Leaving the mountains behind, the highway parallels the Salt River channel for several miles. Now usually dry, the river's bed is quarried for sand and gravel. Water is removed at upstream dams and distributed by a network of canals and irrigation ditches to Phoenix Basin farms and residential areas.

The Superstition Mountains, seen here across the sloping pediment of the Mazatzal Range, are the remains of a volcanic complex that includes several calderas.

IV
Scenic Wonderland — The Colorado Plateau

Stretching almost across Arizona, the Colorado Plateau is a land of contrast: sun and shadow, mesa and cliff, desert and forest, rocks red and gray, green and purple and white. This is horizontal country, offering wide vistas and scenic and colorful landscapes, a series of flat-topped plateaus 4000 to 9000 feet above sea level. Each with an Indian name — the Hualapai, Coconino, Kaibab, and others — the plateaus are collectively named for the great river that courses among them: the Colorado.

Here on the Colorado Plateau a great block of the earth's crust has remained coherent and recognizable through 600 million years and more, while blocks around it have been tilted and squeezed and broken. The difference between the Plateau (when used alone, capitalized, I refer to the whole Colorado Plateau) and the Basin and Range deserts to the west and south is not just in today's scenery. There seems to be some sort of underlying difference here, a difference that has lasted through much of earth history, as if the more or less circular patch of the earth's crust that makes up the Plateau obeyed a different set of geologic rules. Below the Plateau the crust is thicker, and heat flow from the interior of the earth is lower, than in surrounding regions. A belt of sporadic but minor earthquakes runs along the Plateau's west boundary, a boundary that is continuous in Utah with the western limit of the Wasatch-Rocky Mountain system. And geologists have found that measurements of the earth's magnetism and gravity change here, too. It is as if a raft, strong and sturdy, floated in a sea of flotsam. The logs of the raft shift from time to time, but it drifts on, retaining its identity through seas alternately stormy and calm.

Compared with that of the Basin and Range deserts and the rugged Central Highlands, the geology of the Colorado Plateau seems straightforward and easy to understand. The rocks are abundantly exposed to view. In cliffs and ledges, flat-lying sedimentary strata come in set sequences of oldest to youngest — as they were deposited. Marked by sinuous lines of cliffs and dramatic bends of naked rock, an assortment of faults and monoclines edges the individual plateaus. Mesas and buttes

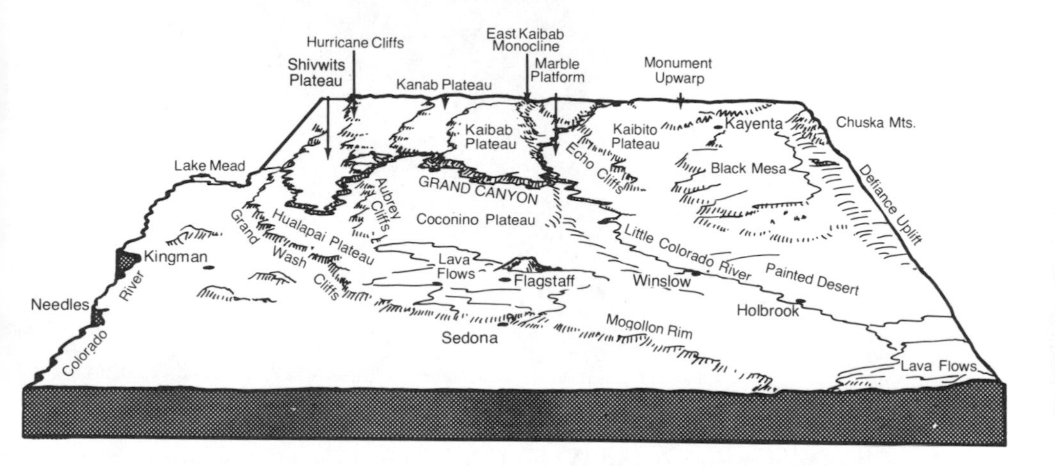

The Arizona part of the Colorado Plateau is divided by faults, monoclines, and the Grand Canyon into a number of lesser plateaus and mesas.

capped by resistant rock layers stand silhouetted against the sky. Stately stratovolcanoes look down upon the volcanic fields that surround them, speckled with cinder cones and streaked with dark lava flows that appear to have erupted only yesterday. Myriads of canyons, large and small, cut through the flat plateau surfaces, exposing to view in cliffs and badlands the layered strata below. And in the deepest canyon of all are exposed the ancient Precambrian crystalline rocks that underlie the sedimentary sequence.

Many parts of the Plateau are arid; few plants are able to sustain life there. In some areas the very rocks are inhospitable, and plants will not grow at all. Rains may be abrupt; flash floods occur on the high deserts during the summer rainy season. From about 5000 to 7000 feet, pygmy forests of juniper and piñon trees clothe the plateau surfaces. From 7000 to 8000 feet, rainfall and snowfall are sufficient to support forests of pine, Douglas fir, and aspen; a little higher are other pines and spruce and fir. The summits of the highest peaks reach above timberline, their rocky soils harboring only alpine plants.

A flash flood surges down a dry wash near Echo Cliffs, east of the Grand Canyon.
Tad Nichols photo.

The west edge of the Plateau is well defined by Grand Wash Cliffs along the Grand Wash Fault. The elevation change here is about 2000 feet. Most of the southern edge of the Plateau is equally well marked out by an erosional rampart, the Mogollon Rim (pronounced MUGgy-own). "The Rim," as it is often called, extends southeastward into New Mexico, but is covered in eastern Arizona by the many lava flows of the White Mountains. All in all the Arizona part of it is about 200 miles long as the crow flies, and 100 to 2000 feet high.

On the north the Colorado Plateau extends into Utah; eastward and northeastward it overlaps into New Mexico and Colorado, where its flat-lying rocks eventually turn up along the edge of the Rocky Mountains.

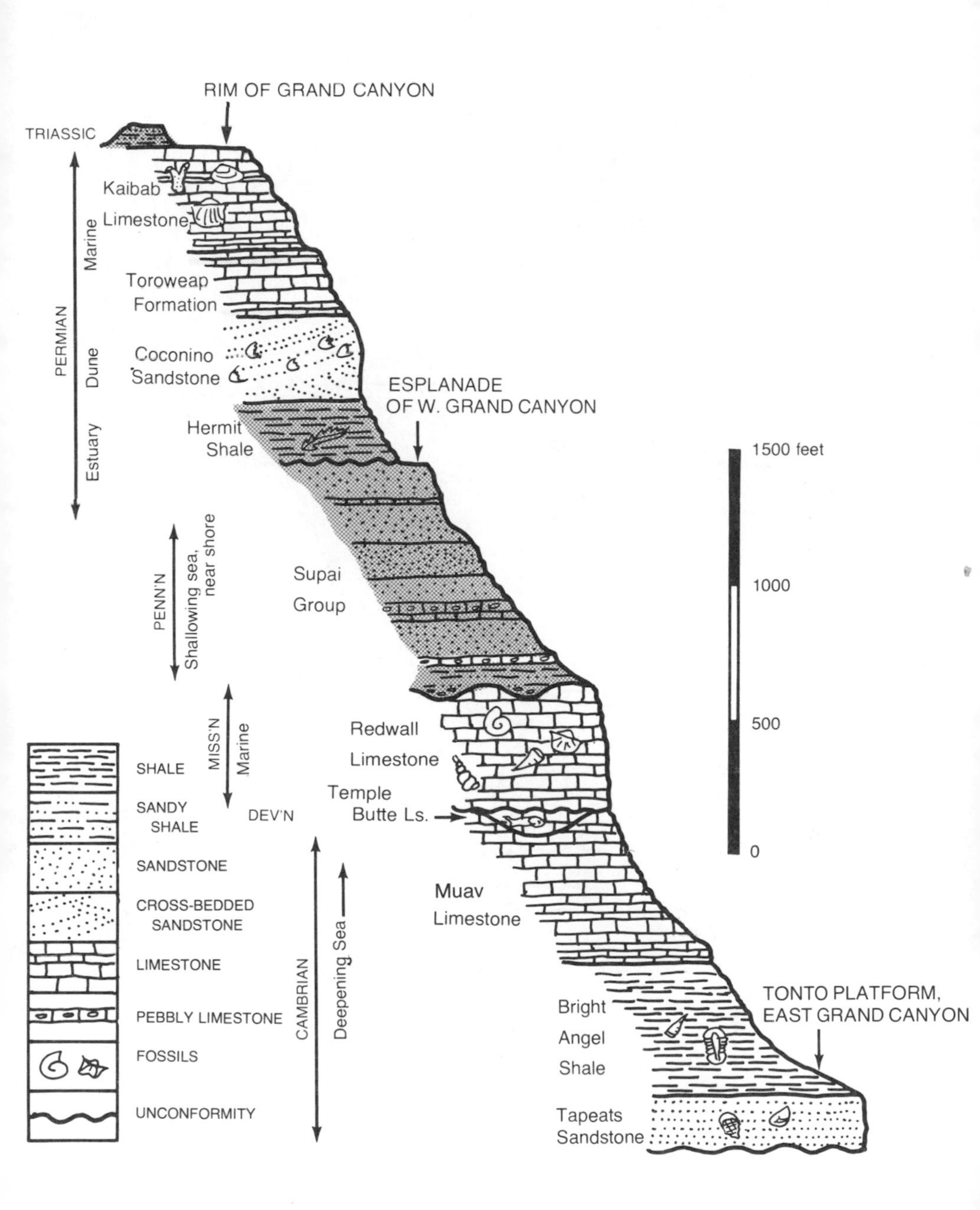

Stratigraphic section of Paleozoic formations of the Colorado Plateau

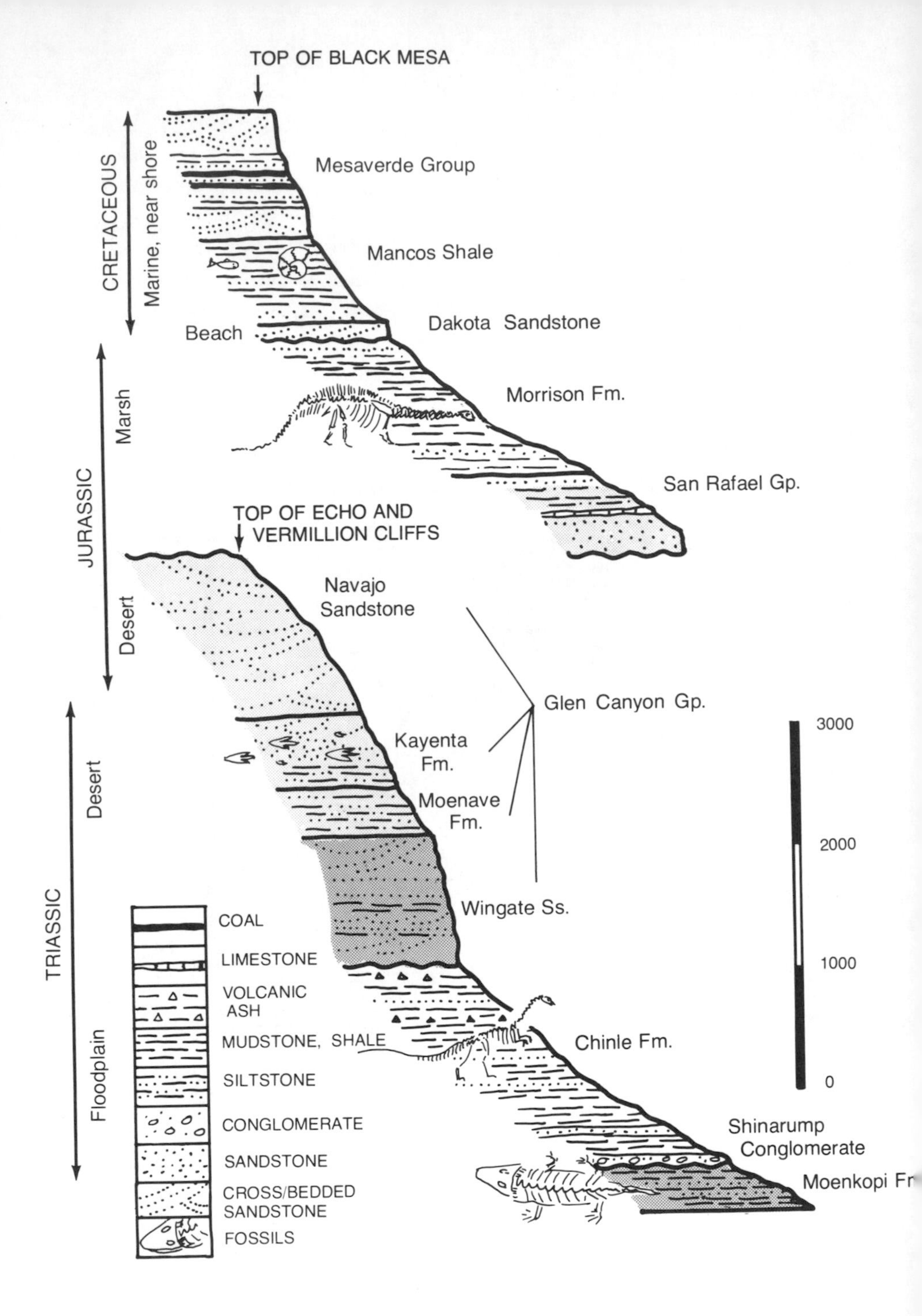

Stratigraphic section of Mesozoic formations of the Colorado Plateau

In the colorful plateau country long dry spells — a desert climate — leave many rock surfaces nearly devoid of vegetation. And the rocks come in all tints and shades: deep red, pale pink, white or buff-colored or yellow; the gray-purple of basalt; the green, purple, red, white and yellow of the Painted Desert. Good exposures, horizontality, and vivid colors make the sedimentary rock layers easy to recognize here, so in this chapter I have brought in names of **formations** and **groups** to an extent unmatched in the preceding chapters.

The diagrams given here for both Paleozoic and Mesozoic strata illustrate the alternations of rock types and the corresponding changes in environment that brought about these alternations. (Note that scales differ in the two diagrams.) They show also the way these formations weather — as slopes or cliffs or alternating slopes and ledges. Though the individual rock units — the formations and their subdivisions — may change in thickness and character across the state, most of the recognizable units once extended across most of northern Arizona.

Well exposed outcrops continuous over many miles have enabled geologists to construe the conditions under which the various rock units were deposited. We know, for instance, what Cambrian seas were like and how they advanced across the land. We know about the floodplains and deltas of Permian time, even about a large sandbody — perhaps an offshore bar — that developed in them in the Oak Creek Canyon area south of Flagstaff. We know that Kaibab seas of late Permian time came also from the west; we know how far they extended, where the marine limestones interlayer and change to sandy near-shore deposits, and where they finally give way to dune sandstones closely resembling those of the underlying Coconino Sandstone. We learn from the Triassic rocks of the Painted Desert what sorts of animals splashed in Triassic swamps or lumbered or scrambled on the land, what kinds of trees lived and died, what floods occurred. We can in our imagination stand on Triassic or Jurassic dunes, or walk a Cretaceous beach, or wander on the shore of a Pliocene lake. We can look on while a mighty river breaks through a divide and changes its course. We can watch a volcano pour hot lava into that river's deep-cut canyon. And as did early inhabitants of the Plateau, we can stand in awe or run in fear when a volcano erupts at our doorstep, filling the sky with fireworks and covering our home with cinders.

Most of the vivid colors in the rocks of the Plateau come from iron oxides — the minerals **hematite** and **limonite**, the material of ordinary red and yellow rust. The iron is derived from biotite, hornblende, and other common iron compounds in igneous or metamorphic rock. It is passed on to the sedimentary rock directly or in solution in groundwater. Varying amounts of the iron oxides give tones from yellow and pink to dark purplish red. Where the iron becomes part of the sedimentary rocks in a reducing environment, that is, without the possibility of combining with oxygen, it usually lends a gray-green cast to the rock. Reducing environments occur on sea floors, where many sedimentary rocks are deposited, and on land where there is an abundance of decaying animal and plant matter, as in swamps and marshes or near dead and decaying organisms where the decay bacteria use up most or all of the oxygen.

Many of the chalky yellow or buff-colored rocks of the plateaus are limestones deposited in shallow seas. Others — notably the Coconino Sandstone — are wind-winnowed dune sand almost lacking in iron minerals. Not all marine limestones are light-colored, though; some such as the Redwall Limestone of Grand Canyon are gray.* And not all windblown sand is white or yellow; the beautiful pink and red cliffs of northeastern Arizona are made of dune sand, too.

In terms of geologic history the Plateau itself is a late developer. But by descending into the Grand Canyon we can glimpse a much earlier history of this area, a history in many ways similar to that of central and southern Arizona. The oldest rocks here, found deep in Grand Canyon's Inner Gorge, are gneiss and schist nearly 2 billion years old. Crushed and partly melted, distorted, folded by collision between crustal plates, they were the roots of ancient mountains. Later in Precambrian time, in another type of mountain-building, they were intruded by masses of granite and webbed with light-colored dikes. Then — erosion.

---

*Why is the Redwall gray? Despite its name, the rock of this formation is uniformly gray on freshly broken surfaces. But in the Grand Canyon it is washed with fine red silt and clay from the formations directly above it, so it forms a "red wall.")

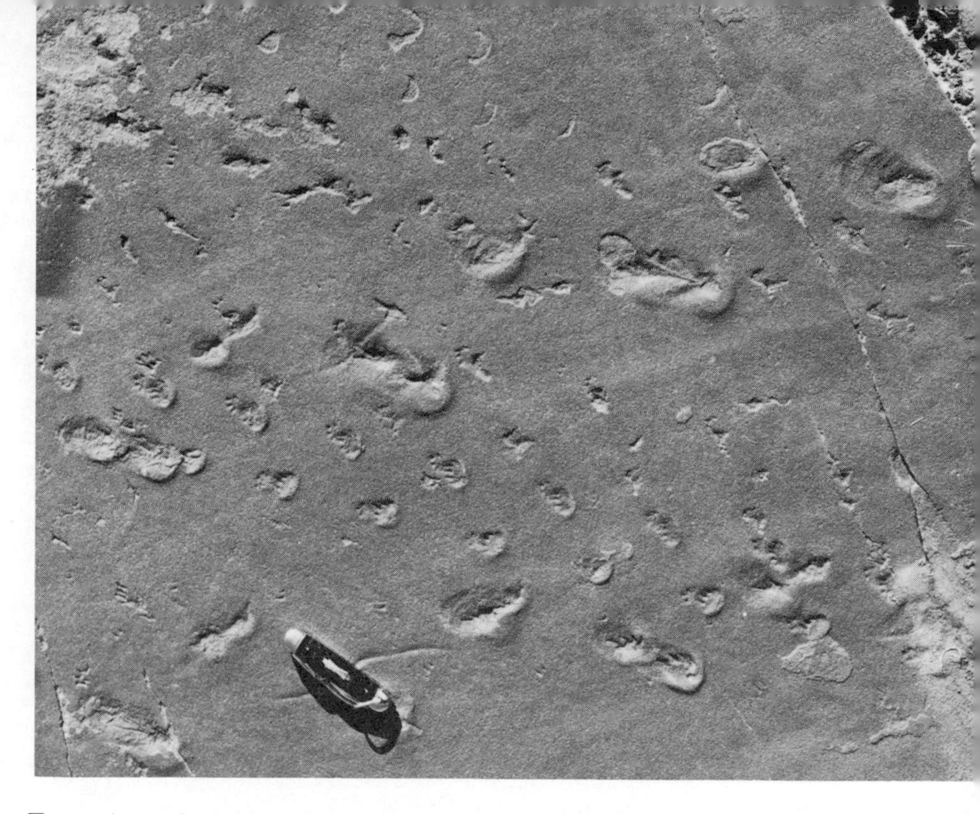

Footprints of animals long extinct give exciting glimpses into the past. These are Jurassic Navajo Sandstone. Tad Nichols photo.

After millions of years of erosion the beveled mountains were covered with sedimentary rocks interlayered with lava flows. Such deposits probably covered a vast area, for they occur in scattered outcrops in central and southern Arizona, too, though not all are of identical age. Here in the plateau country, as elsewhere, they bear the remains of some of the earliest known life — primitive algae and traces of simple, shell-less marine organisms.

Cut and tilted by fault-block mountain-building and then beveled once again, only wedges of these Precambrian sedimentary rocks remain. Seas creeping in from the west across the beveled surface heralded the onset of the Paleozoic Era. Layers of marine sandstone and shale and limestone, most of them with fossils, show that northern Arizona was shelf-like, immersed in a shallow western sea. At times the land rose, deposits were washed away, streams cut shallow channels, and rivers brought floodplain and delta deposits. At other times advancing dunes swept across barren deserts, or erosion

Barren hills of the Painted Desert develop in rock layers that contain bentonite formed from decomposing volcanic ash. This clay swells when it is wet, shrinks as it dries, preventing plants from gaining a foothold. Tad Nichols photo.

reigned. A remarkable feature of the Paleozoic Era was the *lack* of mountain-building, the *lack* of earth movements except for the slight downs and ups that permitted the sea to advance and retreat, and the *absence* of even the slightest hint of volcanism or igneous intrusion.

The Mesozoic Era brought a return to continental types of deposits: floodplain-delta sandstones and siltstones, and desert sands. Newborn mountains in Colorado, New Mexico, and central Arizona played strong roles in contributing to these continental sediments. Though we know of no actual Mesozoic volcanoes in this area, it is apparent that violent volcanic outbursts not far afield spread layers of ash over the land, ash that today contributes to the colorful badlands of the Painted Desert and to equally colorful rocks in northeastern Arizona. Late in Mesozoic time the sea returned — from the east, this time — to deposit thick gray Mancos Shale and later the beach and swamp deposits of the Mesaverde Group.

This sea retreated as the Central Arizona Highlands, and to the northeast the Rocky Mountains, rose at the end of Mesozoic time. Debris from the Highlands was carried northeastward across the sloping former shore and into Utah, Colorado, and Wyoming.

The Cenozoic story of the Colorado Plateau centers around the history of the Colorado River. For years it was supposed that a pre-established Colorado River cut Grand Canyon gradually as the present Kaibab-Coconino Plateau slowly rose. However new evidence from many parts of Arizona suggests a different story, as shown on the accompanying diagram. The story is tentative and only partly unraveled; many intriguing questions remain. But as it concerns not just the Grand Canyon but the Plateau as a whole we should look in outline form at the major geologic happenings:*

- At the close of Laramide mountain-building, streams and rivers still drained northeast across what is now the Colorado Plateau, carrying with them pebbles and boulders from the newborn highlands.
- Normal faulting and uplift of various segments of the Plateau began in Eocene or Oligocene time and continued through Miocene and Pliocene time. (Some faults remained active into Recent time.) As individual plateaus were raised, a long dome, the Kaibab Arch, developed in north-central Arizona.
- On the individual plateaus, and depending to some extent on how high they were raised, erosion stripped away many of the sedimentary rock layers. On the Kaibab Arch, the highest, almost all of the Mesozoic strata were removed, leaving a surface of Permian limestone thinly covered with red Triassic sedimentary rocks.
- On top of these strata, and particularly in northeast-directed stream channels, was deposited gravel containing pebbles of hard igneous, metamorphic, and sedimentary rocks derived from the Central Arizona Highlands and perhaps more distant areas. Remnants of these deposits, the Rim Gravels along the Mogollon Rim, demonstrate that uplift in the Central Highlands kept pace with or surpassed that of the Kaibab Arch.
- For a long time a large river with headwaters in the Colorado and Wyoming Rockies was blocked by the Kaibab Arch, forced to turn south into what is now the Painted Desert — the valley of the little Colorado River. In Pliocene

* Recent studies show that the scenario presented here is unlikely, and that the Colorado River must formerly have drained northwestward into inland lakes in Utah or Nevada.

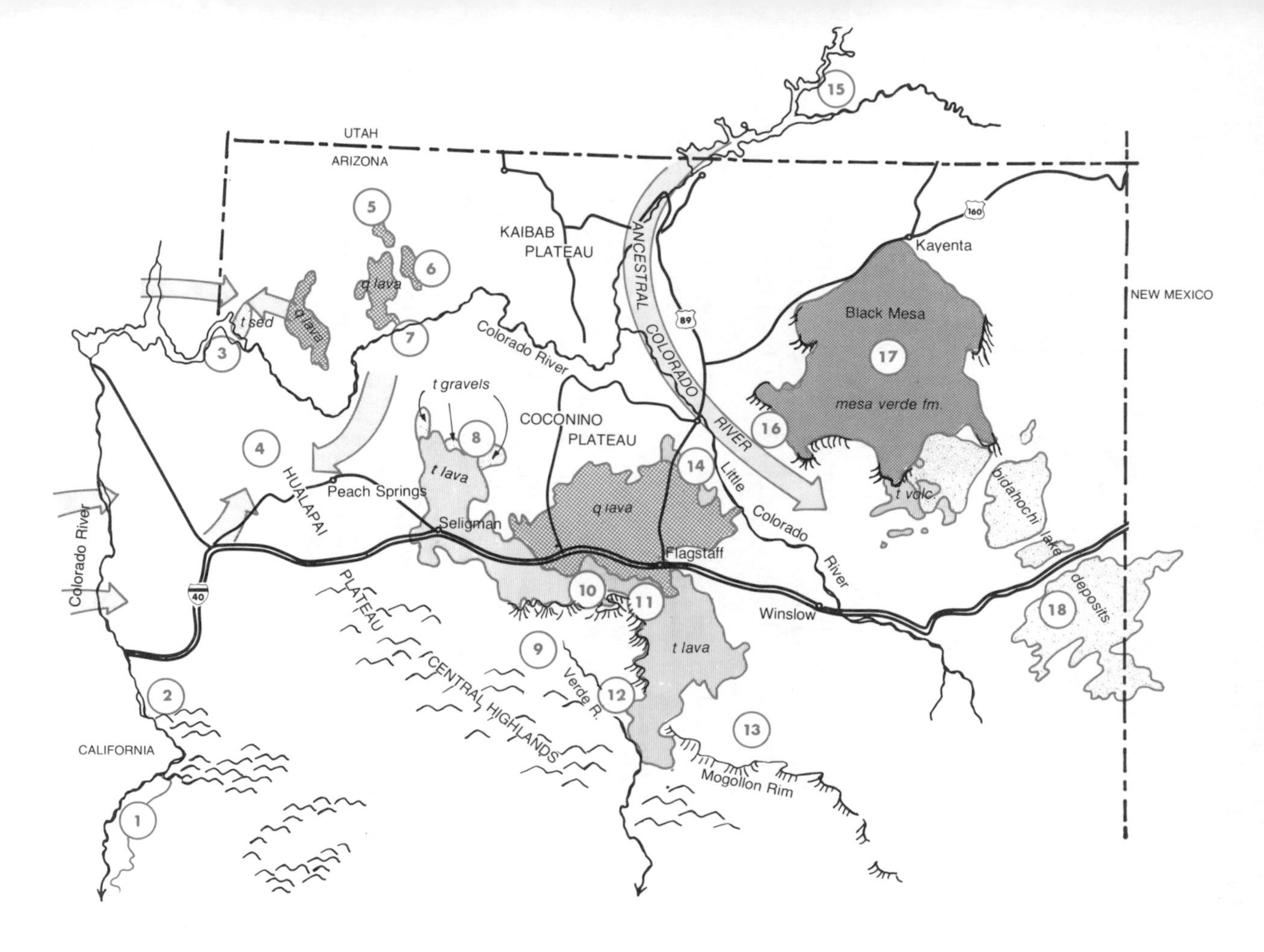
UTAH
ARIZONA
NEW MEXICO
CALIFORNIA
KAIBAB PLATEAU
ANCESTRAL COLORADO RIVER
89
160
Kayenta
Black Mesa
17
mesa verde fm.
15
5
6
q lava
t sed
q lava
3
7
Colorado River
t gravels
8
COCONINO PLATEAU
16
14
t lava
4
HUALAPAI
Peach Springs
Seligman
q lava
Little Colorado River
t volc.
bidahochi lake deposits
18
Flagstaff
Colorado River
40
PLATEAU
10
11
Winslow
9
t lava
CENTRAL HIGHLANDS
Verde R.
12
2
13
Mogollon Rim
1

Evidence for Pliocene to Pleistocene age of Grand Canyon comes from widely scattered localities. Colored arrows show probable direction of drainage before stream and river reversal.

1. Here and farther south, brackish-water Tertiary sediments on both sides of the river show that this area was an estuary of the Gulf of California through Miocene time.

2. Here the present Colorado River flows across the grain of the Basin and Range mountains and valleys, possible only if the river first became entrenched in thick sedimentary layers that covered the ranges completely. Terraces bordering the river show alternate erosion and deposition in Pleistocene time.

3. Until after the end of Miocene time, the trough formed by Grand Wash Fault received sediments from both east and west. The Colorado River could not have been here at that time.

4. Southern Grand Wash Cliffs show northeast-tilted Paleozoic strata beveled by erosion, indicating uplift of the southern edge of the Plateau country. Miocene gravels of the Hualapai Plateau contain pebbles that came from the west and southwest.

5. Recurrent movement along Hurricane Fault displaced Cenozoic lava flows. Though movement began in Eocene and continued to Recent time, 75% of the displacement occurred between outpourings of older (Miocene) and younger (Quaternary) lava.

6. Products of four volcanic episodes register displacement by successive movements of Hurricane and Toroweap Faults.

7. Just 1.2 million years ago, lava flows dammed the Colorado River. Its canyon at that time was only 50 feet less deep than at present.

8. Pre-lava gravel deposits occur beneath lava flows in channels cut into the surface of the present plateau by north-flowing streams. Volcanism was followed by faulting and deposition of post-lava gravels.

9. This region was the highest part of the late Cretaceous coastal plain. Intermittent uplift kept it higher than the rest of the region through most of Cenozoic time. Before excavation of the Chino and Verde valleys, north-flowing streams carried pebbles and cobbles derived from this southern source.

10. A pre-lava pediment exposed in Sycamore Canyon shows a north-sloping surface mantled with gravels of the Precambrian and Paleozoic rocks from high mountains to the south.

11. Remnants of gravel-filled valleys show a north-trending drainage system later disrupted by volcanic flows.

12. Prior to volcanism and uplift of the Black Hills to the south, a north-trending drainage system also existed here. The present south-flowing Verde River is a Pleistocene development.

13. Pre-lava gravels from a southern source are preserved beneath lava flows that edge the Mogollon Rim.

14. The eastern edge of the San Francisco volcanic field impinges on the Little Colorado River valley, making it possible to date successive stages in the development of the Little Colorado drainage basin.

15. River deposits, erosion surfaces, and deeply entrenched meanders of the Colorado and San Juan Rivers record episodes of downcutting separated by periods of valley filling.

16. Four terrace levels indicate four alterations of deposition and erosion in the Little Colorado River valley.

17. Sedimentary strata on Black Mesa record the eastward retreat of the Cretaceous sea.

18. Extensive Bidahochi lake sediments overlap early lavas and show that in Pliocene time a lake filled what is now the upper basin of the Little Colorado River. The Ancestral Colorado probably deposited these sediments.

In Tertiary time San Francisco Mountain and other volcanoes of the San Francisco volcanic field covered the sedimentary rock surface of the southern Kaibab Upwarp. Tad Nichols photo.

of the little Colorado River. In Pliocene time sediments of the great river, the Ancestral Colorado, accumulated in a large interior lake, Lake Bidahochi or Hopi Lake, near the Arizona-New Mexico border. Volcanism along the lake shores contributed volcanic ash to the lake sediments, and interlayered them with lava flows.

- Minor faulting and major erosion south of what is now the Colorado Plateau established a 2000-foot escarpment, the Mogollon Rim, and the deep trough that now separates the Plateau from regions to the south.
- On the southern part of the Kaibab Arch, north of this rim, the San Francisco Mountain volcano developed, along with several of its small neighbors.
- Sometime between 10 and 2 million years ago a stream on the west side of the Kaibab Arch cut headward and penetrated the arch, "capturing" the waters of the Ancestral Colorado. This great river abandoned its southward course in favor of the westward route through the Kaibab Arch.
- The Little Colorado River, coming from new-formed volcanic highlands in east-central Arizona, took possession of the abandoned valley and flowed north where its predecessor had flowed south. To keep up with the downcutting of the Colorado

River in Grand Canyon, the Little Colorado cut a deep, narrow canyon through the lower, northernmost part of its route.

• The poweful Colorado, aided and abetted by the onset of glaciation in its headwaters area, undermined strong rock layers at the expense of weaker ones, so that its canyon walls took on a stepped cliff-slope-cliff-slope appearance. Gradually the walls receded, widening the canyon. Eventually the river cut down and into the very hard Precambrian rocks that underlie the Paleozoic strata.

• Volcanic outbursts continued right into Recent time. New lava spread from the San Francisco volcanic field, surrounding its central stratovolcano. More than once lava flows cascaded into western Grand Canyon and into the little Colorado River's channel south of Cameron, damming the river flow. Eventually the dams were breached by erosion or, in the case of the Little Colorado, bypassed by the river.

* * *

Many distinctive geologic processes are well displayed in the plateau country. You will see plentiful examples of desert pavement and rocks shiny and blue-black with coatings of desert varnish. Badlands develop where volcanic ash swells with every rain. Rockfalls and landslides are easy to spot. Canyon-cutting takes many forms but seems commonly to be due to downcutting by already established streams with already deeply incised meanders. Differential erosion shapes large-scale features: cliffs of resistant sandstone above slopes of soft shale, knobs and domes of resistant rock, hard volcanic plugs left standing as surrounding softer rock washes away. The wind plays a major role here (as in all deserts), sweeping away loosened material so that cliffs and pinnacles meet the ground at right angles. Erosion by both wind and water undermines such cliffs, so that giant blocks break away and tumble down the shale slopes. On a smaller scale differential weathering accents fine but firmly cemented layers of limestone, mudstone, and sandstone. On the Permian Coconino Sandstone of Grand Canyon and its equivalents farther east, and on thick, massive spectacular pink and red Jurassic sandstones of northeastern Arizona, weathering brings out wide, sweeping cross-bedding of ancient sand dunes. Elsewhere cross-bedding on a small scale defines ancient stream channels or the foreset beds of small deltas.

Unlike central and southern Arizona, northern Arizona does not contain known large copper deposits. But small amounts of copper have been mined here and there; there were even pick-and-burro mines within the Grand Canyon. Jurassic sedimentary rocks of northeastern Arizona, particularly the Morrison Formation, contain a good deal of uranium; smaller amounts occur in Triassic rocks of the Painted Desert area. Arizona's only oil production comes from the far northeast corner of the state, near the edge of the true Plateau country. On Black Mesa, the one part of the Plateau where Cretaceous rocks remain, there are thick beds of coal formed in near-shore swamps, coal that now fuels power plants on nearby Kaibito Plateau or travels by slurry pipeline across the state to Nevada.

Northern Arizona's great problem is water. The Colorado River's water is all spoken for; Glen Canyon Dam furnishes hydroelectric power but not irrigation or domestic water. In and around the volcanic centers, where clouds gather above the mountains, rainfall and snowmelt are plentiful and water supplies are adequate for the present population. On the Coconino and Kaibab Plateaus rainfall and snowmelt are greater than on the lower plateaus, but both rain and melted snow sink into solution cavities in the Kaibab Limestone. Much of this water goes on sinking until it meets with the fine, impermeable shales that underlie the Redwall Limestone, some 2000 feet down. It then travels sideways in solution channels in the Redwall — channels that supply good well water when and if they can be pinpointed from the surface. In Grand Canyon, in the Verde Valley, in the Tonto Basin, and along some plateau-edging faults water emerges at the base of the Redwall Limestone as springs. Springs occur also in northeast Arizona at the base of thick, porous Triassic and Jurassic sandstone layers, above the shales which underlie them. As more and more groundwater is pumped to supply the growing towns, coal mines, and slurry pipelines, though, many northeastern Arizona springs are endangered, leaving less and less water for the also growing needs of the Indian residents of this area.

Interstate 17
Camp Verde — Flagstaff
(53 miles)

The Verde Valley is a sort of no man's land between the mountainous highlands of central Arizona and the flat-topped Colorado Plateau to the north. Early Tertiary gravel deposits found today along the Mogollon Rim, which here edges the Coconino Plateau, contain pebbles of rock types that could only have been washed from the Central Highlands south of the Verde Valley, so we know that early drainage in this region was from south to north, across what is now a deep valley.

In Oligocene time, however, a trough seems to have developed here, an erosional valley that closely paralleled the present one, and that by mid-Tertiary time cut through or undermined the Permian cliff-formers — the Kaibab, Toroweap, and Coconino formations that edge the Plateau. This trough was later deepened by uplift south of the Verde Fault, which today marks the southwest side of the Verde Valley.

Over a period of many millions of years, in Miocene and Pliocene time, the high, flat-topped ridge of the Black Hills was elevated. The uplift eventually totaled 4,000 feet or more, and brought Precambrian rocks to their present elevations of 5000 to 7000 feet. Tertiary lava flows that form horizontal bands at the top of the Black Hills were lifted about 3000 feet; their counterparts lie beneath the Verde Valley's floor, on top of gravels deposited in the Oligocene trough. Fault movement played further havoc with drainage patterns, about 6 million years ago impounding the waters of the Verde River, inaugurating a succession of lakes.

The Pliocene-Pleistocene lakes, like the present Verde River, received tributary streams that drained the limestone surface of the Coconino Plateau, and that were as a result highly charged with calcium carbonate. Much of the calcium carbonate ($CaCO_3$) accumulated on the lake floor as the Verde Formation, now forming the chalky-looking white-banded hills of the Verde Valley. The chalk-like limestone layers are interlayered with pink silt and sand brought in by streams, and with volcanic ash from long-continued volcanism in this region.

I-17
Camp Verde to Flagstaff
US 89a
Flagstaff to Sedona

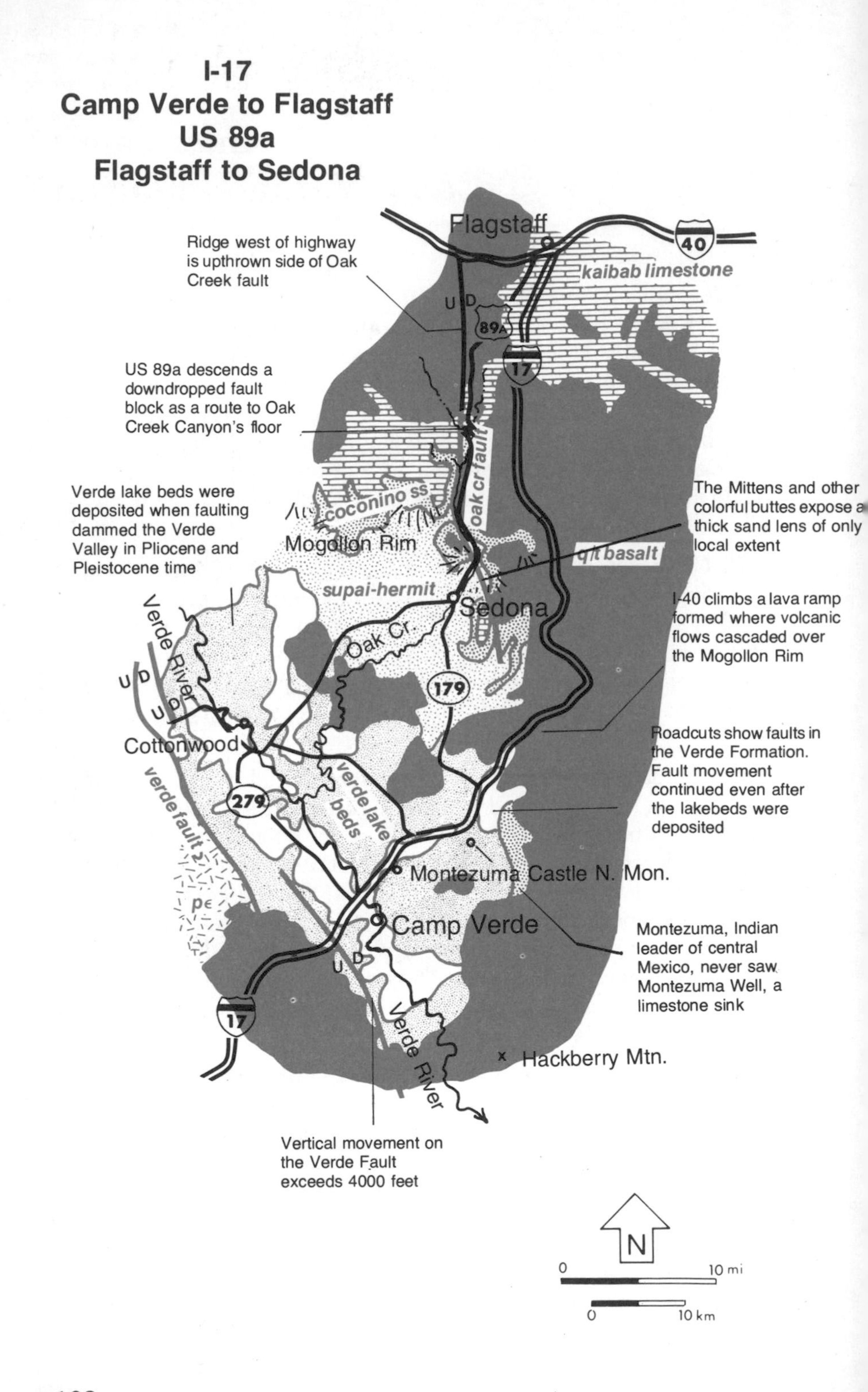

The Verde Formation lake beds total about 3000 feet in thickness. They contain fossil freshwater mollusk shells — both snails and clams —, plant stems and seeds and pollen, bones of mastodons, horses, rodents, and turtles, and tracks of mastodons, camels, lions, bears, and tapirs. The successive lakes may have been quite shallow, not over 10 to 20 feet deep, and at times they were brackish or salty. There is evidence that they were edged with marshes fed by sluggish, winding streams. Salt crystals that developed as lake waters dried suggest an arid climate.

Eventually a channel was worn through the mountains at the southeast end of the Verde Valley, and the last of the lakes drained away. The Verde River and its tributaries carved into the chalky limestone and pink siltstone and layered tuff of the Verde Formation,

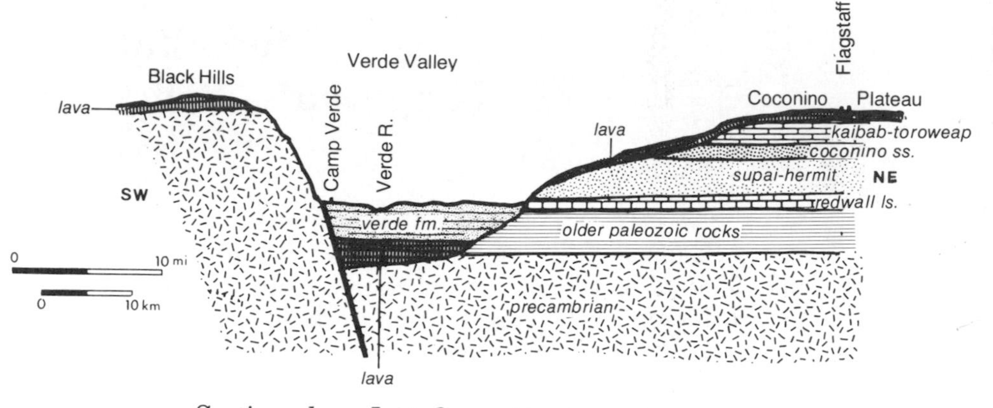

Section along I-17 Camp Verde to Flagstaff.

shaping them into hills and valleys now traversed by the highway. Deepening drainage brought about **solution caverns** in the lake limestones, stream-eroded caves such as that at Montezuma Castle National Monument, and **sinks** such as that at Montezuma Well (see Chapter V).

North of Montezuma Well the highway climbs a lava-flow ramp to the surface of the Coconino Plateau. Red sandstone showing through some of the basalt is part of the Supai (Permian) Group. Overlying buff-colored rocks with well developed, very large-scale, dune type cross-bedding make up the Coconino Sandstone, relic of a Permian desert. At the top of the plateau are occasional exposures of Kaibab Limestone, the resistant light tan marine limestone that surfaced this part of the plateau before the outpouring of the lavas. It, too, is Permian in age. Along the edge of the plateau, the Mogollon Rim, lava flows cascaded down from the plateau surface at such an angle that in places they drained out from under their own cooling, hardening skin, leaving long, vaulted lava tunnels. Some of these tunnels were uncovered during highway building, but closed by gratings to discourage Tom Sawyers. Be on the lookout for other volcanic features as the road climbs: lava flows up to 50 feet thick, **columnar jointing** caused as the lava cooled and shrank, thick brick-colored baked soil zones, and today's dark brown soil developed from basalt lava and volcanic **cinders**. Most of the rim lava flows are 6 to 8 million years old; the cinders are younger.

The high, straight hill to the west marks the extension of the high west rim of Oak Creek Canyon. San Francisco Mountain, a tall stratovolcano discussed under I-40 Seligman to Flagstaff, comes into view as the highway nears Flagstaff.

Interstate 40
Kingman — Seligman
(73 miles)

Leaving Kingman, this highway skirts the north end of the Hualapai Mountains, like the Cerbat Range north of Kingman composed almost entirely of Precambrian rock — granite, strongly banded gneiss, and metasedimentary rocks that still bespeak their origin in clearly marked sedimentary layers.

As the highway rises east of town you can see Grand Wash Cliffs in the distance to the northeast, the well defined west edge of the Colorado Plateau. The cliffs mark the position of Grand Wash Fault, a north-south normal fault with the east side lifted several thousand

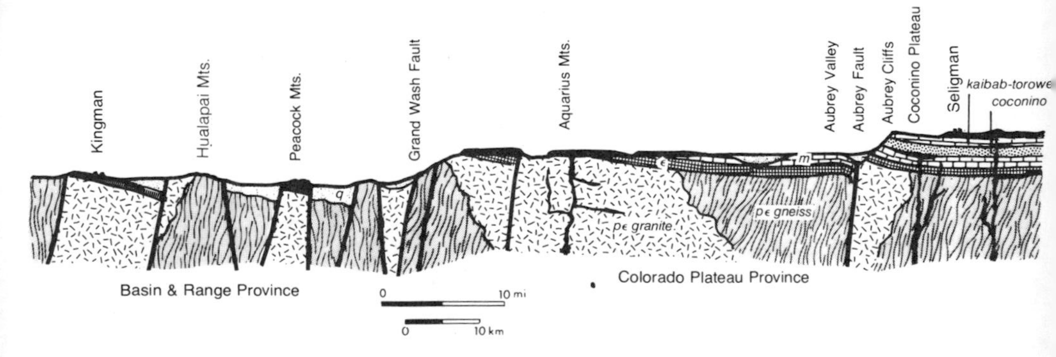

Section along I-40 Kingman to Seligman.

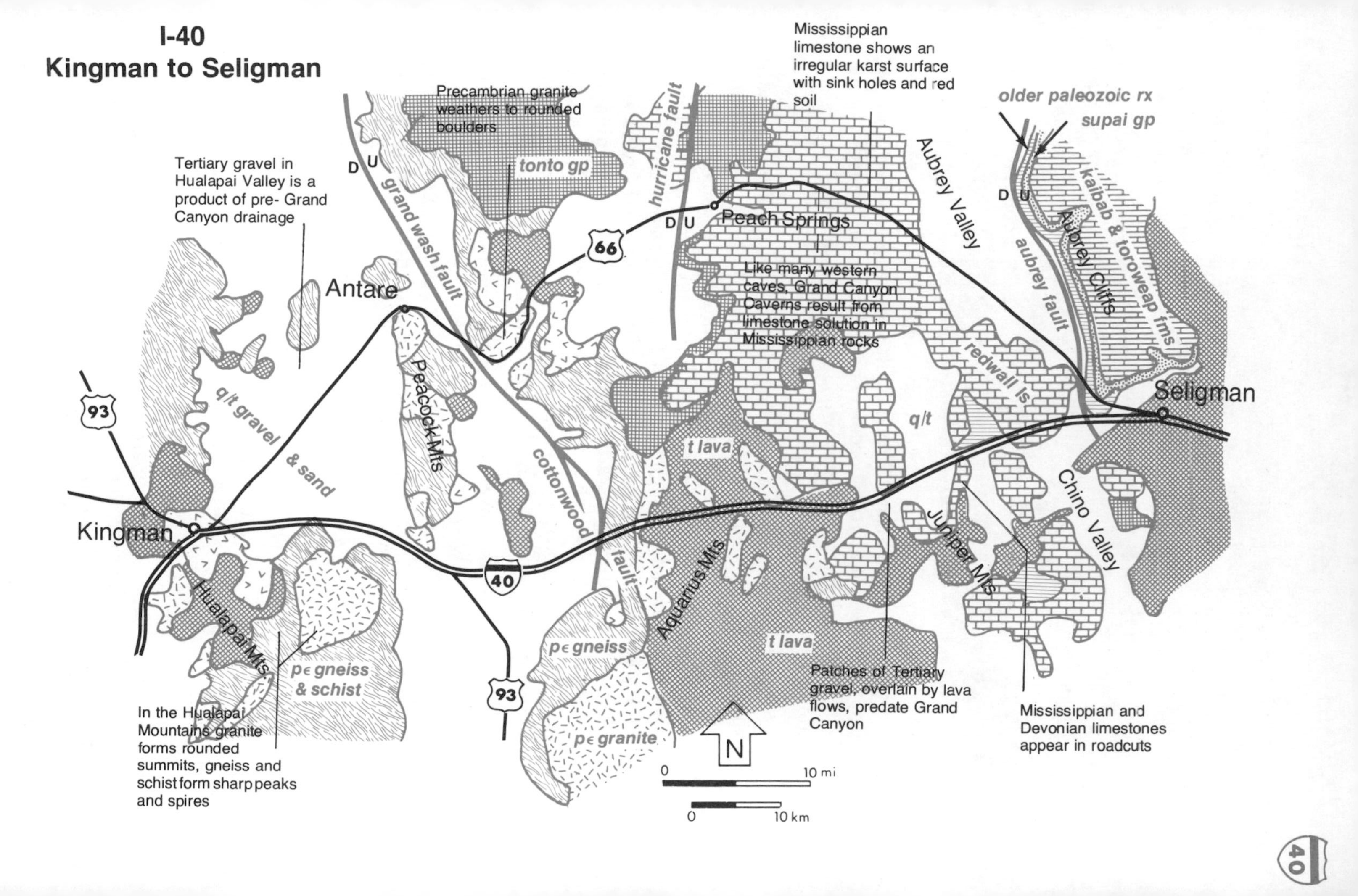

I-40
Kingman to Seligman
Precambrian granite weathers to rounded boulders
Mississippian limestone shows an irregular karst surface with sink holes and red soil
older paleozoic rx
supai gp
Tertiary gravel in Hualapai Valley is a product of pre- Grand Canyon drainage
D U
grand wash fault
tonto gp
hurricane fault
D U
Peach Springs
66
Aubrey Valley
D U
kaibab & toroweap fms
Aubrey Cliffs
aubrey fault
Like many western caves, Grand Canyon Caverns result from limestone solution in Mississippian rocks
Antare
redwall ls
Seligman
q/t
93
q/t gravel
& sand
Peacock Mts
t lava
cottonwood
Kingman
Chino Valley
fault
Juniper Mts
Aquarius Mts
40
Hualapai Mts.
t lava
pє gneiss
pє gneiss & schist
93
Patches of Tertiary gravel, overlain by lava flows, predate Grand Canyon
In the Hualapai Mountains granite forms rounded summits, gneiss and schist form sharp peaks and spires
pє granite
N
Mississippian and Devonian limestones appear in roadcuts
0
10 mi
0
10 km
40

Grand Wash Cliffs mar the west edge of the Colorado Plateau. Dark lower rocks are Precambrian gneiss; stratified Cambrian, Devonian, and Mississippian layers lie above.

J.R. Stacy photo, courtesy of USG

feet above corresponding rocks to the west. With many related but smaller faults, the Grand Wash fault zone extends north to and beyond the Arizona-Utah border. The southern end meets Cottonwood Cliffs, which mark Cottonwood Fault, crossed by I-40 at about mile 78.

Precambrian rocks on the east side of Cottonwood Fault are well exposed in large roadcuts. Here again are Precambrian rocks — both sedimentary and volcanic. Even at Interstate speeds you can glimpse hard quartzite formed from sandstone interlayered with shale, and greenstone derived from ancient lava flows. Reddened zones along joints and faults and in some places along individual layers show where groundwater penetration caused oxidation of iron minerals. Rocks such as these are clues to the mysteries of Precambrian landscapes in that they bear evidence of origin in shallow seas and on wide tide-swept flats along gently shelving shores.

Near the top of the roadcuts the Precambrian sedimentary rocks are cut off abruptly by a granite intrusion. The highway winds through granite country for some distance, among rounded, piled boulders characteristic of granite terrain and due in large part to spheroidal weathering.

The serrated ranges and sediment-filled valleys of the desert are now left behind as we enter a country of cliff-edged tablelands where flat-lying rock layers provide a predominantly horizontal landscape.

Beyond mile 107, and especially near miles 113 and 114, are patches of gray Devonian and Mississippian limestone. Some of this rock displays an irregular karst surface characterized by caverns, sinkholes, and light red soils. These features develop today in tropical and semitropical regions where there is abundant precipitation — not a good description of the region now. That the karst surface developed at the end of Mississippian time is indicated by dark red Pennsylvanian mudstone and siltstone that fill in some karst depressions.

A number of isolated peaks in view to the east are volcanoes, discussed in the next section.

The highway comes to another north-south rampart at mile 118: Aubrey Cliffs, the edge of the Coconino Plateau. The cliffs lie along the Toroweap Fault, though they have eroded back from the actual fault line, which in most places is covered with sand and gravel. The fault continues north, crosses western Grand Canyon at Toroweap Point, and runs northeast through Pipe Springs National Monument. Then it joins the Hurricane Fault to create Hurricane Cliffs, the steep west margin of Utah's Wasatch Mountains.

Rock layers exposed along Aubrey Cliffs are much younger than the Precambrian rocks of Cottonwood Cliffs. These are Paleozoic sedimentary rocks — the same strata that lend shape and color to Grand Canyon's upper ledges, slopes, and cliffs. At the top are bands of buff and light gray limestone and limy sandstone of the Kaibab and Toroweap Formations, each representing an incursion of the sea in Permian time. Below them the Coconino Sandstone, which in Grand Canyon forms a prominent 300-foot white cliff, here is slope-forming and barely visible. Below it in turn are red sandstone and siltstone of the Hermit Formation and Supai Group. Mississippian strata like those west of Aubrey Valley are below the surface here.

West of the fault, the cliff-forming Kaibab and Toroweap Limestones are not just covered with gravel and sand; they are missing. Not what one would expect in the down-dropped and therefore the more protected block. The Quaternary gravel and soil in the Aubrey Valley cover older gravels that lie in an older valley eroded in the Hermit and Supai redbeds — a **paleovalley** that seems to pre-date the faulting. A similar picture exists southeast of here, where the Aubrey Cliffs give way to the Mogollon Rim, south edge of the Plateau. In all likelihood the paleovalley, and therefore the forerunner of the Mogollon Rim, was in Oligocene time continuous across this part of Arizona.

North of Seligman the Coconino Plateau is surfaced with gray-white limestone of the Toroweap and Kaibab Formations. East of Seligman these same layers are covered by a blanket of volcanic rocks, on which the highway will remain to Flagstaff and beyond.

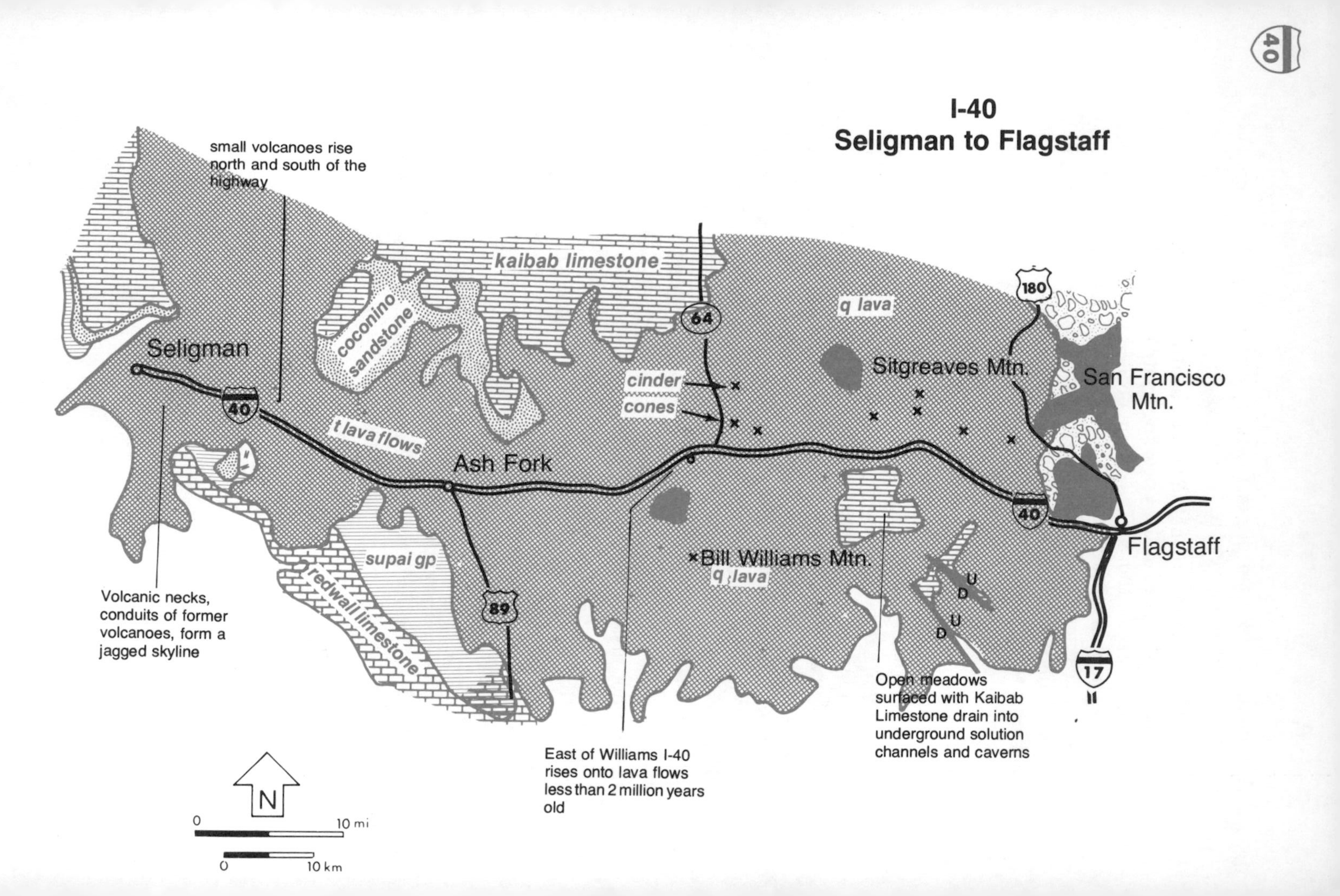
I-40
Seligman to Flagstaff
small volcanoes rise north and south of the highway
kaibab limestone
64
q lava
180
Seligman
coconino sandstone
cinder
cones
Sitgreaves Mtn.
San Francisco Mtn.
40
t lava flows
Ash Fork
40
Flagstaff
supai gp
Bill Williams Mtn.
q lava
U
D
U
D
Volcanic necks, conduits of former volcanoes, form a jagged skyline
redwall limestone
89
17
Open meadows surfaced with Kaibab Limestone drain into underground solution channels and caverns
East of Williams I-40 rises onto lava flows less than 2 million years old
N
0
10 mi
0
10 km

Interstate 40
Seligman — Flagstaff
(74 miles)

East of Seligman and the Aubrey fault, the Coconino Plateau is veneered with lava flows. Flow surfaces hide beneath a meager soil, but the lava is well exposed in roadcuts and natural embankments. Eastward the highway rises onto successive flat-lying flows, each one younger than the one below.

What goes on, geology-wise, beneath the veneer of lava flows that extends from Seligman to Flagstaff? The flows conceal flat-lying Permian limestone and, locally, Triassic red sandstone and shale similar to those around the margins of the volcanic field, strata that here and there are slightly warped, cut by faults, or bent sharply as monoclines. In a few places, perhaps along hidden fault zones, masses of intrusive igneous rock penetrate or dome up the horizontal layers. Elsewhere slender volcanic conduits pierce the lava; their remnants appear as volcanic necks near Seligman.

The oldest lavas, west of Seligman, are roughly 15 million years old. Between Seligman and Ashfork the flows, as well as the volcanic necks, are younger: 15 to 6 million years old. Lava east of Ashfork, part of the San Francisco Volcanic Field, has been dated as less than 6 million years old. And some lava flows near Flagstaff are less than 1 million years old.

Bill Williams Mountain near the town of Williams is one of several centers where, in this sea of dark gray basalt, lighter, silicic volcanic rocks occur, rocks such as dacite, andesite, and rhyolite. Silicic magma does not flow as readily as basalt, but squeezes up into volcanic domes or short, thick lava flows. Bill Williams Mountain consists of a central cluster of conduit-like spires, visible from mile 157, surrounded by 12 volcanic domes and several compact flows. Sitgreaves and Kendricks Mountains farther east and north are similar but smaller centers. When silicic magma erupts over and over again from the same vent, sometimes explosively, sometimes as short, stubby lava flows, a **stratovolcano** develops. San Francisco Mountain is a full-scale example.

An interesting feature of these centers of silicic volcanic activity is that they, like the basalt flows, become younger eastward. Bill Williams Mountain dates from about 4 million years, Sitgreaves from 2.5 million years, and Kendricks Peak from less than 2 million years. San Francisco Mountain's eruptions lasted from 2.8 million to about 200,000 years ago. East of Flagstaff, Sugarloaf and O'Leary Peaks erupted 220,000 and 200,000 years ago.

The first **cinder cones** appear just east of Williams. They are true volcanoes, produced from frothy, gas-rich magma, and they come in various sizes from 100 to 1000 feet high. Many have craters at their summits, or did at one time. Lava flows issued from the bases of some, evidence that the blow-out of frothy cinder-forming magma from the foaming top of the magma column was a preamble to eruption of fluid, relatively gas-poor magma. The cinder cones, too, tend to be progressively younger eastward. The most famous, and youngest, is Sunset Crater northeast of Flagstaff, which erupted less than 1000 years ago. Cinder is quite light in weight, and is quarried as aggregate for railroad beds, road material, and concrete blocks. Several cinder cones in this region have vanished completely, and as you will see east of Flagstaff others will soon meet the same fate.

Most of the lava-covered part of the Coconino Plateau near Williams and Flagstaff is clothed in a forest of ponderosa pine. Many open meadows are lava-free areas underlain by Kaibab Limestone, the Permian rock that surfaces much of the Coconino Plateau and most of the Kaibab Plateau north of Grand Canyon. Because limestone dissolves easily, an underground network of solution cavities and passages steals water from the surface, so many of these meadows do not show a stream drainage pattern. Some, though, have lines of sinks where limestone caverns have collapsed. Water that percolates down through the Kaibab Limestone and the underlying Toroweap Formation eventually ends up in the porous Coconino Sandstone, one of the main aquifers providing water for the city of Flagstaff.

As the highway approaches Flagstaff, San Francisco Mountain, locally called San Francisco Peaks, dominates the scenery, its upper slopes and multiple summits wreathed in clouds or standing stark against the blue plateau-country sky. It is surrounded by younger basalt and volcanic cinders that surface the plateau here.

The volcano's once classic Fujiyama shape was modified at the summit and on the northeast side by collapse into a partly emptied magma chamber, probably after a Mt. St. Helens-like sideward explosion, roughing out an interior valley — the Inner Basin. A succession of Pleistocene glaciers later smoothed this valley and deposited moraines across its mouth. The sculpturing by fire and ice left the volcano with three summits: Humphreys Peak on the north (12,670 feet, the highest point in Arizona) Agassiz Peak, and Fremont Peak.

Elden Mountain, between San Francisco Mountain and Flagstaff, is a smaller volcanic center only a million years old. Lava flows form its southern slope, which can be seen from the highway. Note how columnar jointing cuts the volcanic flows, and the way the lava appears to have plunged down the steep mountain flanks. Columnar jointing results as lava cools and shrinks. The north and east sides of Elden Mountain are less like a volcano and more like a laccolith, with two large blocks of Paleozoic sedimentary rocks apparently tilted up by magma forcing its way toward the surface.

Below the volcanic rocks the plateau is surfaced with red Triassic sandstone and mudstone, the Moenkopi Formation. Visible in roadcuts east of the center of town, these rocks in turn lie on the Kaibab Limestone.

Interstate 40
Flagstaff — Winslow
(58 miles)

For the first few miles east of Flagstaff, lava flows and volcanic cinder cones hide the underlying rock. Some of the cinder cones are being quarried for lightweight aggregate and road material. Erosion, operating even as quarrying proceeds, has gullied the soft material, but the original slope of the cinder layers, away from a central conduit, is still apparent.

The highway soon leaves the volcanic rocks and continues across the uppermost part of the Kaibab Limestone, the youngest Paleozoic unit of this region. In roadcuts, at Walnut Canyon National Monument (see Chapter V), and farther east in the walls of Padre Canyon and Canyon Diablo, at miles 219 and 225, you will see that this formation is made up of layers of yellowish gray limestone. The Kaibab Limestone is progressively sandier eastward, an overall measure of the approach to the one-time shore of the Kaibab sea.

The east edge of the Coconino Plateau is not sharply defined along this route The Kaibab surface slopes eastward gently, with small swales and ridges reflecting the slight folds and faults in the rock

I-40
Flagstaff to Winslow

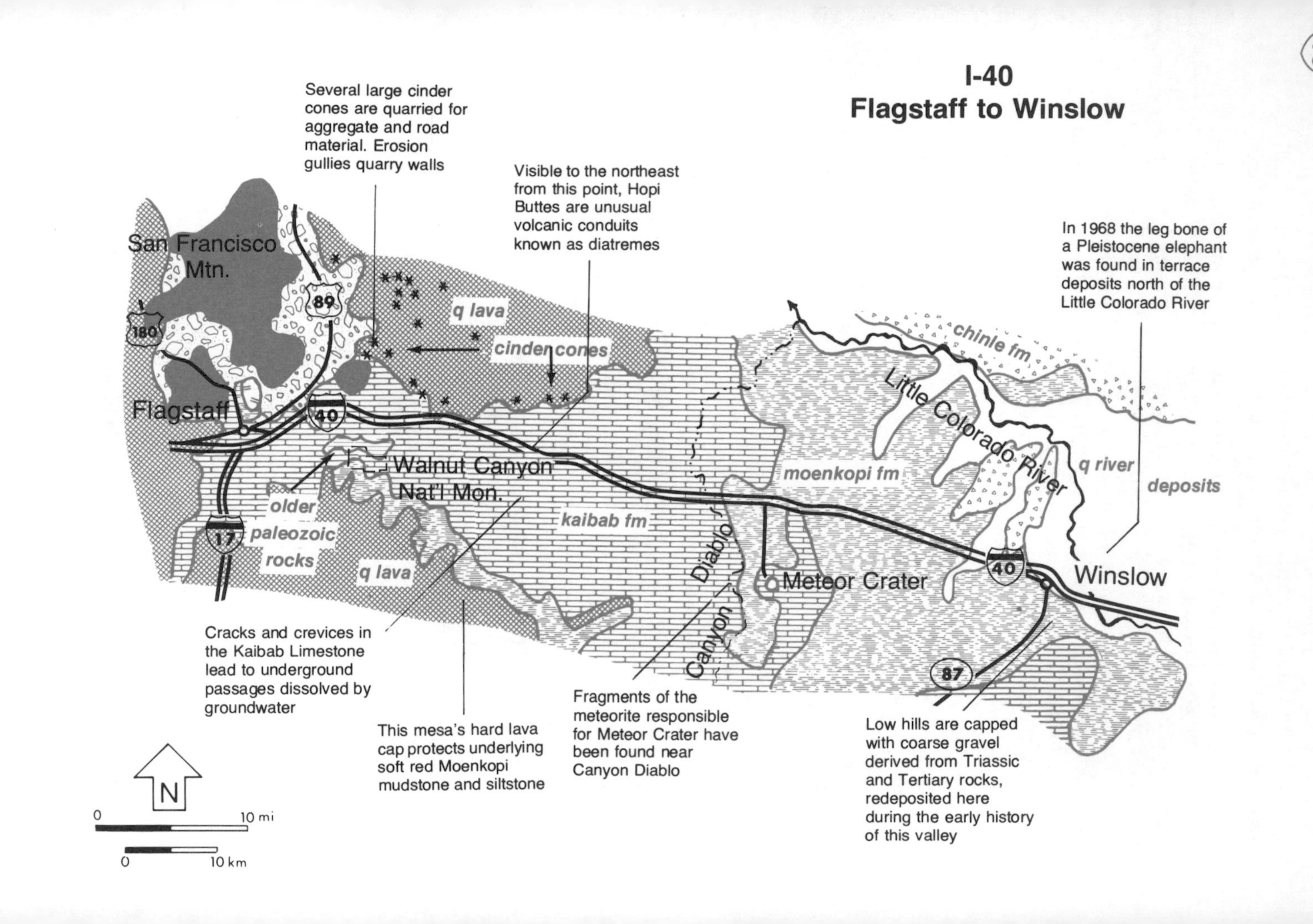

Skirting the lava flow that dammed it, the Little Colorado's muddy stream cascades over its own former canyon wall at Grand Falls northeast of Flagstaff. Tad Nichols photo.

layers. In places the Kaibab Limestone is riddled with cracks and crevices that allow surface water from rain and snowmelt to percolate downward, enlarging the cracks as it does so. The water ends up in the Coconino Sandstone, a thick wind-blown sandstone that underlies the Kaibab Formation here and that is the main aquifer of this region. Springs and wells tap this aquifer, but in places the water is too highly mineralized for domestic use.

Late in Permian time this region was lifted above the sea and eroded into a low-lying plain cut by shallow stream channels. On this surface in Triassic time the Moenkopi Formation was deposited.

Beyond mile 217 the Hopi Buttes Volcanic Field is in sight in the distance to the northeast. This is a region of many small volcanic necks unusual in that magma surging up and down through the conduits broke off and carried downward with it fragments of rock from the conduit walls, preserving sedimentary layers for which there is no longer any nearby surface record. Called **diatremes** (their counterparts in South Africa contain diamonds), the conduits are surrounded by lake deposits and the terraces below them tell of the large lake that filled this valley in Pliocene time and that may have received its sustenance from the Ancestral Colorado River.

South of mile 230, about 10 miles away, the rim of Meteor Crater rises above the plateau surface. This crater is well worth seeing, for it

Honeycomb weathering commonly decorates sandstone of the Moenkopi Formation. Water and wind deepen holes etched in less resistant parts of the dark red rock. Tad Nichols photo.

is the classic example of the type of crater caused by impact of a meteorite. It was the first recognized meteor crater, and is one of the best preserved ones on earth. Similar craters pock the surface of the moon, Mars, and the moons of many of our Solar System neighbors, and there are other meteor craters, less well preserved, scattered over the face of the earth. The crater resulted from the impact and explosion of a mass of almost pure iron weighing some 63,000 tons and having a diameter of 80 feet or so, a fiery ball that plunged through the atmosphere about 22,000 years ago. Because Meteor Crater resembles craters on the moon, it was used as a training ground for Apollo astronauts. Thousands of once-molten fragments of the meteorite have been found, especially near Canyon Diablo, confirming the explosive finale of its earthward plunge. In the crater rim, Kaibab Limestone and the underlying Coconino Sandstone were pushed upward and outward by the force of the explosion.

At about mile 230 the Kaibab Formation disappears under red mesas and buttes of the Moenkopi Formation. Near Winslow this Triassic unit is about 250 feet thick, and appears as layers of cliff-forming sandstone, ledge-forming siltstone, and easily crumbled slope-forming mudstone cut by fine and very numerous veins of gypsum. These deposits formed on a river floodplain or tidal flat, a place of alternate wetting and drying, where current ripple marks, mudcracks, raindrop pits, and salt crystal molds marked the once-soft surfaces. Fossil amphibians have been found near Meteor Crater and elsewhere in this area — strong, thick-skulled creatures as much as 10 feet long, a far cry from today's small frogs, newts, and salamanders. Their short, weak legs suggest that these creatures were adapted to life in shallow rivers and embayments. The Moenkopi contains in addition many footprints and trackways of a four-footed reptile called *Chirotherium*, named more than a hundred years ago from similar tracks found in Germany. Cross-bedded stream deposits of coarse silt and sand become more common eastward — shoreward in Triassic time — as do layers of pebbly conglomerate. It is not difficult to envisage the Triassic landscape, a plain shelving seaward, gradually becoming covered with layers of mud and sand and gravel brought in by rivers from higher areas to the east. The silt and mud were probably not red at that time; the redness came later as their iron minerals oxidized below the surface.

The rest area at mile 235 offers a distant view of the Painted Desert, the valley of the Little Colorado River. Triassic formations — the predominantly red Moenkopi Formation and the varicolored Chinle Formation — lend the Painted Desert its famous colors. Across the valley, dim with distance, Hopi Buttes rise above the horizontal surface of old lake deposits.

A few ledges of the lowest part of the Chinle Formation appear north of the highway near Winslow, ledges made up of a hard, resistant, pebbly unit called the Shinarump Conglomerate. The contact between the Moenkopi and Shinarump is irregular and wavy, and represents a time of erosion. Gravels in the conglomerate contain stream-rounded pebbles of durable rocks such as quartzite, chert, jasper, and bits of petrified wood, all but the last thought to come from Triassic mountains in central Arizona. Soft, colorful sandstone and siltstone strata of the Chinle Formation, many of them rich in volcanic ash, have yielded many types of fossils. The most renowned are of course the tree trunks of the Petrified Forest.

I-40
Winslow to Petrified Forest

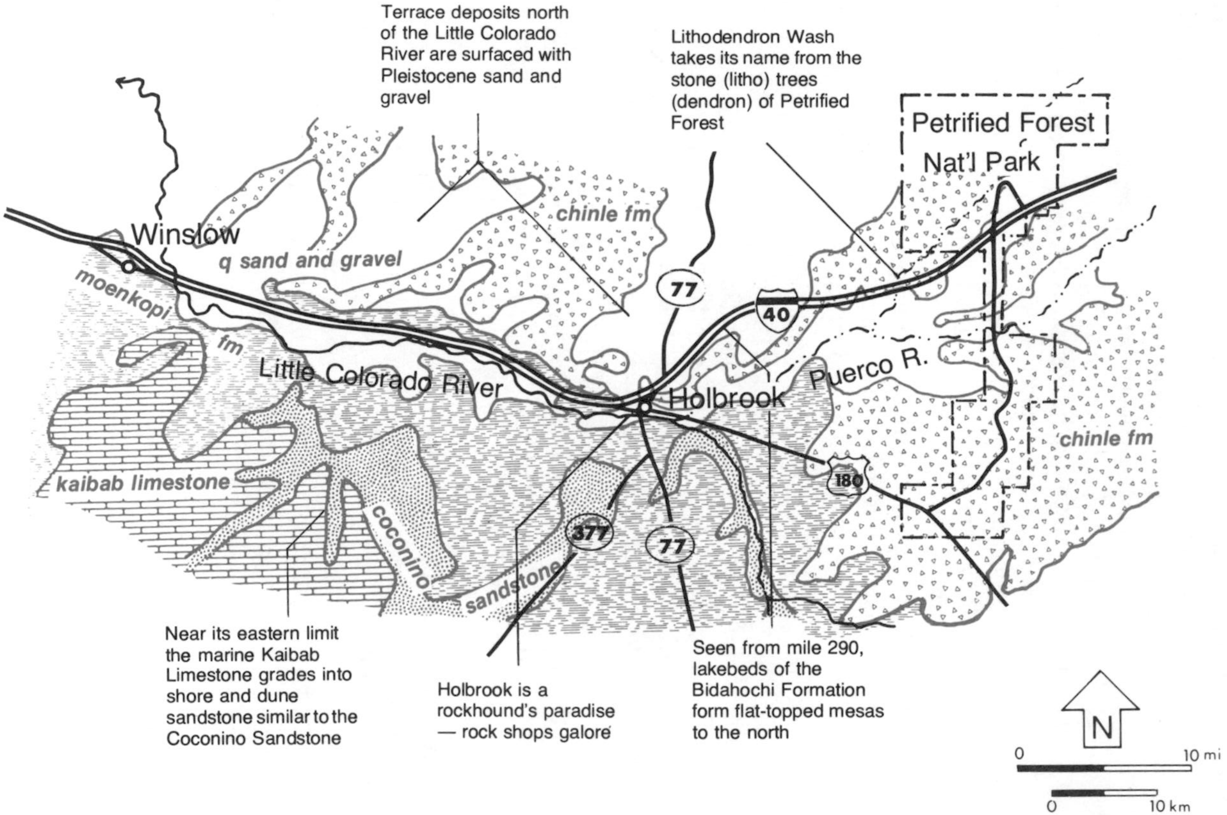

Interstate 40
Winslow — Petrified Forest
(61 miles)

Three miles east of Winslow, I-40 crosses the Little Colorado River, a river that drains 20,000 square miles and yet usually looks bone dry! Its sandy bed conceals an underground river that percolates through porous sand and gravel, nourishing the cottonwood trees and other vegetation along its banks. During heavy summer thunderstorms, muddy red-brown torrents inundate the river bed, sweeping downstream all the way to the big Colorado.

Near the highway as it approaches Holbrook rise red ledges and small mesas of Moenkopi Formation sandstone and siltstone. Behind them stretch barren hills eroded in soft mudstone and volcanic ash of the Chinle Formation. Both of these units are Triassic. The base of the Chinle is marked by a resistant ledge of Shinarump Conglomerate, a unit that contains hard quartzite pebbles derived from highlands that existed in central Arizona in Triassic time.

Although there are scattered outcrops of the Moenkopi Formation north of the highway as far east as Holbrook, many-hued hills of the

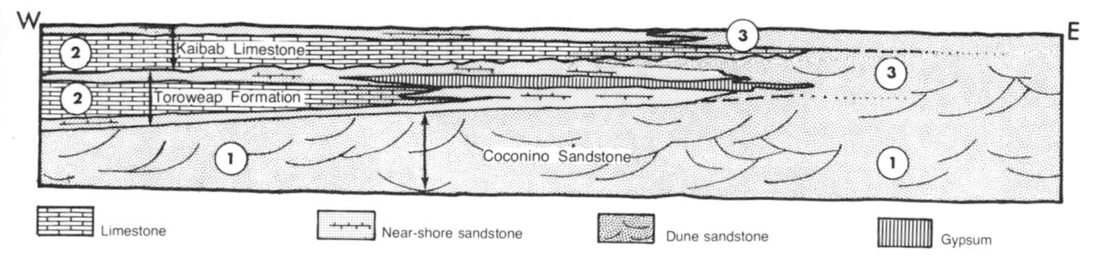

In mid-Permian time, a desert of windblown sand swept northern Arizona, creating the Coconino Sandstone (1). Then shallow Kaibab and Toroweap seas advanced from the west, each depositing a wedge of marine limestone bordered with nearshore sandstone, siltstone, and gypsum (2). Desert conditions still prevailed on the shores, so dune sandstone there is of the same age as the marine limestones and nearshore deposits.

Chinle Formation are progressively more common. This unit contains large amounts of bentonite, a type of clay formed from volcanic ash. Bentonite puffs up when it is wet, shrinks when it dries, a feature that discourages plant growth and contributes greatly to development of badland topography. The Chinle Formation lends its palette of color — blue, gray, deep red, yellow, white and green — to the Painted Desert, and a little farther east contains the petrified wood of the Petrified Forest.

The rise to the south, almost paralleling the highway, is the Holbrook Anticline. The north flank of the anticline, the one that slopes toward I-40, seems to be just a continuation of the very gentle northward dip of all the strata in this region. The south flank, out of sight, is steeper. Most of the anticline is surfaced with sandy gray-white layers of Kaibab Limestone. Here, near the strand line of the Permian sea that gave it birth, the Kaibab is more standstone than limestone. In fact as you go eastward it finally becomes indistinguishable from the dune-deposited Coconino Sandstone that in the Grand Canyon lies well below it. The anticline surface is pitted with depressions, especially along the steep south flank. These are caused by solution of salt and gypsum known to be present in the Supai Group about 1000 feet below the Coconino Sandstone. These minerals probably accumulated in a shallow bay where sea water evaporated, and then were covered with other strata. Solution of salt and gypsum is responsible, too, for the Holbrook Anticline itself, with collapse of overlying formations causing the steep south slope.

The Holbrook Anticline is about 60 miles long. Anticlines are favorable sites for accumulation of oil and gas, which migrate upward through porous rock until they are trapped by impermeable layers. This anticline has been drilled repeatedly, but without much success. Drilling associated with oil exploration shows that underneath the Kaibab and Coconino Formations is an almost complete Grand Canyon-like sequence of Paleozoic sedimentary rock units. Precambrian rocks like those in Grand Canyon's depths lie 3800 feet below the surface.

Terraces north of the river date back through Pleistocene time. They have been traced throughout the Little Colorado River's drainage basin, from Grand Canyon to the New Mexico border. Each terrace mirrors two stages in development of this valley, a stage of deposition followed by one of erosion. Altogether they show five such alternations. Unfortunately for geologists, the five terrace levels do not all occur together anywhere, and only painstaking mapping and study of their surface characteristics have enabled geologists to piece together their story. Some of the younger terrace deposits have yielded bones of fossil mammals, among them horse bones and one leg

bone of an extinct elephant. The youngest terraces bear artifacts and other relics of preColumbian man.

The drainage history of this area is closely tied in with that of the Ancestral Colorado River, which until about 5 million years ago may have flowed southeast, the opposite direction from that now taken by the Little Colorado. In this broad basin, in Pliocene time, Bidahochi Lake (also called Hopi Lake) developed. And in the lake the lower part of the Bidahochi Formation was deposited, a limy, gravelly formation that forms level mesas north and east of the Painted Desert. The lake may have been fed with waters from the Ancestral Colorado River. Some of the Bidahochi Formation's sand and gravel and clay, however, came from the northeast, deposited by streams that drained the Defiance Plateau and the Chuska Mountains.

Looking north from mile 290 you can see the Hopi Buttes jutting above the flat surface of the Bidahochi Formation. This cluster of nearly 200 small volcanoes erupted in Pliocene time, while the lake was here. Some rose as domes of viscous lava; others discharged flows of more fluid lava or clouds of incandescent ash. Still others really exploded, producing broad, shallow **maar** or steam explosion craters half filled now with fragments of volcanic rock. Maar explosions require water, furnished in this case by Lake Bidahochi. All of the volcanic centers are surrounded by the Bidahochi deposits, and the center part of the formation records the volcanic activity of the buttes in layered ash beds. Stream deposits make up the upper third of the formation.

Interstate 40
Petrified Forest to New Mexico
(46 miles)

East of the Petrified Forest this highway continues to climb toward Lupton, going upstream along the Puerco River, traveling on river gravels and Triassic rocks of the Moenkopi and Chinle Formations. Immediately north and south of the highway the tan sandstone and siltstone layers of the Bidahochi Formation lie horizontally across these older, somewhat tilted rocks, in an angular unconformity that can be seen from several places along the route.

The Bidahochi Formation, which is Pliocene in age, now covers parts of an area that measures about 112 miles north-south and 100 miles east-west. Bidahochi or Hopi Lake, in which the lower part of it was deposited, seems to have been larger than this, extending west over what is now the valley of the Little Colorado River. The lake may have been dammed by resistant beds in the Chinle Formation, such as those that today cap Newberry Mesa and Ives Mesa north of Winslow, by volcanic rocks long washed away, or by the northward rise in the Kaibab Limestone or some other resistant formation. In any case, the lake seems to have been roughly circular and nearly as large as Lake Erie.

When the lake was here the climate was certainly wetter than it is now, and water flowing into the lake may have exceeded the present flow of the Colorado River. There is a possiblity that it was fed by the Ancestral Colorado, which may have flowed southward through the valley now occupied by the Little Colorado, as described in the introduction to this chapter. There is no for-sure record where the lake's outlet was; perhaps water left the lake via channels now occupied by the Puerco River or Carrizo Wash; perhaps it escaped southward through a channel since concealed by volcanic rocks of the White Mountains. The lake's ultimate demise probably relates to infilling with volcanic ash and lava flows and to the shift in flow direction and rapid downcutting of the Colorado River as it was diverted westward through what was to become Grand Canyon. Downcutting in Grand Canyon would have steepened the gradient of the newborn Little Colorado River, and any time a river gradient is increased, water flow accelerates and erosion speeds up, erosion that in this case would have cut down through the hypothetical western dam that once contained the lake waters.

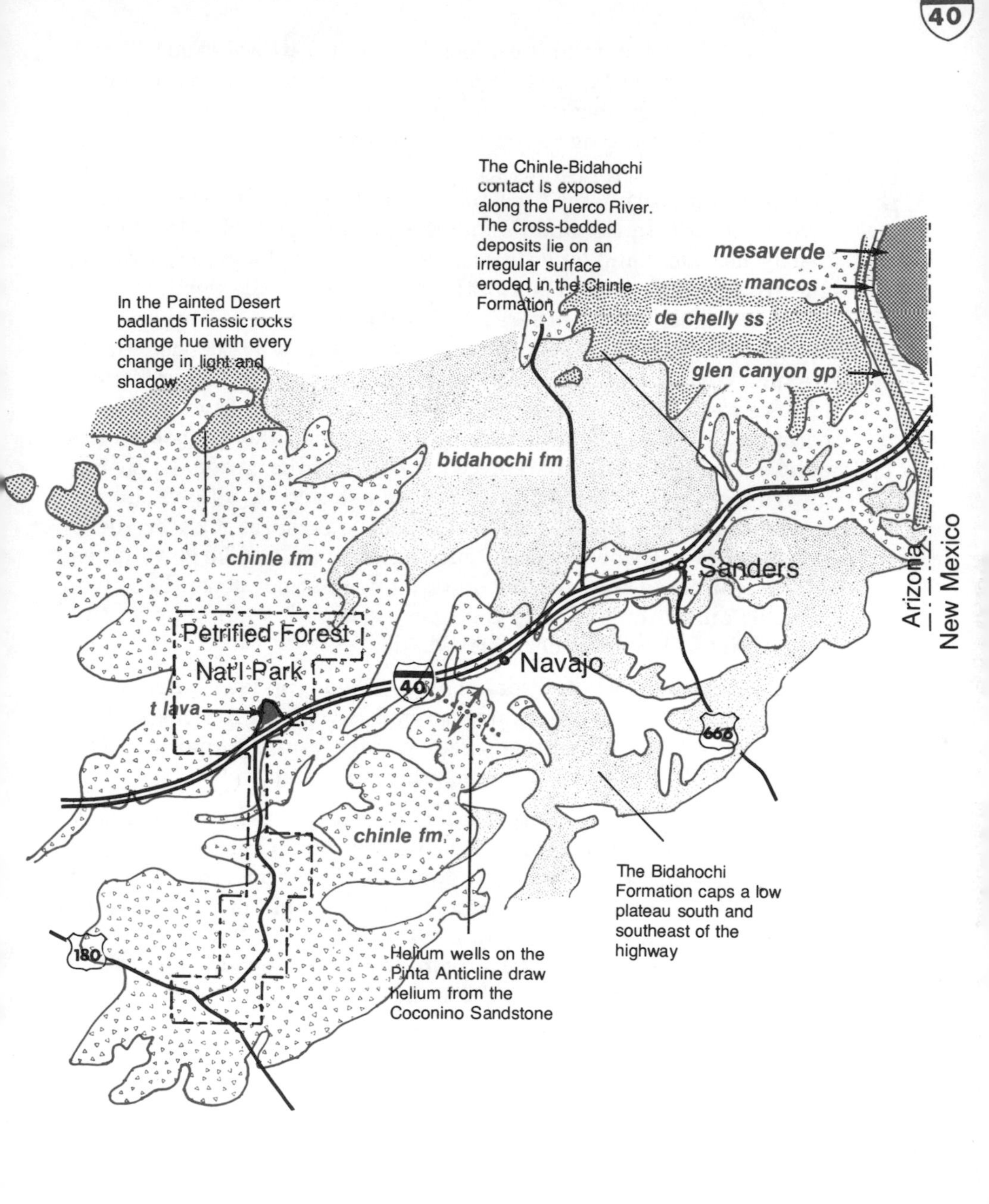

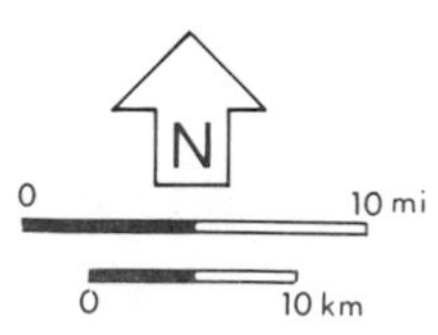

I-40
Petrified Forest to New Mexico

Bidahochi sediments contain fossils: small freshwater snails and clams, well preserved fish skeletons, tracks of water birds, the skeleton of a beaver. The formation has been divided into three members that reflect the changing environments of the lake valley. The oldest, lowest member consists of flat-lying lake beds — sand and conglomerate and clay, with some limestone — and then interspersed layers of volcanic ash. In the middle member, the volume of volcanic ash greatly increases and lava flows reach out toward the center of the lake from the Hopi Buttes volcanic area just to the north. In the upper, youngest member there are few if any true lake beds; most of the material is cross-bedded sandstone probably deposited by streams. We can assume that by the time the upper member formed the lake was almost entirely filled with volcanic ash, lava, and lake sediments. So we have a picture of a highland lake, probably surrounded for much of its existence with wooded hills, gradually filling in with sand and silt and then volcanic ash and lava flows. Later, when the lake was deprived of its once generous source of water, these deposits were trenched by down-cutting streams, and large volumes of lake sediments and volcanic deposits were swept down the Little Colorado and through Grand Canyon, eventually ending up on the delta of the Colorado in northern Mexico.

East of Sanders, where Bidahochi lake beds cap mesas north and south of the highway, there are arched caves and alcoves in this relatively poorly consolidated, geologically young formation. The fine cross-bedding of the upper member shows up well.

Almost at the New Mexico border, colorful Triassic sedimentary rock layers bend down boldly, as do Jurassic and Cretaceous formations, in a thick rock sequence similar to that in other parts of northeastern Arizona. This bend is the east side of the Defiance Plateau, a 75-mile-long anticline complicated by faulting, which brings to the surface in the Canyon de Chelly region some of the Paleozoic rocks that underlie these Mesozoic formations.

U.S. 60
New Mexico — Show Low
(60 miles)

US 60 enters Arizona in a fascinating area where lava flows coming from the White Mountains, and pink Tertiary sediments rather obviously derived from the red Mesozoic rocks that underlie them, conceal the southern rim of the Colorado Plateau region. Just what goes on beneath the lava blanket we do not know; we see underlying rocks only in a few natural "windows" that were never covered with volcanic rock or were covered with volcanics that were later eroded away. Hidden from our eyes there may be clues to drainage patterns of the Ancestral Colorado River and the lake into which it flowed, as well as to the origin of the volcanic rocks themselves. We can only roughly guess the position of the Colorado Plateau margin, which must swing southeastward toward and into New Mexico.

A few miles west of the state line, north of the highway, a small block of some of the Mesozoic sedimentary rocks of this region has hinged downward along a fold and a fault, like a trap door, thereby gaining some protection from subsequent erosion. The rocks are colorful gray and purple mudstone, easily eroded because they are rich with volcanic ash, and easily recognized as the Triassic Chinle Formation of Painted Desert fame. Lighter-colored rocks at the edge of the trap door are Cretaceous.

Fifteen miles south of the highway Escudilla Mountain rises to nearly 11,000 feet. This mountain is a small, steep-sided volcano whose flows protect a mysterious patch of river deposits. Like many other lava flows along the margins of the White Mountain volcanic field, these flows seem to have filled a paleovalley, protecting it from the erosion that wore away surrounding unprotected paleoridges, so that now what was a valley has become a ridge, and what were ridges have become valleys. The highway crosses a similar but younger flow between milepost 395 and Springerville, and as the road climbs onto it and descends from it the former surface can be seen at the base of the flow, well baked and terra-cotta red. A white deposit of caliche, calcium carbonate precipitated as lime-rich water evaporated at or near the surface, shows up about 8 inches below the baked zone.

US 60
New Mexico to Show Low

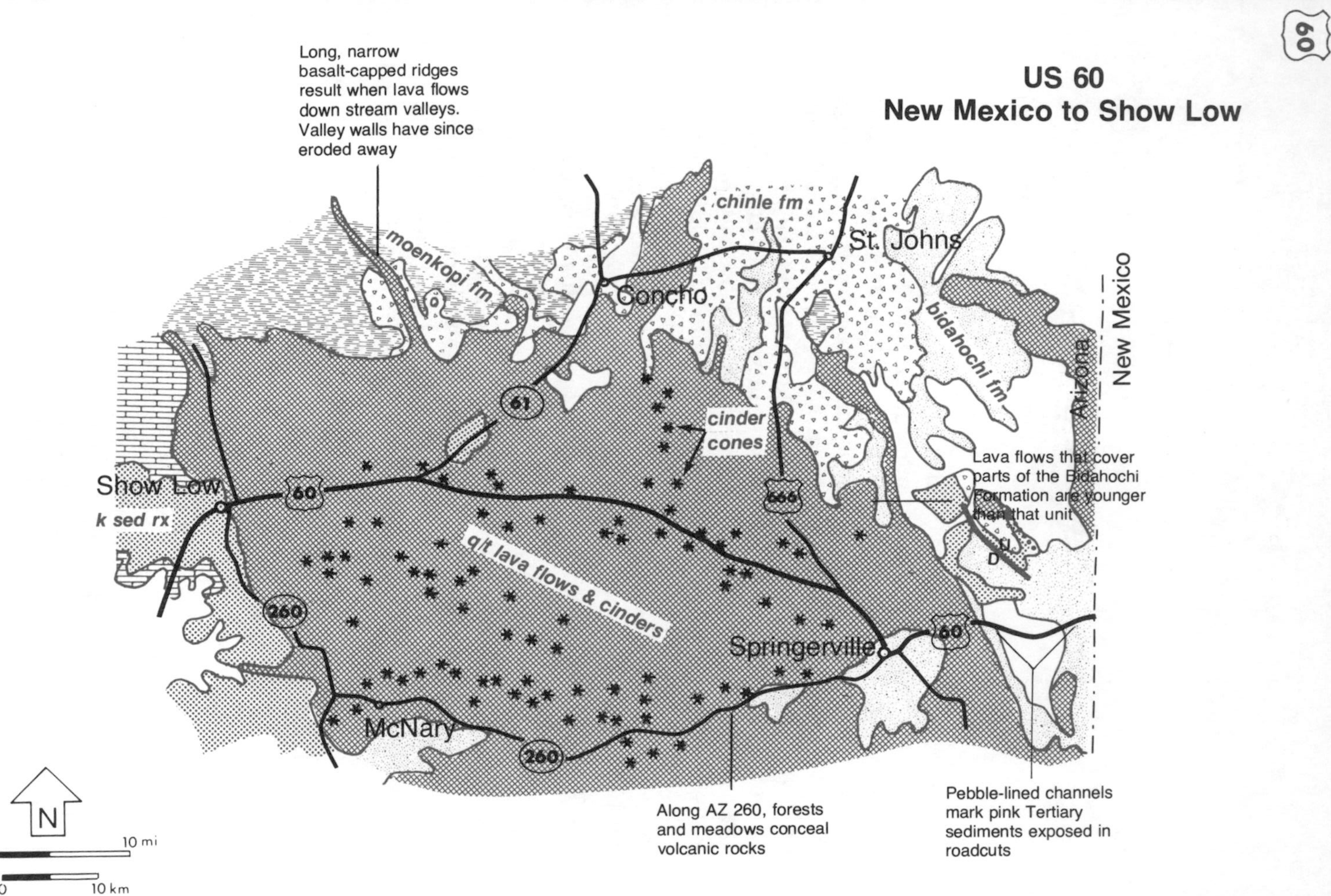

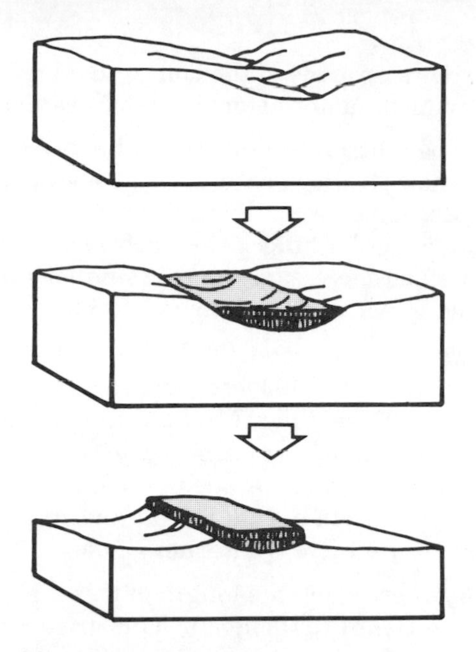

Lava is more durable than most types of sedimentary rock. When it flows down a valley, hardening there, bordering ridges may wear away, leaving a new ridge to mark the position of the former valley.

Headwaters of the Little Colorado River lie south of Springerville in the White Mountains. One of the few large tributaries flowing into the Colorado River from the south, this river is also one of the few permanent streams in this region. But by the time it gets to Holbrook most of its waters have sunk into the river sands. They continue to flow underground, however, surfacing again at Cameron (see US 89 Flagstaff to Cameron) more than 200 river miles downstream. During spring thaw in the White Mountains, or in the summer rainy season, the river may flow or even flood for its entire length.

West of Springerville the highway crosses a broad, rolling sea of lava flows, with increasingly numerous cinder cones on both sides. There are some 200 such cones here, small volcanoes formed when spurts of frothy lava hurtled skyward, cooled quickly, and fell to earth around the vent from which they came. Some have still-preserved craters at their summits; some do not. Many craters are clearly breached by erosion. On others, a resistant rim edges the crater, a rim of lava spatter that fell back around the vent while it was still molten enough to stick together. Large blocks and twisted scraps of lava, called **volcanic bombs**, lie around the base of some of the cinder slopes. Lava flows, many pimpled with rounded blisters or accented with sharp ridges called **squeeze-ups**, emerge from the bases of some cones. A lava flow also emerges from a cleft in the rim of a cone near milepost 373.

Cinder quarries scar many of the cones; the light-weight cinder is attractive ready-made material for road surfaces and concrete blocks.

Flowing lava pays little attention to established drainage patterns, except to flow down valleys as described above. So lava surfaces characteristically have no really organized drainage patterns, but are pocked with irregular hillocks and poorly drained depressions of all sizes. Many of the depressions contain small ponds or lakes; others are partly filled in with soil and vegetation.

The White Mountain volcanic field rises southward toward a central volcano, Mt. Baldy (11,490 feet), where the pile of volcanic rocks is probably a full 4000 feet thick. Mt. Baldy's slopes and the rest of the volcanic region that extends across east central Arizona were built by Tertiary eruptions and partly eroded by Quaternary glaciers. Only here on the north margin of this region and in a small area much farther southwest are there Quaternary flows.

Almost everywhere along the length of this highway one sees and feels the Painted Desert to the north. This great expanse of colorful badlands stretches from the small scrap of Triassic rock in the trapdoor graben near the New Mexico border to the Colorado River, a distance of about 220 miles. The Painted Desert is more fully discussed under I-40, US 89, and Petrified Forest National Park.

U.S 89
Flagstaff — Cameron
(51 miles)

Just north of the eastern part of Flagstaff is Elden Mountain, a small silicic volcanic center that is partly extrusive and partly intrusive. The southern part of the mountain, nearest the highway, displays steep columnar-jointed dacite lava flows. The intrusive part, on the north and northeast, is like a laccolith in that intruding magma domed up overlying Paleozoic sedimentary rocks, which still outcrop on its flanks.

From its junction with I-40, US 89 heads north across the cinder-covered apron of landslides and glacial deposits that surrounds San Francisco Mountain. From about milepost 421 you can look into the Interior Valley of this beautiful volcano, the view partly obscured by humpy glacial moraines and by Sugarloaf Peak, a small volcano in its own right. San Francisco Mountain and its story are discussed under I-40 Seligman to Flagstaff. The Interior Valley, or Inner Basin, was long thought to have been carved by Ice-Age glaciers. There is no doubt that glaciers helped, but the size of the Inner Basin is far too large to be explained by glaciers alone, of the size that could have existed this far south and at this altitude. Sugarloaf Peak has been dated radiometrically as 220,000 years old, and quite obviously predates the glaciers that deposited rock debris around and against it. So the Inner Basin must have been there before the glaciers. San Francisco Mountain itself is 2.8 million to 200,000 years old — its youngest volcanic rock about the same age as Sugarloaf. The basin may have come into being as the central part of the volcano collapsed into its partly emptied magma chamber, or it may be a product of a violent sideways explosion like the 1980 eruption of Mt. St. Helens. Certainly the horseshoe shape of the remainder of the mountain resembles Mt. St. Helens now.

The black cinders that border the highway are the freshest cinders in Arizona. They were emitted by Sunset Crater in eruptions that began in 1064 or 1065 A.D. and that probably were still going full blast a year or so later when William the Conqueror set out to conquer England. The highway loop through Sunset Crater and Wupatki National Monuments, described in Chapter V, rejoins US 89, adding about 20 miles to the distance to Cameron.

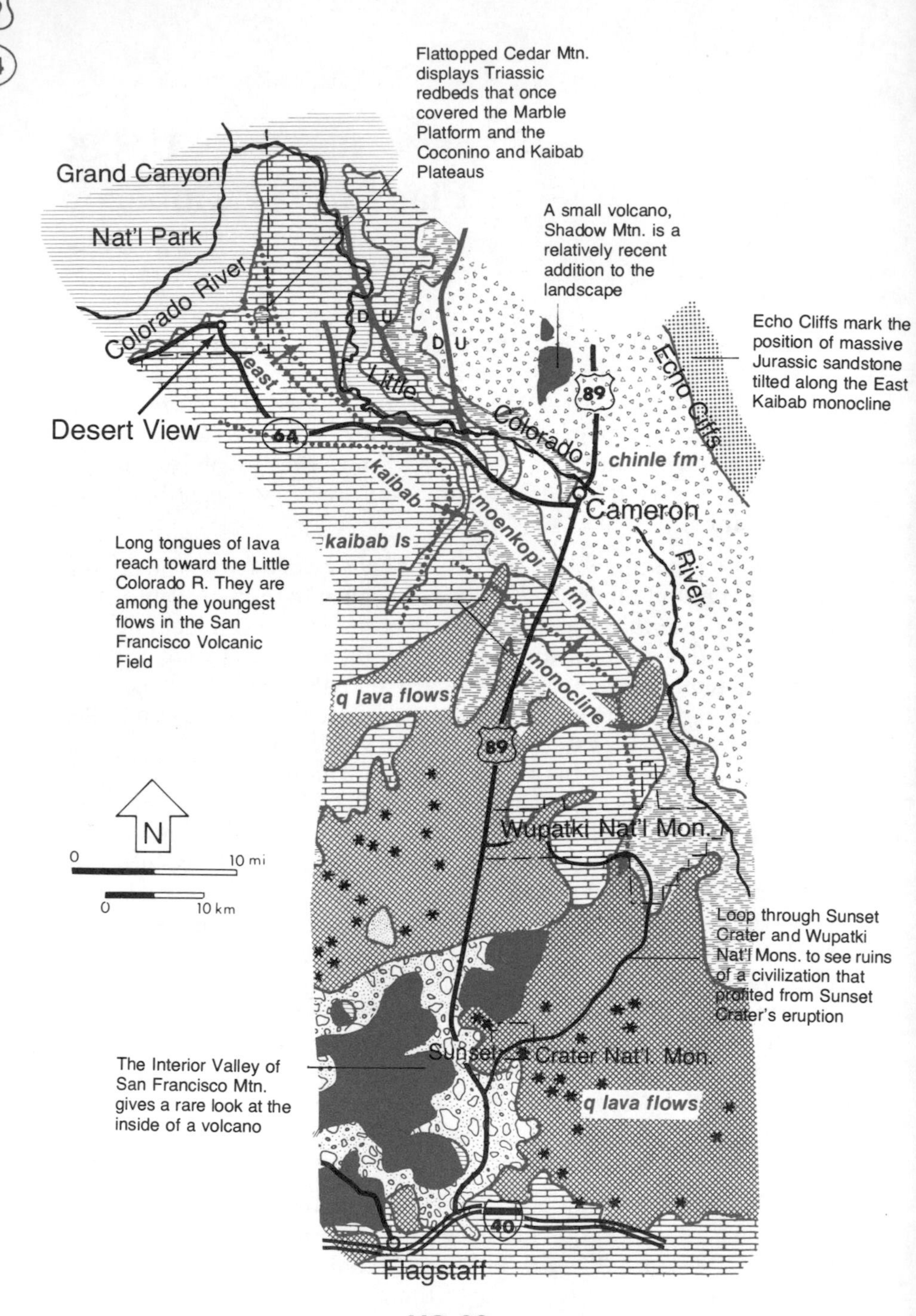

US 89
Flagstaff to Cameron
AZ 64
Cameron to Desert View

Lava flowing from a fissure at the base of a cinder cone spread out over part of the Painted Desert. The crater's rim is edged with spattered lava. Tad Nichols photo.

Just north of Sunset Crater is O'Leary Peak, easy to identify by its zigzag road and fire lookout. O'Leary is not a cinder cone like its small but famous neighbor Sunset Crater, but is a cluster of volcanic domes that pushed up 170,000 to 240,000 years ago, bulged out at the surface, and hardened in place. Most volcanoes of this type wear a skirt of slope-forming fallen rock, and O'Leary is no exception.

A very foamy volcanic rock called **pumice**, mined near the highway north of O'Leary Peak, was used as a lightweight additive in concrete for Glen Canyon Dam.

Not far north of O'Leary Peak the vista changes abruptly as occasional glimpses of the Painted Desert become reality. Suddenly you leave the cinder cones and volcanic peaks, as well as the pine and piñon of elevations above 6000 feet, and look out over the vastness of the valley of the Little Colorado River. Echo Cliffs rise in the distance almost straight ahead, and Hopi Buttes show as dark spires far to the east. Northwest, out of sight, is Grand Canyon.

The highway stays on a lava surface for some distance, then drops onto the buff limestone of the Kaibab Formation, brought to the surface by an anticline. After that it continues to Cameron among colorful Triassic rocks of the Moenkopi Formation. Most of this side of the valley is surfaced with these dark red sandstone, shale, and mudstone layers; the other side is surfaced with slightly younger rocks, the Chinle Formation. The rock layers slant fairly steeply here, defining the eastern edge of the Coconino Plateau. To the north you'll see the even more pronounced dip on Coconino Point, where flat-lying layers suddenly nose-dive into the desert as the East Kaibab Monocline. The Moenkopi Formation has been studied by measuring and recording its rock layers at discrete sites in southern Utah and northern Arizona, then diagramming and comparing the kinds of rock encountered. Many of the sites studied are strung out along the valley of the Little Colorado. Matching up the diagrammed sections with what is known about the origin of the rock types shows that the Moenkopi was laid down on a vast west-sloping plain across which the Triassic sea advanced and retreated several times.

Skeletons of fossil amphibians have been recovered from the Moenkopi Formation near Cameron, and some overhanging sandstone ledges show, on their lower surfaces, casts of reptile footprints. More than half of the tracks were made by a four-footed reptile called *Chirotherium;* they look pretty much like chubby human handprints. The creatures that made them seem to have ranged far and wide, for similar trackways are known from Triassic rocks in Germany, England, and elsewhere. Many of the fossil amphibians are also known from Europe, Africa, India, and Australia. These fossils remind us that when Triassic rocks were deposited, the continents we know today were still attached to one another, so that species of land animals met with no ocean barriers to prevent their dispersal.

Gypsum occurs in the Moenkopi Formation, filling veins and veinlets or making up thin, flat-lying, sometimes nodular beds. This white mineral (a common form is alabaster) forms in lagoons where evaporation of sea water causes concentration and precipitation of gypsum and other salts.

If you wander among some of the Moenkopi ridges, you'll find that many siltstone layers are decorated with ripple marks. Evenly spaced, parallel ripples result from moderate currents that swept across Triassic mudflats. Larger, more irregular ripples in coarse sand formed in stream channels where currents were stronger. Polygonal mud cracks, tiny circular craters formed by raindrops, and casts of salt crystal cubes testify to periodic drying of the surface. It is not difficult to picture, from this evidence, the original Moenkopi environment: broad, river-swept or sea-swept mudflats, partly tidal partly deltaic, with periodic inundation by sand-bearing floodwaters. The climate was warm enough and dry enough to allow evaporation of ponded sea water.

Nearly horizontal terraces edge some parts of this valley, but they are hard to distinguish from rock-capped ledges. Lava flows cover some of them, and can be dated radiometrically, adding to the various lines of evidence that reveal the story of the Little Colorado Valley, showing that it may once have been the route of the south-flowing ancestor of the present Colorado River.

At Cameron the Little Colorado River is entrenched in a canyon cut through solid rock. Upstream the river may be dry — it often is. But here where its channel begins to plunge through bedrock, the water hidden in its sandy upstream bed is forced to the surface. There is usually at least a trickle here, long known to Indian and early settler. Summer thunderstorms in the region to the south can turn the trickle quickly into a muddy brown torrent. When this happens, it is fascinating to watch the tumbling stream front — the "wall of water" of Western fame — churning into this narrow ravine. Here the Little Colorado is just beginning the long descent toward its confluence with the main Colorado River some thirty miles to the northwest and nearly 2000 feet lower. If you are heading toward Grand Canyon you will have an opportunity to see part of the deep, narrow cleft through which it flows.

When it is flowing, the Little Colorado carries much more sediment than most rivers, for the rocks of the Painted Desert region are easily eroded. Its average annual load — 14 million tons — may be carried past Cameron in just a few days. Its waters also contain dissolved salt and gypsum, which make them undesirable for household, irrigation, or stock use.

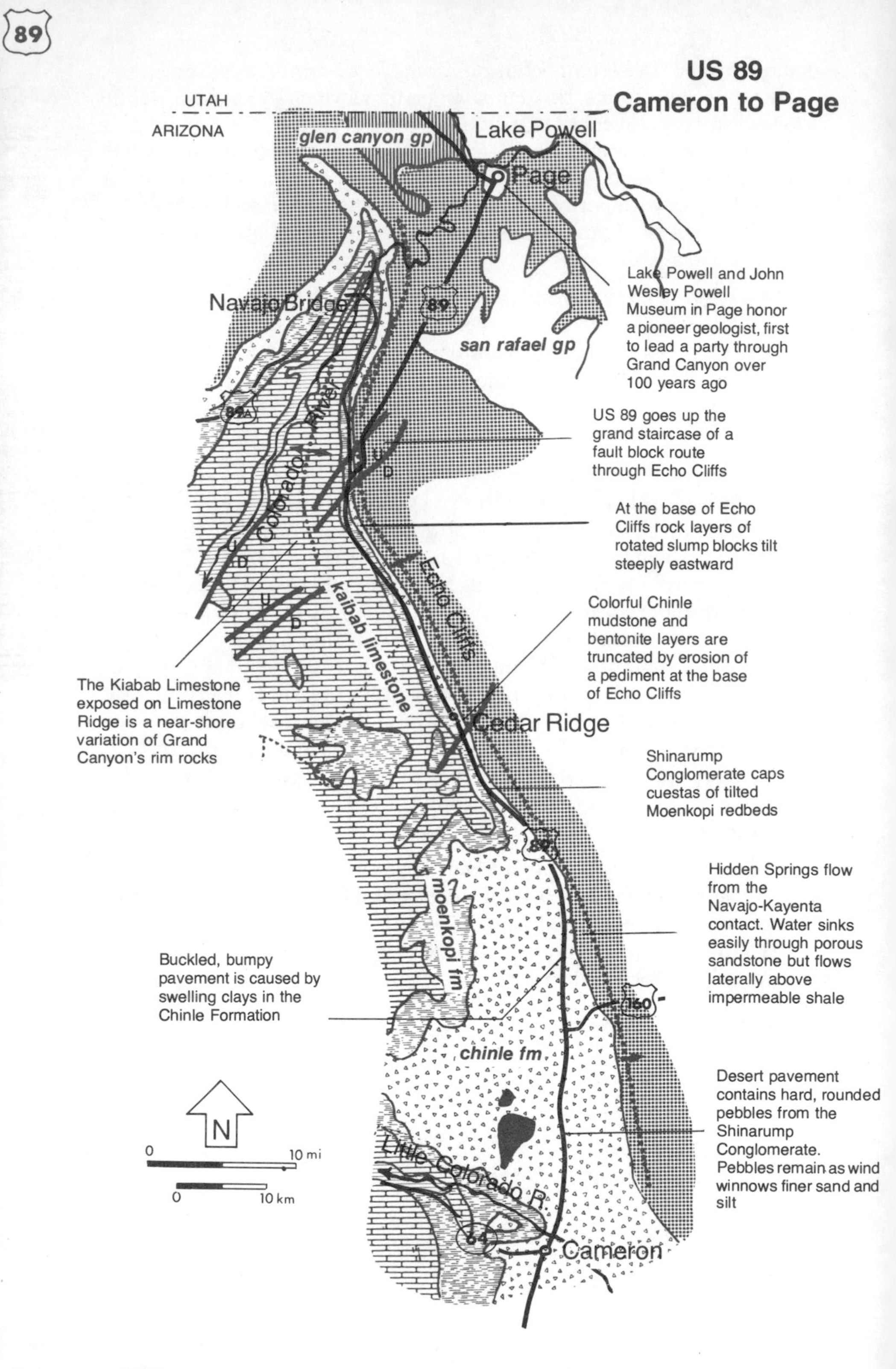

89
US 89
Cameron to Page
UTAH
ARIZONA
glen canyon gp
Lake Powell
Page
Lake Powell and John Wesley Powell Museum in Page honor a pioneer geologist, first to lead a party through Grand Canyon over 100 years ago
Navajo Bridge
89
san rafael gp
89A
Colorado River
US 89 goes up the grand staircase of a fault block route through Echo Cliffs
U
D
At the base of Echo Cliffs rock layers of rotated slump blocks tilt steeply eastward
Echo Cliffs
kaibab limestone
Colorful Chinle mudstone and bentonite layers are truncated by erosion of a pediment at the base of Echo Cliffs
The Kiabab Limestone exposed on Limestone Ridge is a near-shore variation of Grand Canyon's rim rocks
Cedar Ridge
Shinarump Conglomerate caps cuestas of tilted Moenkopi redbeds
89
moenkopi fm
Hidden Springs flow from the Navajo-Kayenta contact. Water sinks easily through porous sandstone but flows laterally above impermeable shale
Buckled, bumpy pavement is caused by swelling clays in the Chinle Formation
160
chinle fm
N
Desert pavement contains hard, rounded pebbles from the Shinarump Conglomerate. Pebbles remain as wind winnows finer sand and silt
0
10 mi
0
10 km
Little Colorado R.
64
Cameron

During infrequent rains, rivulets erode deeply into weak shales and volcanic ash of the Chinle Formation. Note modern wind ripples in the foreground.

Tad Nichols photo.

U.S. 89
Cameron — Page
(67 miles)

The Painted Desert is a land of many moods. Whether in sun or shadow, whether wet (which is rarely) or dry, its weirdly shaped gray, lavender, green, and blue hills seem to belong to another world — a Land of Oz. Capped with hard ledges of sandstone and conglomerate, these hills are shaped by badland erosion — washed with torrential downpours, blown with desert winds — in soft, limy mudstone layers of the Chinle Formation. The Chinle contains liberal doses of **bentonite**, a clay that forms from volcanic ash and that swells when it is wet and then dries into a scruffy, slaky crust that blows away or washes away quite readily. Because few plants can establish themselves in this rapidly eroding surface, no real soil develops.

Casts of salt crystals formed on ancient mudflats are preserved in the Moenkopi Formation. They suggest a near-shore environment where ripple-marked mudflats sometimes dried out. Tad Nichols photo.

At the bottom of the Chinle Formation in Arizona and Utah, and visible at numerous outcrops along this highway, is a coarse, pebble-filled resistant layer that blankets the Moenkopi Formation, filling channels and pockets in its upper surface. This is the Shinarump Conglomerate. Its pebbles are nearly all of durable rock types such as quartzite, jasper, chert, and in a few places petrified wood. Most of them are well rounded. The pebbles become progressively smaller northward, suggesting that their source was to the south, in central Arizona, where there is known to have been uplift in Triassic time. Erosion must have been vigorous to have carried these pebbles far across northern Arizona and into southern Utah.

The Chinle Formation is famous for its fossil trees, the great logs of Petrified Forest, but nearly 40 species of fossil ferns and other plants are known from it also. In addition, the formation contains many fossil animals: freshwater clams, snails, and fish, some large amphibians, and many reptiles — long-snouted phytosaurs and early dinosaurs.

Like the Moenkopi Formation, the Chinle was deposited on a wide, low floodplain, probably a coastal plain. Marshes, ponds, lakes and winding rivers marked the surface, gradually covering it with limy clay and volcanic ash. The many land plants and animals show that the region was never entirely submerged beneath the sea. The colors of the formation, derived mostly from various iron minerals, suggest a low-oxygen environment such as might be found in stagnant swamps and jungle-covered lowlands where there is a good deal of decaying plant material. The climate was almost certainly more humid than during deposition of the Moenkopi's siltstones and mudstones.

West of the highway, hills and cuestas of Moenkopi rocks tilt toward the highway, their upper surfaces protected by the Shinarump Conglomerate. The road runs close to the Moenkopi-Chinle contact, with Moenkopi to the west, Chinle to the east. Farther east rise younger Triassic and Jurassic strata of Echo Cliffs; they, too, dip eastward.

Between mileposts 501 and 504 the highway climbs to an almost level erosion surface at the base of Echo Cliffs. This **pediment** truncates tilted beds of the Chinle Formation, as can be seen at the head of Hamblin Wash near milepost 504.

Twenty miles farther north, US 89 turns east to scale the cliffs. Here the great rampart is broken by upward movement between two parallel faults, and the highway uses the fault block as a stairway. In the jumbled material along the faults, twisted wreckage of colored Chinle strata lends marble-cake swirls to the soft shale layers, which bent plastically, more like mud than rock.

Stop at the parking area at milepost 527 to look at the geologic panorama. The gray-white surface down below is the top of the Permian Kaibab Limestone arching gently across Limestone Ridge. The twisting slit across it is Marble Canyon, a narrow gorge incised by the Colorado River. Southward, you can see where the Little Colorado joins it. Marble Canyon contains no real marble, but where the river cuts through the Redwall Limestone, age-long smoothing by

Along the base of Echo Cliffs, slump blocks sliding along curved slide surfaces rotate so that their strata dip back toward the cliff from which they came.

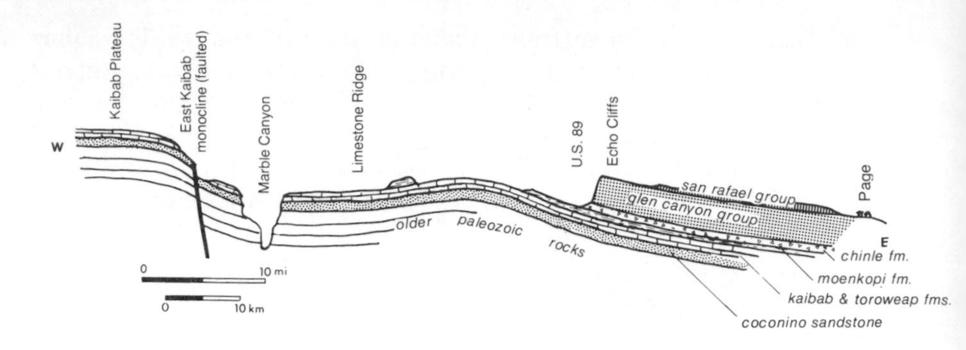

Section across US 89 between Cameron and Page.

muddy, sandy river water has left polished and fluted surfaces that resemble marble.

The Kaibab Plateau rises in the distance beyond Marble Canyon; there the Kaibab Limestone turns up abruptly along the East Kaibab Monocline. The monocline is more complicated than its name implies; in places it is faulted or two-stepped. It can be traced north into Utah and south along the edge of the Coconino Plateau. The immense fold lies above a fault in Precambrian rocks, well exposed in Grand Canyon. On the Kaibab Plateau the Kaibab Limestone tops out at more than 9000 feet elevation, whereas below you and in House Rock Valley north of the river its surface is at about 5000-6000 feet, so you can see that the fault and the monocline displaced the layered rocks about 4000 feet. Strangely, there is evidence that movement along the fault in Precambrian time was in the opposite direction, west side down.

From this vantage point it becomes clear that the highest part of the combined Coconino-Kaibab uplift, called the Kaibab Upwarp, is north of Grand Canyon, and that the Colorado River cuts across the uplift.

In Vermilion Cliffs, north of the river, Triassic and Jurassic rocks are clearly exposed. Using the colorful Chinle Formation as a marker, you can match the formations with the rock layers illustrated earlier. They are the same rocks that make up Echo Cliffs. The horizontal surface above Vermilion Cliffs is the Paria Plateau, which extends north into Utah.

Going up through the big roadcut above the parking area, the highway climbs through these same rocks and through many millions of years of geologic time. The Glen Canyon Group, above the Chinle Formation, has contributed far more than its share to the spectacular scenery in northeast Arizona, southern Utah, and western Colorado. The uppermost formation of this group, the Navajo

Cross-bedding is a prominent feature of the Navajo Sandstone, and with other criteria such as grain size and uniformity indicates that it originated as desert dunes.
Tad Nichols photo.

Sandstone, is particularly easy to recognize because of its pale, salmon-colored blush and the sweeping cross-bedding that shows it was formed of dune deposits. All in all, the Triassic and Jurassic formations record a transition from humid lowlands to vast, windswept deserts that, although they changed character from time to time and place to place, frequently were the sites of rolling seas of windblown sand.

At the top of Echo Cliffs a magnificent panorama spreads out before you, with Lake Powell's blue waters contrasting with soft reds and pinks of sunlit Jurassic rocks. Across the river, these rocks dip in another monocline, defining the east edge of the Paria Plateau. The highway rides on Navajo Sandstone all the way to Page, with barren knobs and domes of younger Jurassic rock — the San Rafael Group — rising above it.

In Page there is a museum devoted to memorabilia of John Wesley Powell, a one-armed Civil War veteran and self-trained geologist who in 1869 led a daring expedition down the uncharted wilds of the Colorado River. Despite many hardships and the ever-present danger of treacherous rapids, he lived to tell the tale, and told it well. Realizing the need for continued exploration of the west, he also was a moving force behind creation of the U.S. Geological Survey, and was its second director. It was he who named many of the geographic features along the Colorado River, including Glen and Marble Canyons and Vermilion Cliffs. Lake Powell honors him in turn.

The Glen Canyon Dam was built between 1960 and 1965. Tours descend inside the dam to the power plant, where the constant roar of the captive river can be heard and felt. Because most of the river's load of sediment is deposited now at the upper end of Lake Powell, the waters of the Colorado, once muddy brown, usually run clear.

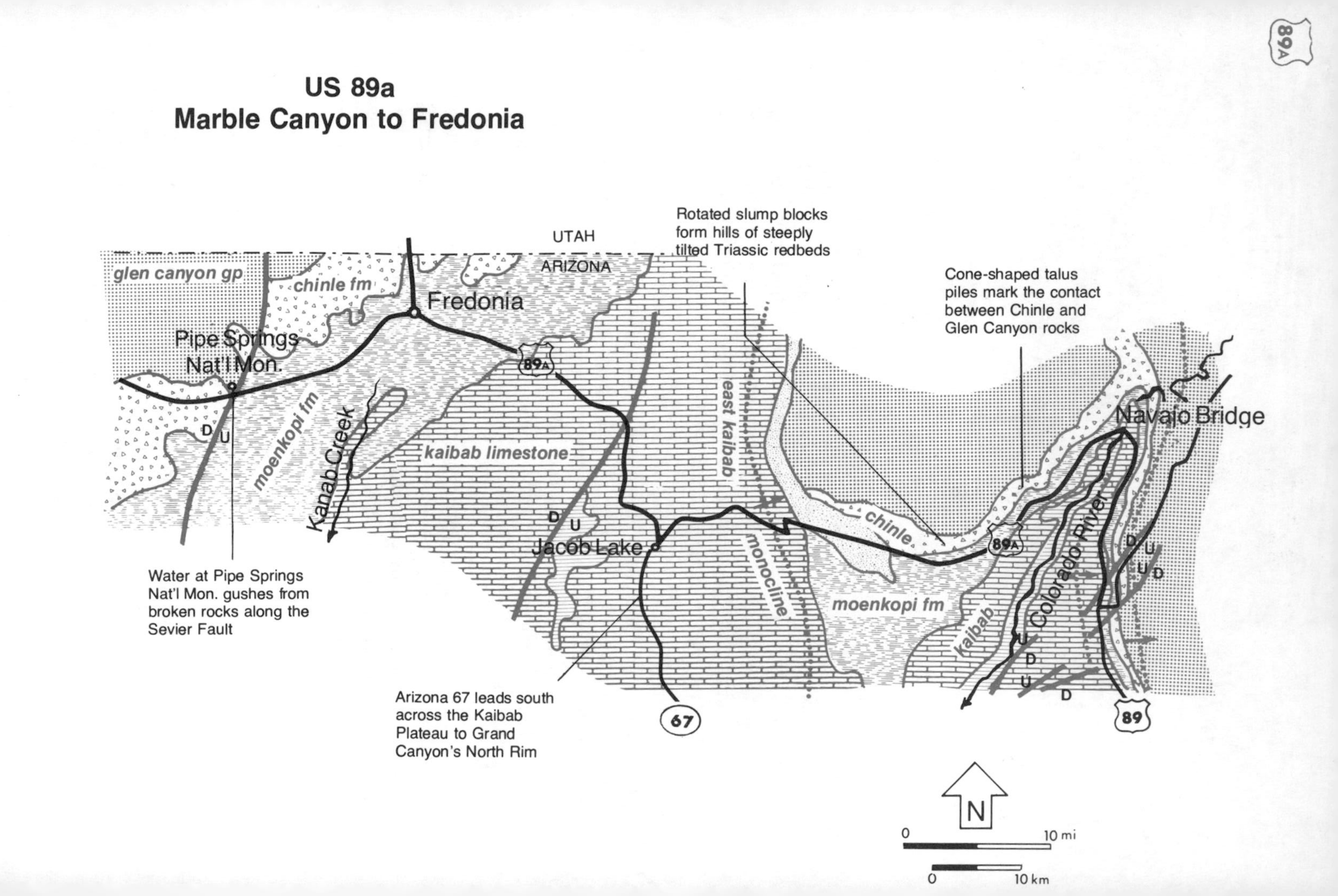

89A
US 89a
Marble Canyon to Fredonia
UTAH
ARIZONA
glen canyon gp
chinle fm
Fredonia
Pipe Springs
Nat'l Mon.
moenkopi fm
Kanab Creek
kaibab limestone
89A
east kaibab
Jacob Lake
monocline
moenkopi fm
chinle
kaibab
Colorado River
Navajo Bridge
89A
67
89
D
U
Rotated slump blocks form hills of steeply tilted Triassic redbeds
Cone-shaped talus piles mark the contact between Chinle and Glen Canyon rocks
Water at Pipe Springs Nat'l Mon. gushes from broken rocks along the Sevier Fault
Arizona 67 leads south across the Kaibab Plateau to Grand Canyon's North Rim
N
0
10 mi
0
10 km

U.S. 89a
Marble Canyon — Fredonia
(83 miles)

As seen from the junction of US 89 and 89a, the jumble of rocks between two faults marks a conspicuous break in Echo Cliffs, the route taken by US 89. A whole unit of the cliffs has been pushed up here. Erosion of soft mudstone and volcanic ash beds in the Chinle Formation of the lifted block provides access to the top of the cliffs.

North of the highway junction, US 89a runs along the contact between gray-white Kaibab Limestone and the deep red sandstone and siltstone layers of the Moenkopi Formation. The Kaibab, youngest of the Paleozoic strata in this region, contains many tiny fossil snails and clams, a sort of "kindergarten fauna" of small or immature shellfish. The fossils, though common, are not very exciting to look at because they occur as tiny hollows in the limestone, molds of the original shells. Casts of the shells can be made by filling the holes with liquid latex, letting it stiffen, and then snapping it out. In some places the calcium carbonate of the original shells has been replaced by silica, and the finely detailed shells can be removed by dissolving away the surrounding limestone with acid.

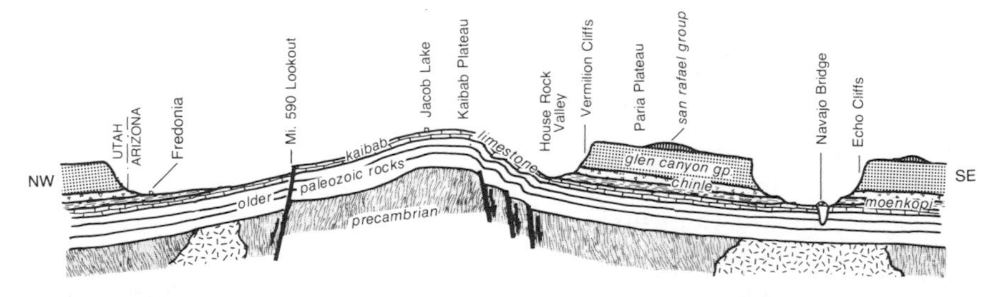

Section along US 89a Marble Canyon to Fredonia.

Triassic and Jurassic rocks appear in section on the steep slopes of Vermilion Cliffs near Lees Ferry. Tad Nichols photo.

In the Vermilion Cliffs, now visible to the north across the incision of Marble Canyon, the sequence of sedimentary rocks is similar to that in Echo Cliffs; it goes from Permian Kaibab Limestone at the bottom to Navajo Sandstone at the top of the cliffs. Higher and farther back from the cliffs are rocks of the San Rafael Group. The Mesozoic formations are not difficult to recognize if you first pick out the blue, gray, and purple shales of the Chinle Formation, which form a wide bench above the cliff-forming Shinarump Conglomerate. Then work up and down through slope-forming and cliff-forming bands of sedimentary rock, comparing them with the diagram in the introduction to this chapter.

Differential erosion — the wearing away of soft rock layers between hard ones — operates on a grand scale here. Erosion pushes the Triassic-Jurassic cliffs inexorably northward and eastward, and bares the resistant, buff-colored surface of the Kaibab Limestone.

At Navajo Bridge the highway crosses the Colorado River where it is entrenched in Marble Canyon. Upriver from the bridge are outcrops of Kaibab Limestone, Moenkopi redbeds, and the variegated Chinle Formation well exposed at the bend of the river. The river's bend is controlled by a monocline, and just there a wider channel has developed. Because flow is slower at this widened spot, with sand bars developing along the banks, it was for many years the site of Lee's Ferry. The story of John Lee is told by a roadside marker at the rest stop. In 1927 the ferry was replaced by Navajo Bridge. Shelving banks that edge the river at the old ferry site are now used for launching rafts and boats going through Marble Canyon and Grand Canyon; they are the last sites accessible by road for some 240 river miles. In the 400-mile width of Arizona, the Colorado can be crossed by car in only three places: Navajo Bridge, Hoover Dam, and Glen Canyon Dam. Lee's Ferry was for many years the only vehicle crossing between southern and northern Arizona! There are now two footbridges in Grand Canyon.

The highway curves along the south side of Vermilion Cliffs, on the bench of the Kaibab Limestone, among large blocks of Shinarump Conglomerate fallen from the cliffs as they were undermined by erosion. Some of these blocks have been here for some time, protecting the mudstone immediately under themselves, so that now they stand on slender mudstone pedestals.

There is a single prominent white bed in the Moenkopi Formation that can be recognized almost everywhere in the Little Colorado Valley and near Vermilion Cliffs. This massive, cross-bedded, fine-grained sandstone forms a prominent cliff with rounded edges, and varies surprisingly little in thickness from place to place. Because its position within the formation rises westward, and because it becomes younger in that direction, it is thought to have been laid down as a sheet of river-borne debris deposited as the delta advanced seaward. As you can perhaps imagine, a delta environment is unstable: Floods, storms at sea, and up and down movements of the land all bring about

In rock layers alternately weak and resistant, differential erosion has created what appear to be stacks of Chinese hats. Desert pavement coats the foreground flat. Tad Nichols photo.

major changes in the kinds of sediment that are deposited. The Moenkopi Formation records many such changes in its succession of sandstone, siltstone, and mudstone beds, and farther west several intertonguing limestone layers record advances and retreats of the sea.

House Rock Valley, where the highway swings northward, is the home of a bison herd. They are occasionally seen near the road. At mile 566 the highway leaves this valley to climb to the top of the Kaibab Plateau. Here again scenery directly reflects the geologic construction of the land. The highway climbs the plateau by stair-stepping up a triple monocline, with nearly level "treads" between the "risers" — a geologic design that helped to engineer the highway. The actual folds of the Kaibab monocline lie directly over one or more prominent faults that displace Precambrian rocks several thousand feet below. The fault zone itself can be seen in the depths of eastern Grand Canyon, even from points on the rim. In roadcuts here you'll see how sedimentary rock responds to the same displacement; thousands of small faults in the Kaibab Limestone and the underlying Toroweap Formation take up much of the movement, accentuating the bend in the rocks. Protruding **concretions**, blobs of gray and white chert, characterize these impure, near-shore limestones.

The Kaibab Plateau is poorly drained in terms of surface streams. Most of the drainage is underground through passageways dissolved in these and other limestone layers. Undrained depressions on the surface are caused by collapse of underground passageways. Jacob Lake, hardly more than a pond, is in a sinkhole formed in this manner by solution in the Kaibab Limestone.

There is only one paved access route to the North Rim of the Grand Canyon: the well traveled road from Jacob Lake to North Rim Village and Cape Royal. This road traverses the highest part of the Kaibab Plateau, passing alternately through dense forest and open, grassy valleys marking areas where the Kaibab Formation has collapsed over caverns and underground passages. A much less traveled, unsurfaced, delightfully geologic route from Fredonia follows the Toroweap Fault south along the west side of the Kanab Plateau, west of the Kaibab Plateau, and reaches Grand Canyon at Toroweap Point, where lava flows from nearby volcanoes spill over the canyon rim. For a run-down on the geology of the Grand Canyon see Chapter V.

U.S. 89a
Flagstaff — Sedona
(44 miles)

For a map of this route see I-17 Camp Verde — Flagstaff.

South of Flagstaff this highway crosses the thickly forested, undulating surface of the Coconino Plateau, descending across successively older lava flows. Because of the protective lava cap, Oak Creek Canyon begins quite suddenly. From the viewpoint at mile 390 the most striking feature is the difference in elevation between the two rims of the canyon. This difference is caused in part by movement along the Oak Creek Fault, and in part by erosion of an earlier, perhaps Oligocene canyon more or less parallel to the present one, a canyon eroded down through the rim-forming limestones and then filled up again with lava flows.

Let's take a look, for a moment, at the rock sequence on the west canyon wall. It tells a tale of Permian time. Except for localized patches of basalt, all the rocks are sedimentary. At the bottom are massive red sandstones and thin-bedded red shales, not very well exposed. They were deposited on a floodplain or delta — a near-shore environment — in late Pennsylvanian and early Permian time. Above them are the yellow-white bluffs of Coconino Sandstone, windblown sand of a great desert that existed here midway in the Permian Period. The environment of the dune sands, recognized by their well sorted, fine, rounded grains as well as by steep, large-scale cross bedding, seems to have alternated for a time with the delta-floodplain environment, for the contact between the two is characterized by alternations of red delta-floodplain deposits and whitish or

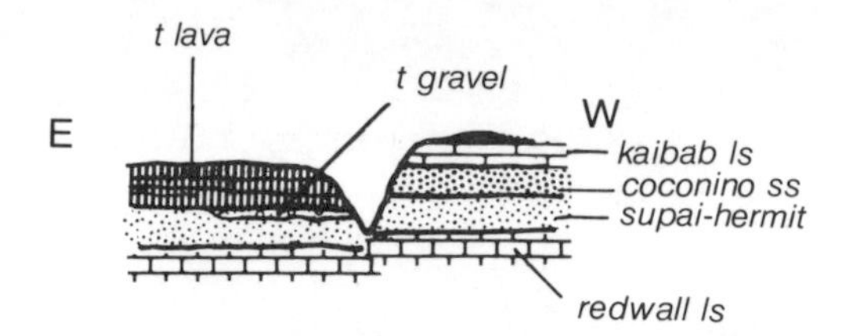

Section across US 89a in Oak Creek Canyon. On the east side of the canyon the normal rim-formers – the Kaibab Limestone, Toroweap Formation, and Coconino Sandstone – eroded away before lava flows accumulated. Movement on the Oak Creek Fault only partly explains the canyon's lopsidedness.

yellowish dune deposits, as you will see farther down the canyon. The scenery here in Permian days may have been much like that of the Tigris-Euphrates or Nile deltas today, with lush, fertile river valleys expanding into deltas, fighting a never-ending war against the relentless incursion of the desert. Though for a time the desert won out and dunes dominated the landscape, the land eventually subsided enough to allow seas to creep in from the west. In them were deposited the tan cross-bedded sandstone of the Toroweap Formation and the also tan Kaibab Limestone, the uppermost two units of the canyon wall. Fossils abundant in these formations tell us that they, too, formed in Permian time.

Beyond the viewpoint, the road descends abruptly into Oak Creek's deep, narrow canyon. It zigzags down a large block of basalt that is tilted and slumped, but nevertheless remains intact, along the Oak Creek Fault. Like many faults this one is a broad belt of crushed and dislocated rock rather than a clean, sharp break. Where the road starts its descent there are really two faults, one much more pronounced than the other, with a narrow down-dropped block between them. The highway cuts through several lava flows which show crude columnar jointing. Brick-colored baked soil zones appear below the flows. A lot of the white and buff-colored sand in the roadcuts comes from Coconino Sandstone crushed along the fault; some fragments escaped total crushing and are large enough to display the cross-bedding characteristic of the Coconino.

The fault and the weakened rock of the fault zone very definitely control the position of Oak Creek Canyon. The creek, a permanent stream, is fed by springs along the fault zone, where water percolating through the Coconino Sandstone and the broken rock along the fault zone intercepts the surface. The creek follows the fault southward in a dog-leg course almost all the way to Sedona. Here near its head Oak Creek Canyon is about 1500 feet deep; at Sedona it is 2500 feet deep and considerably wider.

Much of the east wall of the canyon is concealed beneath slide rock or **talus** made up of broken blocks of basalt filling Oak Creek's older canyon. Because of its rough columnar jointing and evenly spaced horizontal cooling cracks, the basalt breaks into roughly equidimensional blocks. The west wall is rimmed with Kaibab Limestone and doesn't develop such extensive rockslides.

Once down in the canyon you'll find the Coconino Sandstone at roadside level on your left. Several exposures reveal its steep, large-scale cross-bedding. Between miles 383 and 382 it is cut by a gray band of igneous rock, a dike, one of many that fed the seven basalt layers above.

On the west side of the road red shales and sandstones of the Supai Group soon begin to make their appearance. The canyon opens out a bit below milepost 382, and it becomes easier to see the fantastic spires and hoodoos that erosion has carved in the Coconino Sandstone of the west wall.

If you've been to Grand Canyon you may have recognized that the strata here resemble those of the upper walls of the Grand Canyon, though they are not nearly as well exposed. There is, however, a major difference in the red rocks below the Coconino Sandstone, as we shall see.

Near Oak Creek Bridge the lopsidedness of the canyon again is apparent. Deep pools near the bridge — a favorite sunning and bathing place — are shaped by flowing water that stairsteps down ledges of red sandstone, part of the Supai Group-Hermit Shale sequence. Oak Creek Canyon seems to be on the fringes of the Supai and Hermit units; the alternating sandstone and shale layers are cumulatively thinner than their counterparts in Grand Canyon. But

Oak Creek Canyon's striking red buttes reveal a thick lens of massive sandstone deposited as a beach or bar.

above the Hermit Shale is another unit that more than makes up for the thinning, an immense erosion-resisting mass of bright red, rather uniform sandstone that extends east and west of Oak Creek but is not known farther north or south. It is not present at all in Grand Canyon. South of milepost 378, the canyon opens out into a panorama of red buttes, cliffs, slopes, and pinnacles shaped in this massive sandstone, a stunning technicolor landscape long ago discovered by Hollywood. The great sandstone unit seems to have formed in a subsiding basin, but whether it was a coastal sand bar, a beach-and-dune duo, or a high-rising dune field is still uncertain.

Rough-shaped originally by Oak Creek and its tributaries, the red buttes and promontories were later swept clean by rain and wind erosion. Like water, wind uses sand grains as its tools. (Unlike water, it cannot also use pebbles, cobbles, and boulders.) Like a sand-blaster, wind hurls sand grains against bare rock surfaces until they are smooth and rounded; it also etches grooves and caves where the rocks are weak, and leaves projecting rims and ridges of stronger rock.

At Midgely Bridge Oak Creek has carved a narrow inner gorge in Mississippian Redwall Limestone, a formation also known in Grand Canyon. This marine limestone contains fossil horn corals as well as lacy bryozoans, wide-winged brachiopods, and ammonites.

The Mogollon Rim, southern edge of the Colorado Plateau, is very jagged here, cut deeply back by Oak Creek and farther west by Sycamore Creek, a similar deep notch in the plateau rim.

To return to Flagstaff, take any one of the connecting routes shown on the map for I-17, then follow I-17 north as it climbs a lava ramp to the top of the Mogollon Rim. You may want to visit Montezuma Well and Montezuma Castle National Monument on the way.

Badland hills near Cameron owe their ·velopment to an arid climate and ·entonite-rich clays of he Chinle Formation. Struggling vegetation borders ephemeral streamlets.

Tad Nichols photo.

U.S. 160 from U.S. 89 — Kayenta

(84 miles)

From its junction with US 89, 15 miles north of Cameron, all the way to Kayenta, US 160 travels through colorful Mesozoic sedimentary rocks — Triassic and Jurassic delta and floodplain and sand dune deposits, and Cretaceous marine and near-shore deposits. In the northern reaches of the Painted Desert, the valley of the Little Colorado, slopes and benches of soft mudstone and siltstone alternate with ledges and cliffs of resistant sandstone and conglomerate; pink and deep brick red alternate with delicate hues of green, blue, and yellow. The land in many places is devoid of vegetation; elsewhere it is dotted with piñon and juniper or gentled with a blue-green mantle of desert shrubs. The shape of the land is directly a result of hardness, softness, and attitude of rock layers.

As US 160 leaves the valley of the Little Colorado it climbs onto a broad platform capped with resistant sandstone of the Glen Canyon Group. This group includes red ledges of Wingate Sandstone, bright orange-red Moenave Formation, red Kayenta Formation stream-deposited on ancient floodplains, and the Navajo Sandstone, a gray-

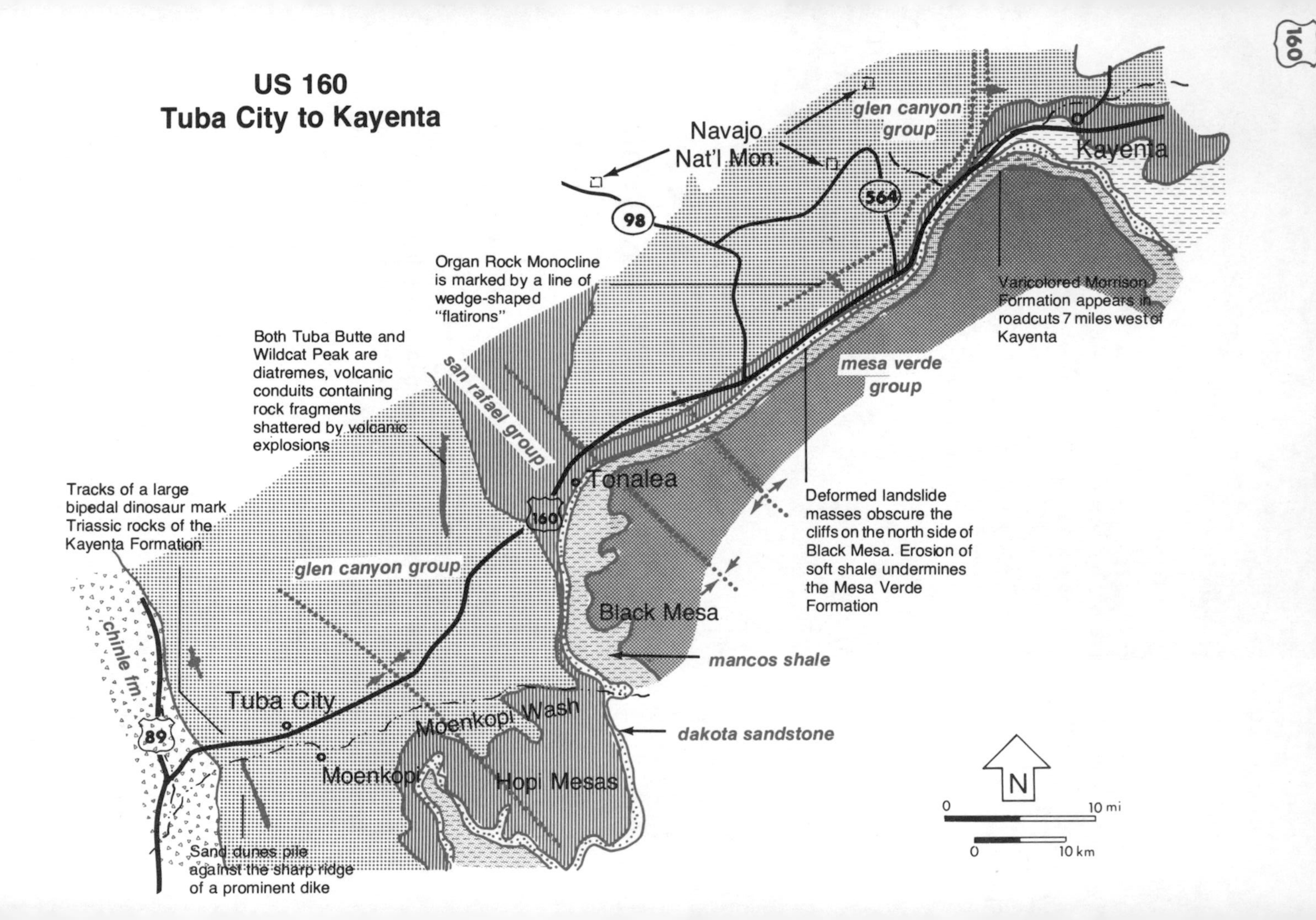
US 160
Tuba City to Kayenta
160
Navajo
Nat'l Mon.
98
564
glen canyon
group
Kayenta
Varicolored Morrison Formation appears in roadcuts 7 miles west of Kayenta
Organ Rock Monocline is marked by a line of wedge-shaped "flatirons"
Both Tuba Butte and Wildcat Peak are diatremes, volcanic conduits containing rock fragments shattered by volcanic explosions
san rafael group
mesa verde
group
Tonalea
160
Tracks of a large bipedal dinosaur mark Triassic rocks of the Kayenta Formation
glen canyon group
Deformed landslide masses obscure the cliffs on the north side of Black Mesa. Erosion of soft shale undermines the Mesa Verde Formation
Black Mesa
chinle fm.
mancos shale
Tuba City
Moenkopi Wash
89
dakota sandstone
Moenkopi
Hopi Mesas
N
0
10 mi
0
10 km
Sand dunes pile against the sharp ridge of a prominent dike

Terrace levels near Moenkopi bevel gently dipping strata. The terraces may have been beveled by the Colorado River before excavation of Grand Canyon.

white or pale pink, cross-bedded, cliff-forming dune sandstone that is responsible for much of the scenic grandeur of the plateau country. The entire group is thought to span late Triassic and early Jurassic time, with the Navajo Sandstone considered to be mostly Jurassic.

At about mile 317, near a side road leading north to Moenave, dinosaur footprints mark a flat rock surface in the Kayenta Formation. Others can be seen near the Hopi Indian village of Moenkopi. The creatures that left these tracks in once-soft mud 200 million years ago were large bipedal dinosaurs with short forelegs suitable only for clasping food. Dinosaur tracks and skeletons have been found at many sites in these Mesozoic rocks, and add to the picture of widespread, possibly lush floodplains swept now and then by flooding rivers.

To the north as the route approaches Tonalea is Wildcat Peak, a volcanic conduit and its associated dikes. Farther north still is White

A dinosaur walked across these once muddy flats, its clawed feet leaving prints that indicate a seven-foot stride.

Tad Nichols photo.

Resistant rock of a basalt dike cuts through sedimentary rocks near Tuba City. Some dikes in this area can be followed for several miles. Tad Nichols photo.

Mesa, formed of late Jurassic sandstone of the San Rafael Group. Between mileposts 355 and 360 the highway crosses the boundary — an unconformity not well exposed here — between the Glen Canyon Group and the San Rafael Group, a distinction marked by a change from the massive, cross-bedded pale pink or grayish white Navajo Sandstone to brownish red, horizontally bedded ledges and slopes of the San Rafael Group.

East of Tonalea is Black Mesa. Though it seems low here, farther eastward this mesa becomes the dominant landscape feature. Its lower slopes expose weak layers of the Morrison Formation, Jurassic deposits of possibly marshy and certainly vast river floodplains. Its upper slopes, above a conspicuous cliff of shallow-water Dakota Sandstone, record in their dark gray shale the incursion and retreat of a Cretaceous sea. This is the Mancos Shale, and it contains fossils of marine shellfish such as oysters and ammonites, and skeletons of fishes (even sharks), turtles, crocodiles, and a large marine reptile

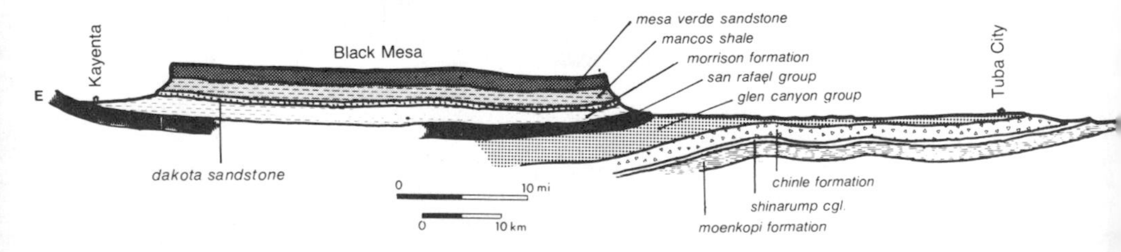

Section south of and parallel to US 160 from Tuba City to Kayenta.

called a plesiosaur. Above the Mancos Shale, shoreline deposits document a fluctuating retreat of the sea: units of beach and bar sandstone, shaly lagoon deposits, and coal-bearing siltstone of the Mesa Verde Group, which tops Black Mesa. The lagoon deposits are near the surface of most of the mesa, and are the source of coal now being strip-mined atop the mesa.

In strip-mining procedures vegetation is removed, topsoil is pushed aside, and as much as 180 feet of overburden is stripped away. Finally, huge power shovels loosen and remove the coal, which is crushed and carried by belt line, railroad and slurry pipeline to power plants near Page, Arizona and in Nevada, 274 miles to the west. As coal is mined, overburden and topsoil are replaced and seeded with pasture grasses and clover, to comply with environmental strictures. About 12 million tons are mined annually, with royalties going to the Navajo and Hopi Indian tribes who jointly own most of Black Mesa.

East of mile 375, and again at mile 389, jumbled terrain and tilted strata south of the highway are landslide debris. Erosion of soft Morrison Formation sandstone and Mancos Shale constantly undermines the mesa's edge, causing such slides, gradually wearing back the walls of this and many other mesas and plateaus — a common form of valley widening, reducing plateaus to mesas, mesas to buttes, and buttes to pinnacles.

At the northernmost end of Black Mesa the highway threads through a narrow defile between the mesa and the Shonto Plateau to the north. Triassic and Jurassic rocks that underlie Black Mesa come

Massive wedges of Navajo Sandstone mark the edge of a monocline, one of many on the Colorado Plateau. Tad Nichols photo.

to the surface here, tilt up steeply in the Organ Rock monocline north of the highway, and then level out abruptly to form the summit of the Shonto Plateau to the north and Skeleton Mesa to the northeast. Where Tsegi Canyon cuts through these rocks (across from mile 382), it exposes older rocks that underlie them.

The prominent pink sandstone of the Organ Rock monocline is, once again, the Navajo Sandstone. Its fine, even grains and steep, sweeping cross-beds are typical of sandstones formed from sand dunes. The same rocks edge mesas east of Kayenta. Where the rock is close to the road it is channeled and pitted by stream rivulets. The Navajo Sandstone is an important aquifer in this region, supplying water for communities on and near Black Mesa, for stock on Indian reservations, and for the slurry pipelines that carry Black Mesa coal. Almost all the water in this aquifer sinks into it on the Shonto Plateau and in the narrow area between that plateau and Black Mesa — a recharge area of only about 100 square miles to furnish an area many times as large. Already the rate of water withdrawal exceeds the rate of renewal.

Fine red-brown or gray-brown, silty stream deposits exposed in gullies between mile 370 and Kayenta are geologically quite young. Careful study has revealed charcoal and Indian relics that give radiocarbon dates within the last 7000 to 10,000 years. Such deposits are favored hunting grounds for archeologists, and tell us something of the drainage history of this area, besides. They show, for instance, that the drainage divide is migrating westward as the stream east of the divide cuts headward.

Kayenta lies on a fairly level valley floor where erosion of the Morrison Formation has widened the valley. Steep slopes of Navajo Sandstone rise north and northeast of town, turned up along the edge of the Monument upwarp (see Monument Valley in Chapter V).

A giant power shovel dwarfs a bulldozer in a Mesa Verde coal mine.

U.S. 160
Kayenta — Four Corners
(79 miles)

Two main structural features combine to form the scenery of the Kayenta-Four Corners area: Monument Upwarp to the north, and many relatively gentle anticlines and synclines, among them the Defiance Uplift, to the east and southeast. These features are punctuated by dark volcanic dikes and necks of the Navajo Buttes.

Comb Ridge begins north and northeast of Kayenta and more or less parallels the highway as far as Dinnehotso, where it curves north into Utah. Geologists have postulated that the Comb Monocline, the bending over of rock layers at the south edge of the Monument Upwarp, overlies a major fault in Precambrian rocks deep below the surface. Some geologists have gone so far as to postulate a southwestward continuation of such a fault, pointing out that other geologic features — faults and small intrusions — lie along the same line.

The main rock unit visible in the Comb Monocline is the Navajo Sandstone, whose pale salmon color, steep, large-scale cross bedding, and tendency to weather into bare rounded surfaces make it easy to recognize. The sand of which it is made is fine and even-textured, sorted by wind like the sand in modern dunes, and some surfaces bear ripple marks formed by winds that blew here in Jurassic time, about 200 million years ago. Studies of the cross bedding in the Navajo Sandstone show that these winds blew from the north and northwest, in terms of present directions.

Between and behind the "teeth" of Comb Ridge are older Triassic rocks, and behind them some more massive windblown sandstone, the Permian De Chelly Sandstone that forms the monuments of Monument Valley north of here and the massive cliffs of Canyon de Chelly to the southeast.

From the top of the hill northeast of Kayenta you'll see several of the volcanic necks that dot the landscape here. Agathla Peak, the largest, and three others line up at right angles to the highway,

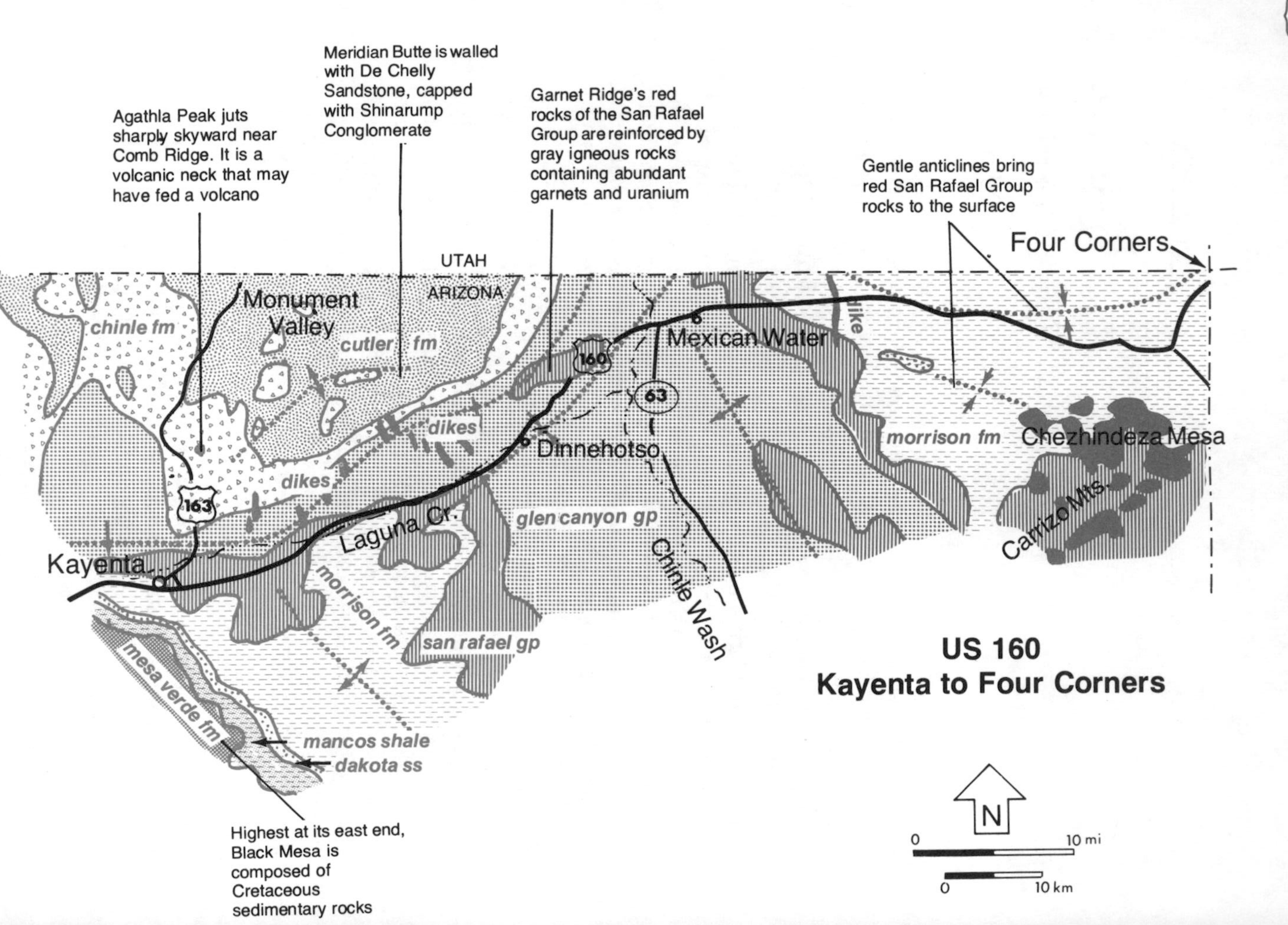

Agathla Peak juts sharply skyward near Comb Ridge. It is a volcanic neck that may have fed a volcano
Meridian Butte is walled with De Chelly Sandstone, capped with Shinarump Conglomerate
Garnet Ridge's red rocks of the San Rafael Group are reinforced by gray igneous rocks containing abundant garnets and uranium
Gentle anticlines bring red San Rafael Group rocks to the surface
Four Corners
UTAH
ARIZONA
Monument Valley
chinle fm
cutler fm
160
Mexican Water
dike
63
dikes
Dinnehotso
morrison fm
Chezhindeza Mesa
Carrizo Mts.
163
Laguna Cr.
glen canyon gp
Chinle Wash
Kayenta
morrison fm
san rafael gp
mesa verde fm
mancos shale
dakota ss
Highest at its east end, Black Mesa is composed of Cretaceous sedimentary rocks
US 160
Kayenta to Four Corners
N
0
10 mi
0
10 km

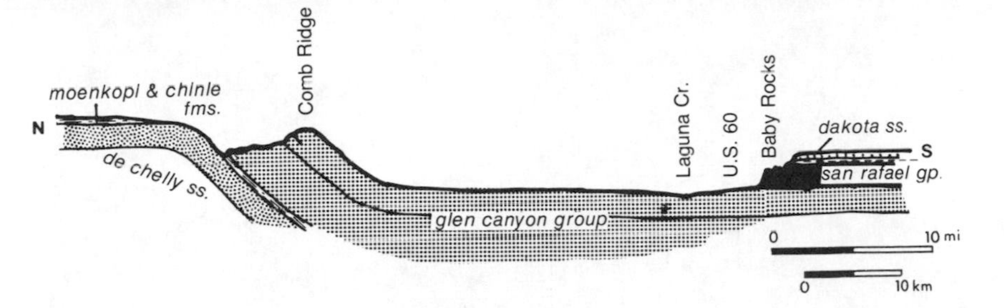

Section across US 160 near Baby Rocks, east of Kayenta.

presumably along a fault. The necks are the conduits, the feeding pipes, of Tertiary volcanoes, some of which erupted explosively, creating broken rock fragments that fell back into their conduits and hardened there as volcanic breccia. There are numerous lava dikes in this region, too, almost all of them with the same northwest-southeast trend.

Baby Rocks, near milepost 408, show well the effects of differential weathering on resistant red sandstone and softer red siltstone of the San Rafael Group. The amusing little pinnacles, as well as the cliffs behind them, are broken by numerous vertical joints and small faults that allow water to penetrate the rock masses, speeding up the weathering process. Note the oddly contorted beds at the base of the cliff. Such deformation occurred as slumping and sliding when the rock was not yet consolidated.

The San Rafael Group is composed of a number of different types of rock — sandstone, siltstone, and claystone. The thick, dark reddish sandstone and siltstone units suggest deposition on river floodplains, as do occasional examples of trough-type cross-bedding indicative of filled-in stream channels. In places some units seem to have been truncated by erosion before others were deposited above them — a situation also compatible with river floodplain origin. Perhaps in Jurassic time the region was somewhat similar to the Nile or Tigris-Euphrates valleys today, with deserts of blown sand competing for dominance with river and floodplain environments.

Roadcuts and natural outcrops between mileposts 426 and 436 display the Navajo Sandstone well, and also demonstrate how precariously thin and sandy the soil layer is, in this arid land. Near the highway and at the edges of mesas, modern dunes build from the sand of old ones. In places you can see on the Navajo Sandstone the type of weathering called **exfoliation** — the peeling off of curving slabs of rock, probably because of rapid heating and cooling in hot desert days and cold, dry desert nights. Curving exfoliation joints develop particularly well in rocks of uniform grain size such as the wind-sorted sandstones here. Very thin horizontal beds in this rock are freshwa-

Fossil animal tracks slowly disappear as this Jurassic sandstone flakes away in a weathering process known as exfoliation. Tad Nichols photo.

ter limestone or fine siltstone deposited between dune areas.

From high vantage points along highway 160 you can see the gentle warping that characterizes much of this area. Basically flat-lying strata south of Comb Ridge are actually very gently bent into low anticlines and shallow synclines. Mexican Water lies on the crest of a far larger structure, the Defiance Uplift, which farther south brings to the surface the Permian rocks displayed in Canyon de Chelly's magnificent cliffs (see Chapter V). At Mexican Water the crest of the anticline is opened up by erosion, so that older rocks are revealed, down to the lower part of the Glen Canyon Group.

At milepost 442 the road goes through a slender ridge-forming dike. The gray rock of the dike is much harder, more resistant, than the surrounding sandstone. Farther north, a volcanic neck punctuates the landscape.

Look carefully at the walls of Red Butte, on the Arizona-Utah line north of mile 453. In places, you can see beveling and **unconformities** in the San Rafael Group strata of which it is composed, as well as slight warping and filling of erosional channels that were part of the ancient Jurassic floodplain. A few small faults can also be discerned. Material from these cliffs spreads out over the valley, giving it a sandy pink coat — poor substitute for soil.

Beyond milepost 457 the highway skirts the north end of the Carrizo Mountains, which rise like a barricade north of the Chuska Mountains. They are the remains of seven intrusive igneous masses called **laccoliths** that in Tertiary time thrust up under the Dakota Sandstone and older sedimentary layers, doming them up. Patches of light-colored Dakota Sandtone are plastered in near-vertical position on the flanks of Chezhindeza Mesa at the west end of the range; the same rocks lie horizontally on top of some of the intrusions.

Arizona officially ends on a small platform of Dakota Sandstone in the vast Four Corners region. To the north, in Utah, is more plateau country. To the northeast in Colorado is Sleeping Ute Mountain, another laccolith, with volcanic conduit toes; Mesa Verde is far beyond, hidden by a near butte. In New Mexico rise the ethereal spires of Shiprock, visible only from the west side of the Four Corners platform. Beyond Mesa Verde and Shiprock, plateau-country scenery gradually disappears, its place taken by the very different scenery of the Southern Rocky Mountains.

U.S. 180
Grand Canyon — Flagstaff
(82 miles)

Leaving the Grand Canyon's rim, this highway runs almost directly south across the same Kaibab Limestone surface that forms both rims of the canyon. Elevation decreases as the highway descends what can be thought of as the broad southern nose of the Kaibab Upwarp, the great geologic flexure that includes both the Kaibab Plateau north of the Grand Canyon and the Coconino Plateau here south of it. Triassic, Jurassic, and Cretaceous rocks once covered the entire upwarp. But except for a few remnants of the Moenkopi Formation (such as that on Red Butte east of mile 225) and of Shinarump Conglomerate, they are gone now.

The change in elevation from about 7000 feet at Grand Canyon Village to 6000 feet at the junction where US 180 turns southeast toward Flagstaff, is reflected in the vegetation. Ponderosa pine forest of the canyon rim yields to piñon and juniper "pygmy forest" and then to open sagebrush flats, a picture identical on every side of the Kaibab Upwarp. Once on this open savannah you will see the San Francisco Peaks to the southeast, and three lesser volcanoes — Kendrick Peak, Sitgreaves Mountain, and Bill Williams Mountain — to the south. These mountains and the volcanic field that surrounds them are discussed under I-40 Seligman to Flagstaff.

Note the lack of well established drainage on the Kaibab surface. Near pilepost 222, for instance, there is an enclosed valley with no outlet — one of many. Rain falling on such individually isolated valleys runs toward central low spots, where it may pond for a time or may sink down through underground solution channels that have gradually widened along joints and fissures. Some of the channels, as

US 180
Grand Canyon to Flagstaff

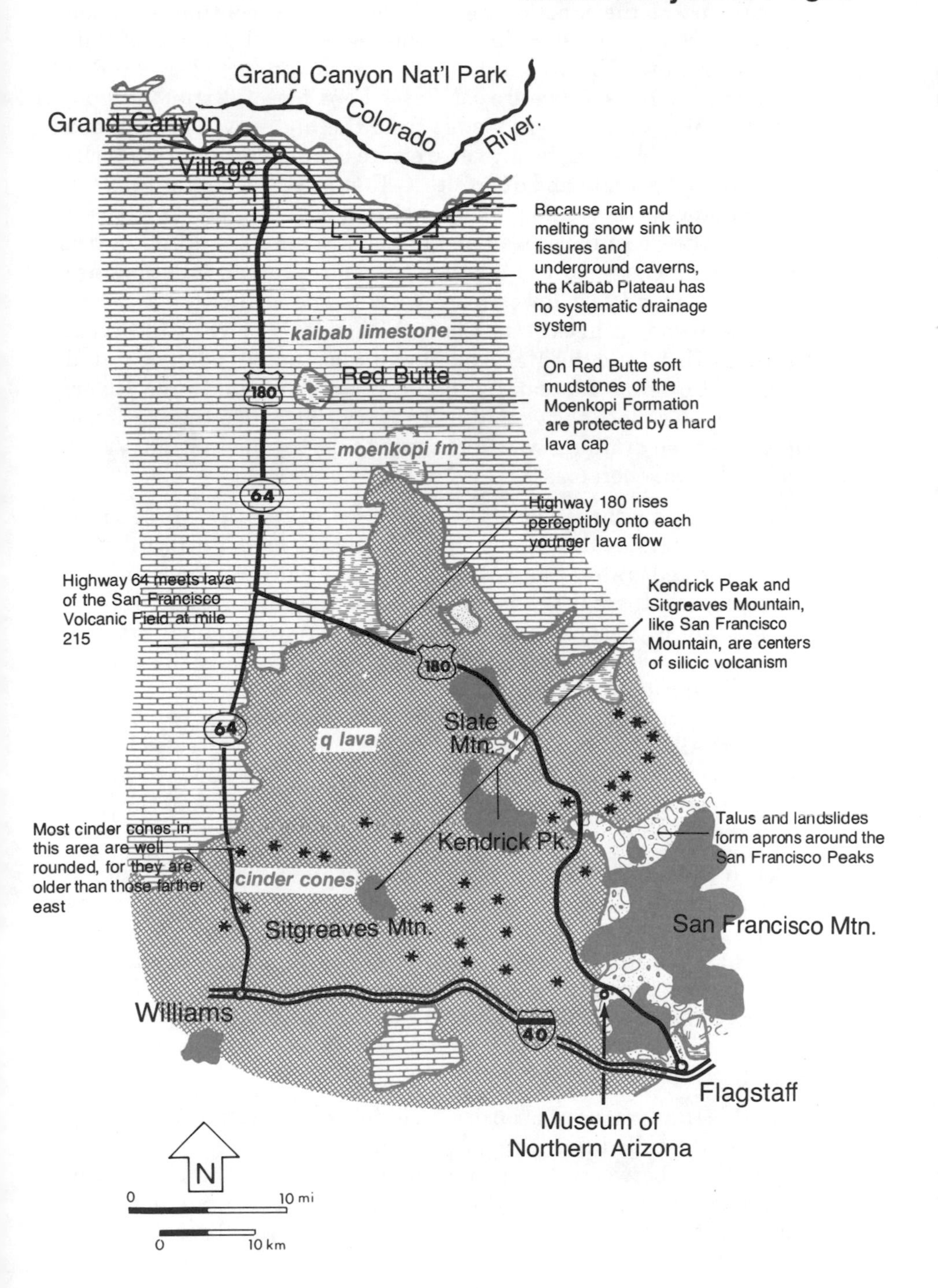

well as the low spots above them, probably date back to the Pleistocene Ice Ages, a time of wetter climatic conditions.

Southeast of the junction near Valle, US 180 continues on the Kaibab surface, then encounters some more small patches of flat-lying, red-brown Triassic sedimentary rock, protected as at Red Butte by lava flows. There are patches in here, too, of Tertiary gravel, and several gravel pits are visible from the highway. The rounded pebbles and cobbles in the gravel are of resistant Precambrian rocks — gneiss and granite and quartzite — that are clues to the Tertiary geologic story of this region. There are no Precambrian rocks exposed on either the Coconino or Kaibab Plateaus. Some show up, of course, in the Inner Gorge in the depths of Grand Canyon, but they are several thousand feet lower in elevation. The search for a source narrows down to ranges of the Central Highlands: the Prescott area, the Black Hills, and the Mazatzal Mountains, sources now separated from the Coconino Plateau by the deep valley of the Verde River. Their presence here shows that in Tertiary time when these gravels were deposited drainage was from south to north, and that the Verde Valley did not then exist.

Proceeding southeastward, US 180 passes just north of three small volcanic centers — Red Mountain with dipping cinder beds exposed in its dissected crater, Slate Mountain, and Kendricks Peak. Awash in a basalt sea, these little mountains formed as lighter-colored silicic magma forced passages toward the surface. At Red Mountain and Kendricks Peak this thick silicic lava broke through to the surface and built up small, steep-sided cones composed of thick, stubby lava flows. At Slate Mountain on the other hand magma never reached the surface at all but domed up overlying Paleozoic and Triassic sedimentary rocks in a blister-like laccolith. Erosion has now opened up the blister, revealing the steeply tilted sedimentary layers and the gray igneous core.

South of mile 233 you will be among the cinder cones of the San Francisco volcanic field, small volcanoes built of foamy, frothy magma propelled from small volcanic vents by rapidly expanding gases. Many of these little cones still have craters at their summits, and a few of the craters contain diminutive crater lakes. In this lava-covered area north of Flagstaff there are also several lava caves — long tunnels formed as still molten lava flowed out from under its already cooled and rigid crust.

All this time you have been drawing closer to San Francisco Mountain, locally called San Francisco Peaks, perhaps recognizing the resemblance between this mountain and famous (or infamous) volcanoes like Fujiyama, Mt. Hood, and Mt. St. Helens. Like them, this mountain is a stratovolcano composed of alternating lava flows and layers of volcanic ash accumulated over several million years of geologic time. Like Mt. St. Helens it may once have exploded sideways, opening a great cavity in its northeast flank. For more complete discussions of this beautiful mountain, see I-40 Seligman — Flagstaff and US 89 Flagstaff — Cameron.

On the outskirts of Flagstaff, this highway passes the Museum of Northern Arizona, where geologic, archeologic, and Indian exhibits tell more of the interesting history of the Plateau country. The main museum building is constructed of dacite from the San Francisco Peaks; it rests on one of the basalt flows which surround the mountain. The research center nearby serves as headquarters for geologists and other scientists working in Northern Arizona.

U.S. 666
Sanders — Springerville
(81 miles)

Leaving the valley of the Puerco River, US 666 climbs onto the surface of the Bidahochi Formation, a loosely knit unit of sand, silt, volcanic ash, and gravel deposited in Pliocene time in Hopi (or Bidahochi) Lake. The route travels for 60 miles and more among mesas capped with these young, poorly consolidated rocks. A large part of the formation is rich in volcanic ash, which decomposes into bentonite — a fine, porous clay. The particular kind of bentonite found here has many industrial uses: as a drying agent or dessicant, in petroleum refining, and in decoloring vegetable oils, for instance. It been mined from the Bidahochi Formation near Chambers and Sanders. Between miles 357 and 354, the normally flat-lying formation is bent into a strong anticline breached along its summit by a tributary of the Puerco River so that it forms two facing scarps separated by the stream valley.

The Bidahochi Formation is geologically young — probably no older than Pliocene — and the fold is obviously even younger. It seems likely to have been caused by solution of salt deposits known to exist in Permian strata well below the surface. For more about this type of anticline see I-40 Winslow — Petrified Forest.

South of the anticline the Bidahochi Formation retains its normal horizontal attitude, capping the many low mesas along the highway. Its tan color and horizontal upper surface are distinctive. The hori-

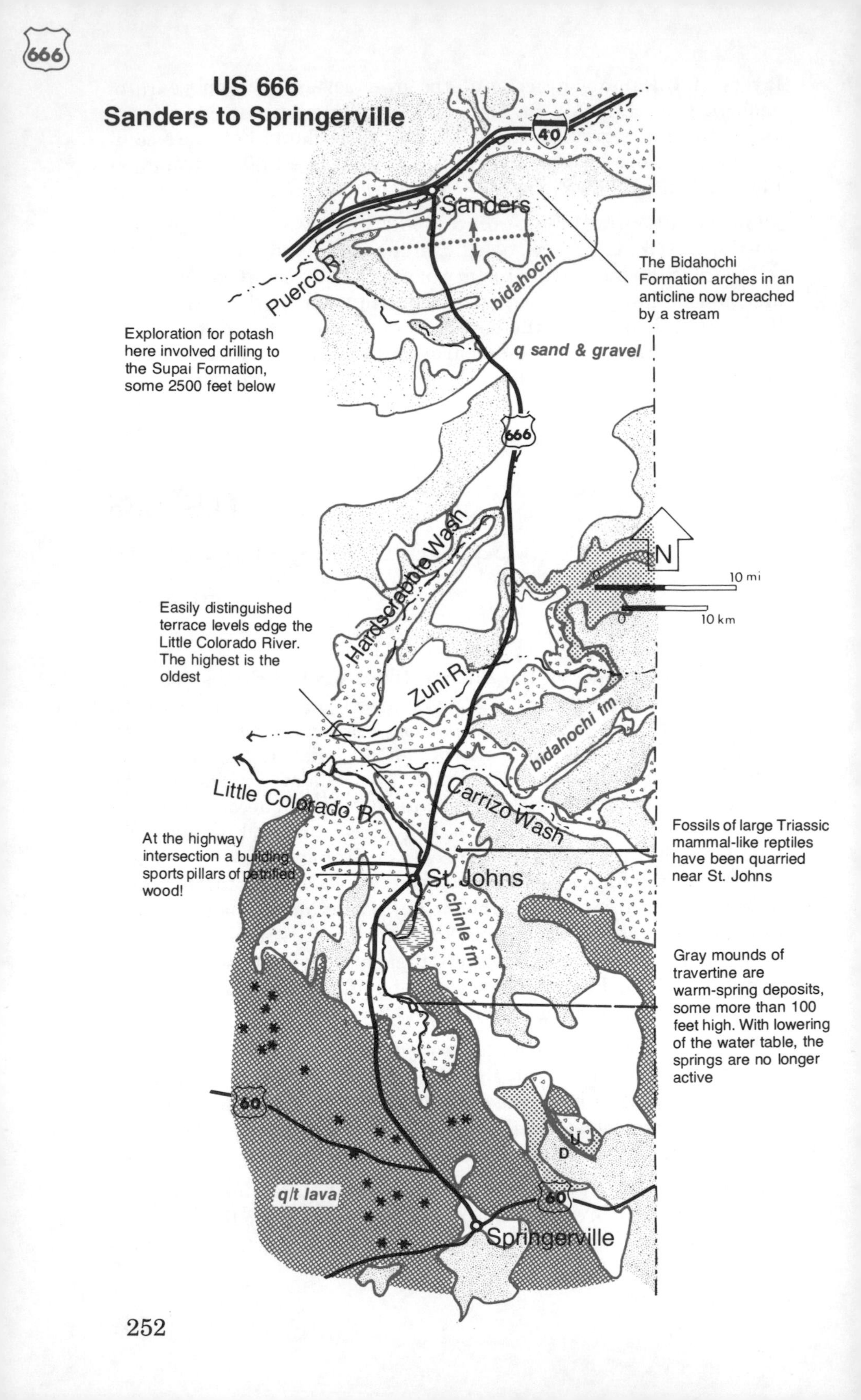
666
US 666
Sanders to Springerville
40
Sanders
Puerco R.
bidahochi
The Bidahochi Formation arches in an anticline now breached by a stream
Exploration for potash here involved drilling to the Supai Formation, some 2500 feet below
q sand & gravel
666
N
0
10 mi
0
10 km
Hardscrabble Wash
Easily distinguished terrace levels edge the Little Colorado River. The highest is the oldest
Zuni R.
bidahochi fm
Little Colorado R.
Carrizo Wash
Fossils of large Triassic mammal-like reptiles have been quarried near St. Johns
At the highway intersection a building sports pillars of petrified wood!
St. Johns
chinle fm
Gray mounds of travertine are warm-spring deposits, some more than 100 feet high. With lowering of the water table, the springs are no longer active
60
U
D
q/t lava
60
Springerville

zontal beds, once continuous over this entire area, have been dissected by the Puerco and Little Colorado rivers and their tributaries. In exposures along the highway, north-dipping cross-bedding shows that these are not the lake deposits or volcanic ash of the middle and lower parts of the formation, which consist of many thin, flat-lying layers, but are the upper member, deposited by flowing streams which carried fine material from nearby mountains out across the flats exposed by the shrinking lake.

East of milepost 336, a lava flow in mid-valley caps remnants of Cretaceous rock brought near the surface by the Zuni Uplift east of here in New Mexico. The lava, younger than the Bidahochi Formation, lies directly on Cretaceous rocks with no intervening Bidahochi; the Cretaceous rocks must have formed a promontory jutting out into Lake Bidahochi. Along Hardscrabble Wash, the Zuni River, and the Little Colorado River, which you will cross at St. Johns, the Bidahochi overlies colorful Triassic strata of the Chinle Formation, visible near the highway from the Zuni River southward beyond St. Johns.

This is the southernmost part of the Painted Desert, and true to form the Chinle rocks weather into badlands banded in shades of green, purple, blue and red. This is the rock, too, of the Petrified Forest, and fragments of petrified wood used to be scattered over the Chinle surface here. Gradually they have been picked up for souvenirs or for sale to tourists. In St. Johns local residents have used the wood as novel ornamental building stone.

Vertebrate fossils also occur in the Chinle Formation, and near St. Johns numerous large, lumbering reptiles called *Placerias* have been found. *Placerias* resembled a rhinoceros (a mammal) more than it did its fellow reptiles and contemporaries, the dinosurs.

South of St. Johns the Chinle Formation and remnants of the Bidahochi Formation are more and more overlain by dark lava flows from the White Mountains. Some of the flows, like the one south of Lyman Reservoir, cap long ridges. One's first reaction is to wonder how the flow accomplished the ridgeline balancing act. However, the lava actually flowed down a long valley and hardened there. Then intervening ridges, which were unprotected, eroded away, leaving the lava-filled valley as a sinuous and resistant new ridge.

The White Mountain volcanic field is composed of a central cluster of large stratovolcanoes fringed on the north by thinner lava flows and about 200 cinder cones. Some of the cinder cones are being quarried now for ready-made lightweight aggregate and road material. In the quarries you can glimpse the internal makeup of these minivolcanoes, their relatively resistant cores and the sloping red and gray layers of which they are composed. Volcanic rocks from the White Mountains cover up all the rest of the geology — Paleozoic and Mesozoic rock that must be present here, the Mogollon Rim that is the southern edge of the Colorado Plateau, the southern boundary of the Bidahochi Formation, and possibly the outlet for the waters of the Bidahochi Lake.

AZ 63
Chambers — Chinle
(77 miles)

After crossing a mile or so of Puerco River sand and gravel, Arizona 63 rises onto a broad tableland made up of Pliocene lake and stream deposits, the Bidahochi Formation. These young, poorly consolidated rocks now cover parts of an area that measures more than 100 miles in diameter. The lake in which they were deposited, called Lake Bidahochi or Hopi Lake, extended westward along the present valley of the Little Colorado River, and must have been comparable in size, though not in shape, with Lake Erie. It is thought to have been fed by the ancestor of the Colorado River, but doubtless it also received the waters of smaller streams draining plateaus and mountains near the eastern border of Arizona, streams that carried an abundance of fine quartz sand derived from the predominantly sandstone strata of this region.

Although lower parts of the Bidahochi Formation are clearly lake deposited, the uppermost unit, the one you will see at roadside here, consists of cross-bedded, tan, river-deposited sand alternating with layers of volcanic ash that drifted over this area as Lake Bidahochi

In a roadcut north of Chambers the characteristic festoonlike cross-bedding of a former stream channel cuts the Bidahochi Formation.

filled with lake deposits. Geologically young, the sandstone weathers into sandy soil commonly blown into dunes. Most of the dunes, still recognizable by their hummocky shape, are now stabilized by vegetation. Here and there among the dune deposits are thin bands of silty white limestone containing fossil freshwater snails and clams, suggesting that a playa or seasonal lake existed here for some time, too. Several roadcuts reveal one-time stream channels filled with cross-bedded, stream-deposited sand, commonly with a layer of pebbles at the bottom of the channel. The dark rich browns of some sandy layers bespeak the accumulation of humus — dead plant matter — and development of ancient soil zones.

From the top of the hill between mileposts 32 and 33 one begins to feel the vastness of the Navajo country, which stretches north and east into Utah, New Mexico, and Colorado. Here one also begins to see the great anticline of the Defiance Uplift, a 100-mile-long dome which crests not far east of the highway. Off to the northwest, red, white, and tawny gray Mesozoic sedimentary rocks dip westward, away from the crest of the anticline, as does the De Chelly Sandstone to the east. Out of sight eastward these same rocks turn down under the Arizona-New Mexico border.

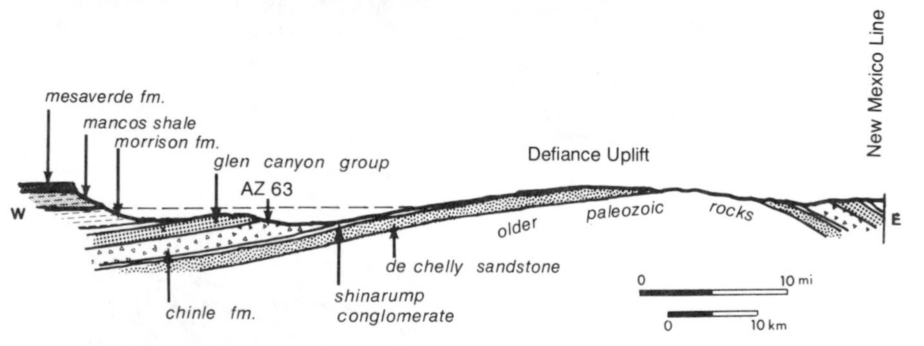

Section across AZ 63 near Ganado. Dashed line shows level of uppermost Bidahochi deposits; north of Ganado only scattered remnants remain.

Near Ganado there is a change in the roadside geology: the sudden disappearance of the Bidahochi Formation, and the entry into the landscape of the Chinle Formation's colorful beds of fine mudstone, volcanic ash, and conglomerate. The west side of the Defiance Upwarp becomes quite evident beyond milepost 53, where the valley of Chinle Wash, called Beautiful Valley, suddenly opens before you and the barren strata display their westward dip well. Chinle Wash is a relative newcomer in this area; satellite photographs show ancient drainage patterns that ran southwest from the Defiance Plateau and the Chuska Mountains across the surface of Black Mesa.

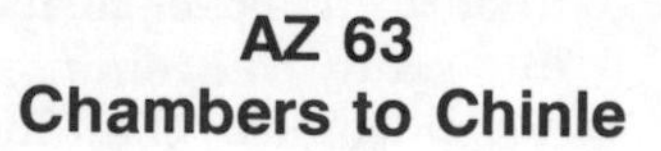

AZ 63
Chambers to Chinle

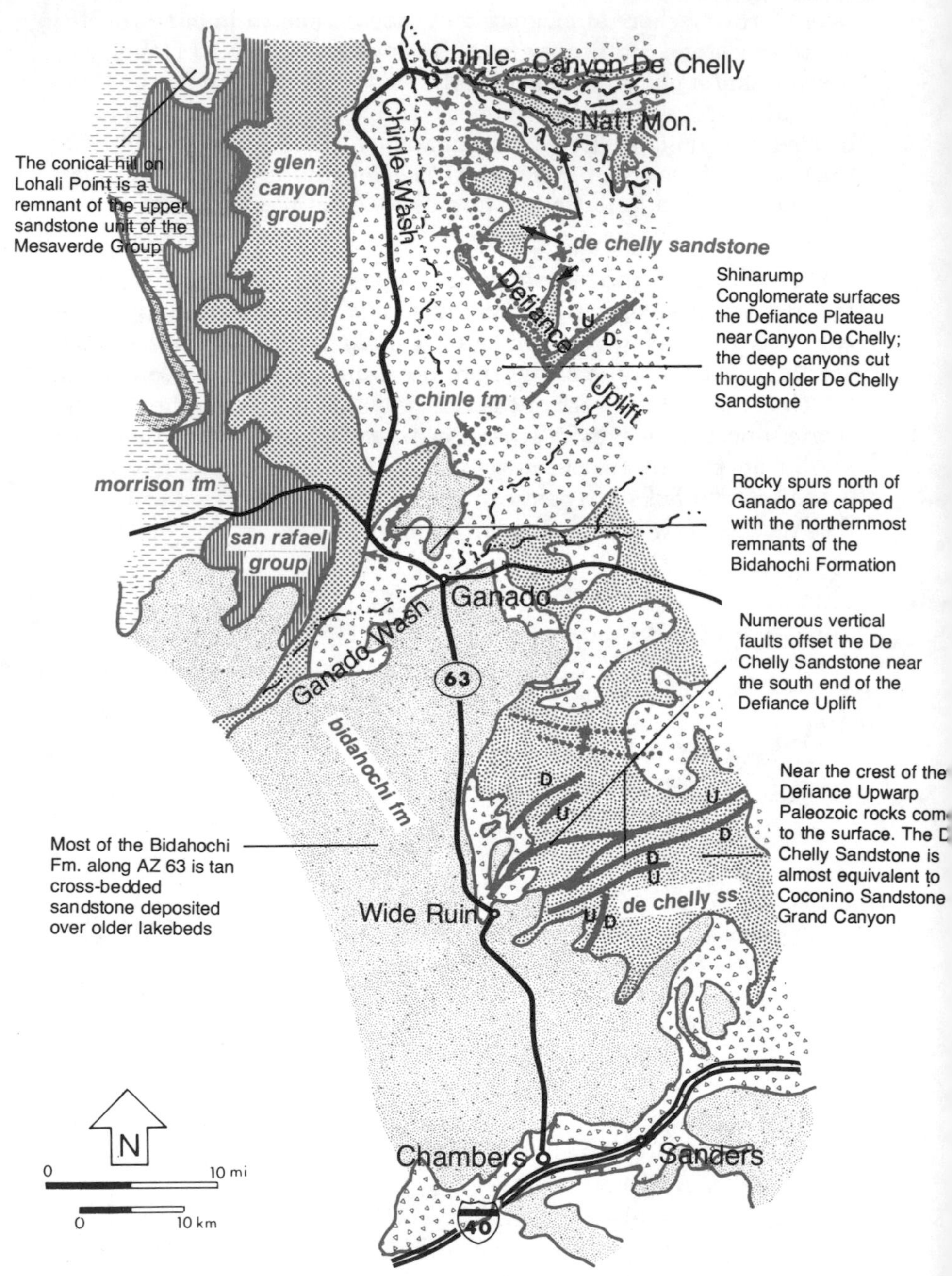

In the distance to the northeast the Chuska Mountains shape the skyline. There, thick layers of sandstone of the same Pliocene age as the Bidahochi lake deposits make up the mountain wall, and are capped with scattered remnants of once extensive lava flows. Just how the Chuska Sandstone relates to the Bidahochi beds is not yet well understood.

West of the highway Black Mesa's Cretaceous strata also dip westward, but more gently than rocks near the highway. The mesa is topped with rocks of the Mesaverde Group, a three-formation group with lagoon mudstone and coal sandwiched between two thick units of near-shore sandstone. Below it is a gray, piñon-covered slope of Mancos Shale, another Cretaceous unit, which contains beautiful, still pearly fossil shells of ammonites and other marine mollusks. A thin ledge of Dakota Sandstone separates the Mancos from a lower slope-former, the Morrison Formation, a widespread Jurassic unit as colorful as the Chinle Formation of the Painted Desert. Near Morrison, Colorado, at Dinosaur National Monument in northeastern Utah, and at a number of other sites this formation has yielded many well preserved fossil dinosaur skeletons.

Just opposite Lohali Point on Black Mesa, with its cone-shaped hill of Mesaverde sandstone, the road drops into a broad valley where Chinle Wash meets Cottonwood Wash and streams coming from Canyon del Muerto and Canyon de Chelly, near the site of the Navajo Indian community of Chinle. For discussion of Canyon de Chelly National Monument see Chapter V.

AZ 63
Chinle — U.S. 160
(62 miles)

North of Chinle, Arizona highway 63 follows the course of Chinle Wash, a typical desert wash in that over much of its course it is dry much of the year. The incline of the Defiance Upwarp rises to the east, and increasingly horizontal rock layers that edge Carson Mesa and Black Mesa border the valley on the west. The valley itself is floored with soft, easily eroded mudstone, claystone, marl (limy mudstone), and bentonite (decomposed volcanic ash) of the Chinle Formation, which is exposed here as in the Painted Desert as multi-colored badlands. Because of their high content of bentonite, surfaces of these rock layers swell with every rain, then shrink and dry to a crackled, easily powdered crust. Such instability of the surface, coupled with rapid erosion and the high gypsum content of the rock, discourages vegetation, as you can see. And the lack of vegetation further enhances erosion, in an endless cycle. The swell-and-shrink characteristics are hard on highway pavement, too, leading to unexpected humps and bumps.

When the Chinle Formation was deposited the landscape here was far different than it is today. The land was low and sloped evenly and gently toward a western sea, much as the present Gulf Coast slopes toward the Gulf of Mexico. The climate was warm and humid, as in southeastern United States today. Silt-laden rivers flowed across the lowland, depositing mud, sand, and occasional layers of pebbles brought from new mountains to the south. Infrequent floods introduced battered logs and other plant fragments. Both vegetation and animal life were semi-tropical, with crocodile-like phytosaurs and other reptiles large and small, fish, amphibians, and fresh water clams.

Flat-topped terraces along Chinle Valley, as well as some of the slopes on the Chinle Formation, are surfaced with desert pavement of small pebbles that become more and more concentrated as fine sand and silt are blown away. Wind erosion is an important desert process. If you have ever collided with a desert "dust devil" you must appreciate its force. Most of the pebbles that make up desert pavement along this route come from conglomerate layers in the Chinle Formation, among them the Shinarump Conglomerate, lowest member of the Chinle, which caps the sloping Defiance Plateau to the east. Scenic Canyon de Chelly and Canyon del Muerto (see Chapter V) cut through it and into the De Chelly Sandstone below. Notice that there are no red siltstone and sandstone layers below the Shinarump as there are west and south of Black Mesa; the Moenkopi Formation is absent here.

West of the highway successive layers of sandstone and shale, all of them of Mesozoic age, form the benches and escarpments of Carson Mesa and, behind it, Black Mesa. These strata, too, are tilted gently

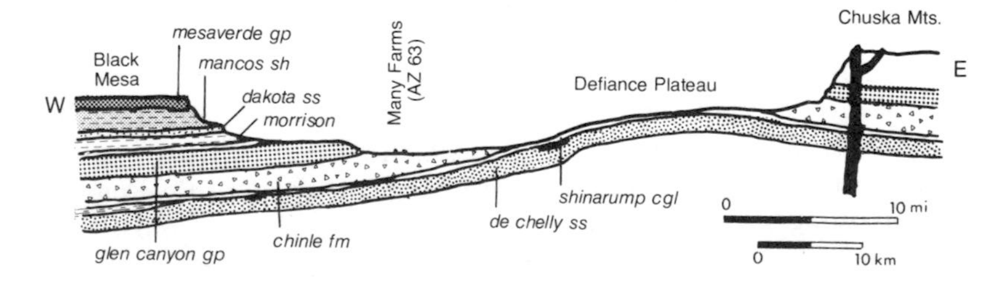

Section across AZ 63 at Many Farms.

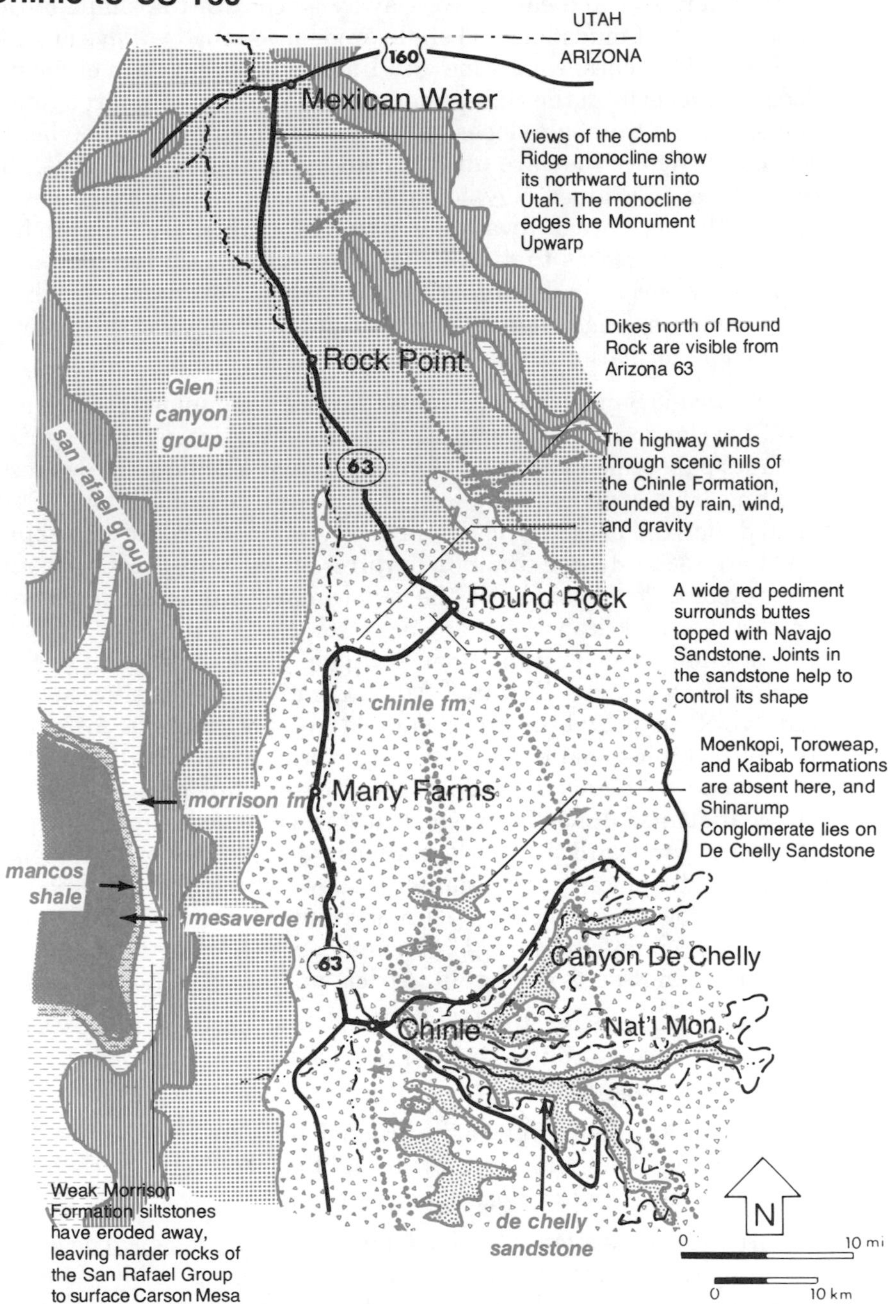
AZ 63
Chinle to US 160
UTAH
ARIZONA
160
Mexican Water
Views of the Comb Ridge monocline show its northward turn into Utah. The monocline edges the Monument Upwarp
Dikes north of Round Rock are visible from Arizona 63
Rock Point
Glen canyon group
san rafael group
63
The highway winds through scenic hills of the Chinle Formation, rounded by rain, wind, and gravity
Round Rock
A wide red pediment surrounds buttes topped with Navajo Sandstone. Joints in the sandstone help to control its shape
chinle fm
Moenkopi, Toroweap, and Kaibab formations are absent here, and Shinarump Conglomerate lies on De Chelly Sandstone
morrison fm
Many Farms
mancos shale
mesaverde fm
63
Canyon De Chelly
Chinle
Nat'l Mon.
Weak Morrison Formation siltstones have eroded away, leaving harder rocks of the San Rafael Group to surface Carson Mesa
de chelly sandstone
N
0
10 mi
0
10 km

westward by the Defiance Uplift. Here you can see exceptionally well the manner in which successive rock layers — some hard and resistant, others soft and easily worn away — control the shape of the land. Layers of mudstone and shale erode into broad valleys or wide slopes and benches; sandstone and limestone layers form cliffs and ledges. Debris from the cliffs, some of it sliding away in large intact blocks, lies on the shale slopes; **talus cones** develop below notches in the cliffs. Where they are undermined by erosion of soft rocks, the massive, even-grained Navajo Sandstone, now coming into view to the north, may break away along curving joints, creating arched recesses and alcoves that in some places were used for shelter or storage by early inhabitants of this region. Conveniently for those inhabitants, springs commonly occur right at the base of this porous, permeable formation.

The Navajo Sandstone is perhaps the most prominent formation in the Glen Canyon Group. Certainly it is one of the easiest to recognize, with its pale pink, salmon, or light gray color and large-scale cross-bedding. This sandstone is one of the greatest accumulations of ancient dune sand in the world. Its massive, rounded pink or white cliffs and bare-rock hills are common features of the Navajo country, and are responsible for many of the outstanding scenic attractions of north-eastern Arizona and adjacent southern Utah: the massive white summits of Capitol Reef National Park, shelters for cliff-dwellings in Navajo National Monument, the high, slim arch of Rainbow Bridge, and the great walled towers of Zion. The formation is 1600 feet thick near Page, Arizona, but only about half that thick in northeastern Arizona. Its high-angle cross-bedding derives from the crescent-shaped dunes from which it formed, where sand blown up the gentle windward slopes of the dunes was deposited on the steeper lee faces. In places, wavy laminae show where sliding sand avalanched down the dune face. Thin horizontal laminae formed as silt and clay accumulated in flat interdune areas similar to those now recognized in deserts of Arabia, Africa, and our own Southwest.

As the highway crosses the Navajo Sandstone north of mile 120, the magnificent cross-bedding and the horizontal interdune deposits are particularly apparent.

The Chuska Mountains east of Round Rock lie close to the New Mexico border. They are capped with light-colored Tertiary sedimentary rocks, the Chuska Formation.

Northeastern Arizona displays many of the geological features normally required for oil and gas accumulation: a possible sedimentary rock source for oil, nearly flat-lying sedimentary rocks of appropriate porosity, which act as oil reservoirs, and several kinds of "traps" — rock structures that prevent oil from migrating out of the

reservoir rocks. As a result, this area has seen quite a bit of dri
most of it unsuccessful. Arizona's only really productive oilfie
the northeast slope of the Chuska Mountains, is unusual in tha
oil has accumulated in igneous sills rather than in sedimentary rocks. The same general area — known as the Lukachukai Mountains — also contains uranium.

North of the Chuska Mountains, the Carrizo Mountains are the remains of a cluster of Tertiary laccoliths.

As Arizona 63 approaches US 160 it drops rapidly off the north end of the Defiance Upwarp. Tall fingers of Agathla Peak, the largest of several volcanic necks near Kayenta, are in view to the west, with the jagged line of Comb Ridge (the Navajo Sandstone again — did you recognize it?) below it. To the north, dim with distance, are the La Sal Mountains of Utah.

AZ 64
Cameron — Desert View
(32 miles)

For a map of this section see US 89 Flagstaff — Cameron.

As this highway leaves US 89 about 2 miles south of Cameron, it heads directly toward the great **monocline** that forms the eastern edge of the Kaibab and Coconino Plateaus. Kaibab Limestone surfacing these plateaus, at elevations around 7000 feet south of Grand Canyon and 8000 to 9000 feet north of it, turns abruptly downward along this monocline, then levels out again to form a lower surface — the Marble Platform — at about 5000 feet. The generally north-south trending monocline branches here, with the more pronounced branch curving southwestward around lofty Coconino Point.

Below the monocline and near Cameron this lower Kaibab surface is hidden by overlying Triassic strata, red sandstone and mudstone of the Moenkopi Formation, in places capped with hard Shinarump Conglomerate.

The East Kaibab Monocline shows to good advantage from mile 291. Its northern extension, the east edge of the Kaibab Plateau, is visible in the distance. Monoclines like this one, similar in structure but with differences in height and orientation, are fairly common across the Colorado Plateau. At Grand Canyon the Colorado River has cut down through several monoclines and into hard, relatively

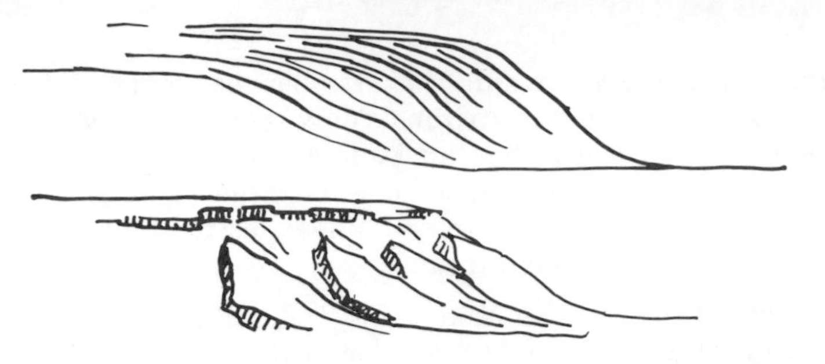

Rocks bent down along the East Kaibab Monocline have eroded into "flatirons" between which older rocks are exposed.

brittle Precambrian rocks below. There geologists find that the sedimentary rocks of the upper canyon walls merely drape across great faults in the rigid Precambrian rocks, like blankets hanging over the edge of a bed and flattening out again on the floor. Many of the faults are complex, especially in eastern Grand Canyon. The folds at the surface seem far simpler, but in reality may be made up of many small breaks and offsets. Erosion along the monocline has carved triangular wedges of Kaibab and Toroweap Limestones, and revealed patches of Coconino Sandstone between these wedges. Some of the monocline-forming faults are extremely ancient, and here again the unsurpassed exposures in Grand Canyon bring evidence to light and reveal what rocks are displaced and which faults are cut by other faults. Careful mapping and study of the faults exposed in the canyon shows that the first movement along many of the large faults occurred in Precambrian time. What is more, Precambrian movement on some, including Butte Fault below the East Kaibab Monocline, was in the opposite direction — with the east side lifted, the west side lowered!

Shadow Mountain, a small volcano northwest of Cameron, lies on the trend of a major fault where broken rock permitted volcanic magma to reach the surface. In the background the East Kaibab Monocline edges the Cononino Plateau.

Tad Nichols photo.

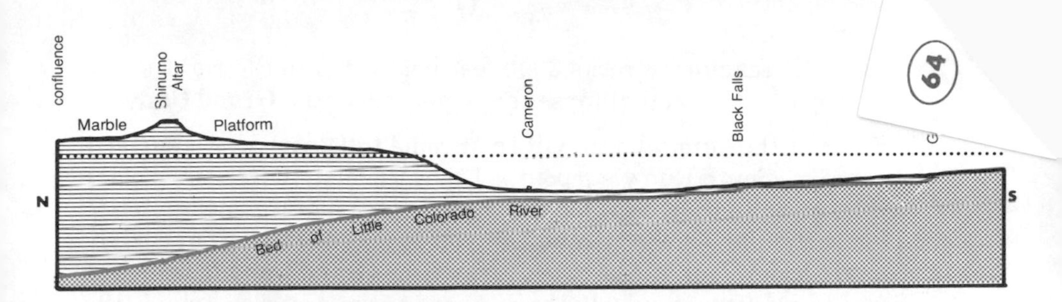

Though the floor of the Little Colorado valley (black) rises northward, the Little Colorado River flows in that direction (color). Ever since drainage reversed, the Little Colorado has had to cut down through the Marble Platform toward its confluence with the Colorado River mainstream.

North of the highway and almost paralleling it between mileposts 285 and 280, the Little Colorado River has carved a narrow, spectacular gorge down through the Paleozoic rocks of the Marble Platform. Several small side roads lead to the perpendicular lip of this gorge — a place to watch your children carefully. The general dip of the Kaibab Limestone surface is to the south; the gorge is the river's way of maintaining its northwestward flow into the Colorado River in Grand Canyon. Kaibab Limestone is the rimrock. Deep in the gorge one can glimpse the massive sandstone of the Coconino Formation, black with lichens; father downstream the river cuts into the same older Paleozoic rocks that wall Grand Canyon. Its gorge remains narrow because this smaller river, with only seasonal flow, must devote its energies to downcutting in order to keep pace with the more efficient erosion of the larger, more energetic Colorado.

Near milepost 280 the highway begins to climb the East Kaibab Monocline to the top of Coconino Plateau. The fold is double here, stairstepping with an extra level, probably reflecting a double fault

Little more than a million years ago, lava flows cascading down the walls of western Grand Canyon dammed the Colorado River. The river eventually cut through the lava dams. Tad Nichols photo

down in Precambrian rocks 3000 feet below. Off to the right as the highway climbs you'll glimpse the upper end of the Grand Canyon.

Once at the summit you will soon come to Desert View, easternmost of the South Rim vista points. Enjoy the matchless scenery for a while, and then for a geological interpretation turn to Chapter V. This particular view of Grand Canyon is magnificent — one of the best. The view looking eastward away from Grand Canyon is magnificent in a different way. In the distance to the east rise Echo Cliffs, about 30 miles away across the Marble Platform and the north end of the Painted Desert. To the southeast is the valley of the Little Colorado, a valley that may have been occupied in mid-Tertiary time by the ancestor of the present Colorado River, flowing southeast rather than west. Beyond the Little Colorado Valley is the horizontal top of Black Mesa, heart of the Navajo country. The rock sections diagrammed in the introduction to this chapter will help you understand the geologic relationships between Permian rim rocks of Grand Canyon, on which you are standing, Triassic rocks of the Painted Desert, Jurassic ones of Echo Cliffs, and Cretaceous ones that make up Black Mesa — a total sequence representing about 200 million years of geologic time.

Looking back into Grand Canyon you see rocks of an even vaster time span; from Permian time 290 to 240 million years ago, back through Paleozoic and Precambrian time, with rocks as old as 2.2 billion years in the depths of the canyon. One of the remarkable features of Grand Canyon is that it covers — uncovers — so much of geologic time. Just think! In your drive today, if you came from Flagstaff, you have seen rocks as young as 1000 years — the products of Sunset Crater's eruption in 1064 or 1065 A.D. You've seen older volcanic rocks of the San Francisco Peaks, back to about 3 million years. You've seen Mesozoic rocks on faraway Black Mesa, on Echo Cliffs, and in the colorful hills of the Painted Desert; they range in age from 65 to 225 million years. Here on the rim of Grand Canyon you are standing on Paleozoic rocks — the Permian Kaibab Formation at the rim and below it successively older layers back to about 570 million years. In this eastern part of the canyon itself are Precambrian sedimentary rocks that are about 1 billion years old. And downstream, where the dark Inner Gorge begins, there are yet older Precambrian rocks, the Vishnu Schist, formed 2.2 million years ago. Even they must have been derived from older rocks.

The record is not complete, of course. There are long gaps in it. As a matter of fact there are undoubtedly more years, more centuries, more millenia missing than there are represented here. But even the gaps tell us about the history of our earth. The Great Unconformity, for instance, above the youngest of the Precambrian rocks and below the oldest Cambrian rocks, tells us of a long, long time of erosion, erosion, and more erosion.That between the older and younger Precambrian rocks tells us how mountains, ranges the size of the Himalayas, gave way to the forces of erosion during an even earlier period when the land was worn low, beveled and smoothed. Lesser unconformities tell of shorter erosional intervals. The unconformity at the top of the Mississippi Redwall Limestone tells of a warm, humid climate when caves and sinks and a rough karst topography developed on the limestone beds. That at the base of the Redwall shows us that Devonian sediments were laid on an irregular surface channeled by streams.

Half the age of the earth is here, in these rocks, spread out for you to see.

AZ 87
Winslow — Pine
(71 miles)

Arizona 87 climbs gradually out of the valley of the Little Colorado River and heads southwest among low hills of deep red Triassic Moenkopi Formation sandstone and siltstone. These sedimentary rocks, deposited on a broad coastal plain near the margin of the ancient supercontinent Pangaea, dip gently — almost imperceptibly — north here, so that even though the highway rises beyond about mile 335 it comes to older and older parts of the formation. Where it crosses Jacks Canyon Creek at mile 330 the contact between the lowest Triassic redbeds and the top of the buff-colored Kaibab Formation below it is well exposed.

The Kaibab Formation is quite sandy here. The Kaibab sea, remember, came from the west across a flat, nearly featureless, slowly sinking surface, and here its marine limestones intermix with near-shore beach and dune sandstones in a pattern very like that found along the Gulf Coast today. Red-tinted sandstone layers between the sandy limestone layers are another near-shore feature.

Drainage here is **dendritic**, branching like a tree, and northbound

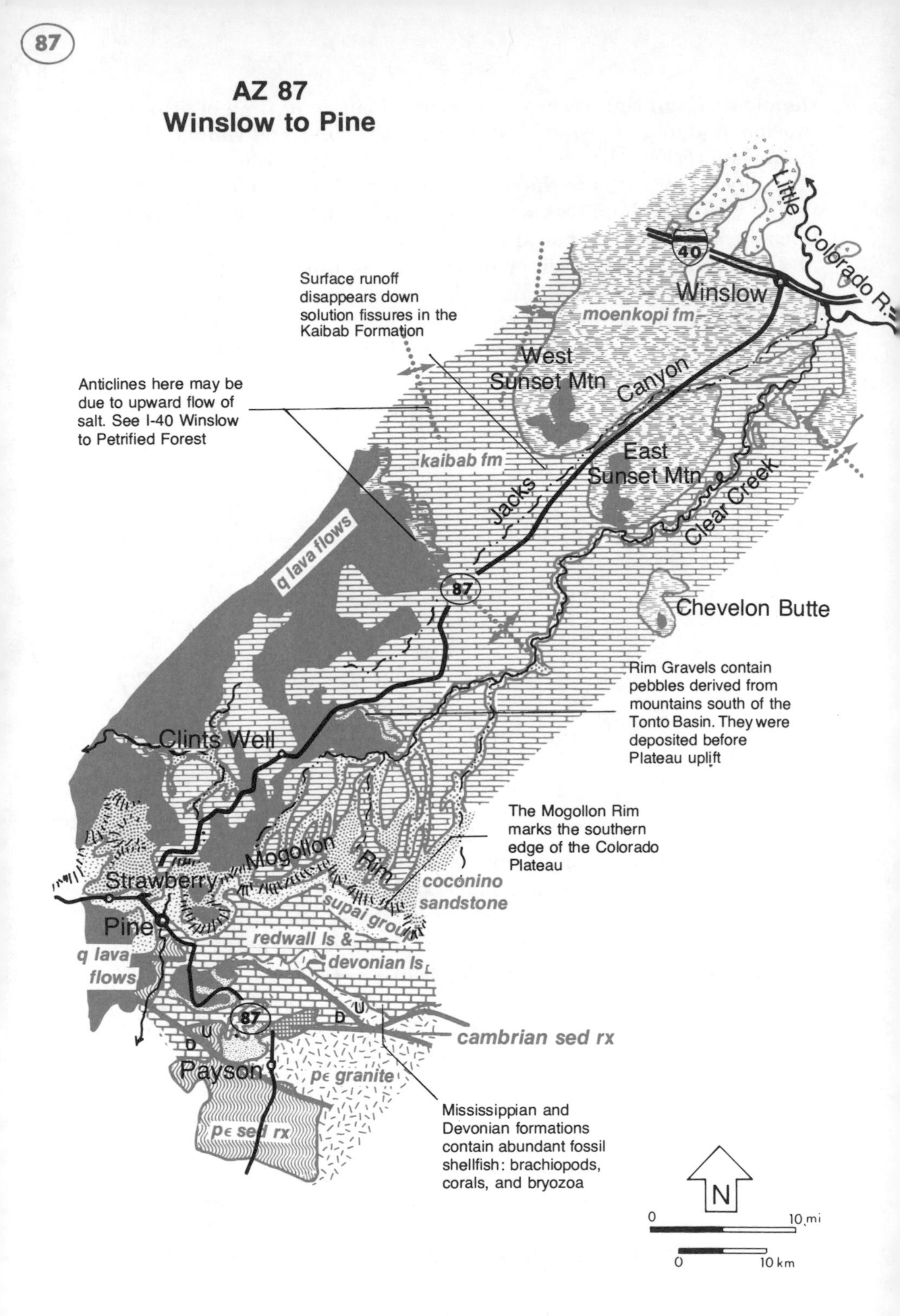
87
AZ 87
Winslow to Pine
Surface runoff disappears down solution fissures in the Kaibab Formation
Anticlines here may be due to upward flow of salt. See I-40 Winslow to Petrified Forest
40
Little Colorado R.
Winslow
moenkopi fm
West Sunset Mtn
Canyon
East Sunset Mtn
kaibab fm
Jacks
Clear Creek
q lava flows
87
Chevelon Butte
Rim Gravels contain pebbles derived from mountains south of the Tonto Basin. They were deposited before Plateau uplift
Clints Well
The Mogollon Rim marks the southern edge of the Colorado Plateau
Mogollon
Rim
coconino sandstone
Strawberry
supai group
Pine
redwall ls &
devonian ls
q lava flows
87
D
U
cambrian sed rx
Payson
pє granite
pє sed rx
Mississippian and Devonian formations contain abundant fossil shellfish: brachiopods, corals, and bryozoa
N
0
10 mi
0
10 km

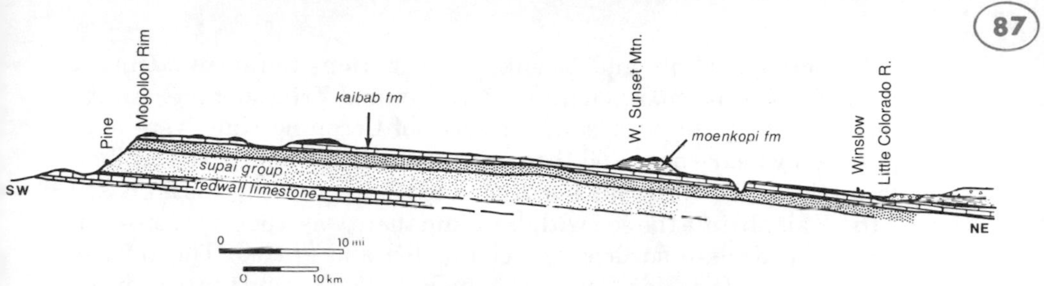

Section along AZ 87 Winslow to Pine.

toward the Little Colorado River. Jacks Canyon and a few other streams have incised twisting clefts through the Kaibab Formation. clefts that may be collapsed underground solution channels in the relatively soluble limestone. Small solution caves can be seen in the walls of Jacks Canyon.

East and West Sunset Mountains, one on each side of the highway near mile 326, preserve with lava caps some outlying remnants of the Moenkopi redbeds. However you'll have to look closely to see the dark red Moenkopi ledges below the lava caps, for they are pretty well covered with basalt debris from the lava above. Watch near Sunset Pass for green banding in the Moenkopi. The formation is rich in iron minerals — hence its predominent dark rust red color where the iron minerals have oxidized to hematite. Where oxygen was not plentiful, particularly where it was used by soil bacteria or by bacterial decay of plant and animal material, the iron minerals give it a greenish color. Paleontologists watch for this color in their search for vertebrate fossils. Large alligatorlike fossil amphibians have been found not far from here in the Moenkopi Formation.

For some distance south of the Sunset Buttes the highway remains on the Kaibab Formation surface. Ahead, more lava flows preserve more Moenkopi redbeds. On the whole, resistant volcanic rocks like basalt, flowing over and protecting underlying strata, played a large part in preserving and maintaining the integrity of the southern margin of the Colorado Plateau, the Mogollon Rim. Lava remnants on East and West Sunset Mountains, once continuous with those farther west, for a long time prevented substantial erosion. Once they were broken through, though, the underlying redbeds, soft and poorly consolidated, made easy going for erosive forces. More resistant than the redbeds, the Kaibab Formation takes on in its turn the job of protecting underlying layers, as it does on the rims of the Grand Canyon and the western part of the Mogollon Rim. Water tends to sink through the Kaibab Formation, dissolving underground channels, so there are few well developed stream channels on the Kaibab surface, and man-made reservoirs on this surface don't hold water well.

Coconino, Kaibab, and Moenkopi Formations tell us of a time of changing environments here in Permian and Triassic time. As the land subsided, the vast sandy deserts of Coconino time were supplanted by near-shore and then marine environments of the Kaibab formation. (The Toroweap Formation sea didn't made it this far east.) After Kaibab time the sea withdrew and there was a period of erosion, when the Kaibab surface was channeled and eroded. The time of erosion spans the end of one era — the Paleozoic — and the beginning of another— the Mesozoic. In Triassic time large rivers flowing from the Ancestral Rocky Mountains in Colorado and northern New Mexico brought in mud and silt and sand, all containing finely broken up iron minerals, and spread them out across the low, flat plain.

Near milepost 297 the road climbs through a lava-protected remnant of the Moenkopi Formation and onto one of the basalt flows. Basalt lava such as you see here is very fluid when it issues from the ground, usually from narrow fissures, and it spreads into a vast, fairly thin sheet, moving out over the surface faster than a man can run. Flows here are virtually continuous with those of the San Francisco volcanic field near Flagstaff, though they probably came from a nearby source. Much younger than San Francisco Mountain, which is visible occasionally far to the west, they are of about the same age as the lava flows and cinders that surround its base.

Pine Creek is one of the few south-draining canyons that cut through the Mogollon Rim. The highway follows it off the lava and down through the buff limestone strata of the Kaibab Formation, thick with lumps and nodules of chert. Then it drops through the light-colored Coconino Sandstone and into orange-red rocks of the Supai Group. The base of the Kaibab is at milepost 273, that of the Coconino at milepost 272. In the Coconino Sandstone notice the long, sweeping cross-bedding which shows, in combination with the formation's fine, rounded, frosted grains, its origin in ancient sand dunes. Watch in highway cuts here for occasional dikes of gray basalt, feeder fissures for the lava flows of the plateau above.

From mile 273 the highway descends to Strawberry and Pine, with views of the forested slopes of the Mazatzal Mountains, highlands of Precambrian granite, schist, and sedimentary rock. The Sierra Ancha farther east exposes great thicknesses of Precambrian sedimentary rocks — quartzite, shale, and limestone interlayered with diabase sills that intruded between the strata in Precambrian time.

The town of Strawberry lies at the base of the Coconino Sandstone, lowest of the cliff-formers along the Mogollon Rim. Light red sandstones and shales between Strawberry and Pine are parts of the Supai Group. Just south of milepost 270 a thin, buff-colored limestone is interlayered with these red deposits. This is the Fort Apache Limestone, a fossil-bearing marine limestone that is thickest in the southeast, near Fort Apache, Arizona and that thins and disappears farther west. Pine lies at the base of the Supai Group and at the top of the Naco Formation, a marine limestone containing abundant Pennsylvanian brachiopods and corals whose calcium carbonate shells have been replaced by silica and coated with a brick red variety of silica called jasper.

V
The National Parks and Monuments

The national parks and monuments need no introduction. Some of those discussed here are primarily geologic — Grand Canyon, Sunset Crater, Petrified Forest, Chiricahua. In others the interest is largely in their records of human history — Navajo, Canyon de Chelly, Walnut Canyon, Montezuma Castle — yet their geology is interesting too, so I have included them here. Organ Pipe Cactus National Monument was established because of its varied flora, but it, too, is worth a geologic look. I have also included Monument Valley, a Navajo Tribal Park on the Arizona-Utah border. Reading applicable chapter introductions will help you understand the park and monument discussions that follow.

Most of the parks and larger monuments have inviting trails that offer good opportunities to look more closely at their geology, biology, and human history. Let me encourage you to leave your car and walk close to — *into* — their natural features. Carry drinking water and allow time and energy for the return trip, especially in the Grand Canyon, where coming up may take three times as much time as going down.

If you don't feel up to strenuous walking, attend some leisurely nature walks led by park personnel, and go to scheduled evening talks. Park guides and trail guide leaflets often discuss geologic details. Most of all, see in the parks and monuments valued natural resources that will add to your enjoyment and understanding of Arizona.

Rocks, fossils, minerals, plants, and animals are protected in the parks and monuments: no collecting is allowed. Photographs make better souvenirs, anyway!

Canyon de Chelly National Monument

Here are colorful canyons, weatherworn cliffs marked with the cross-bedding of Permian dunes, arching recesses sheltering well preserved cliff dwellings, and farms and hogans of pastoral present-day Navajo Indians. All come together in Canyon de Chelly and Canyon del Muerto, both within the national monument.

Many geologic features of these canyons can be seen from the rim drives and numerous vista points along them. However you'd probably enjoy going into the canyons, either walking down the trail to White House Ruin, or touring the canyons with the required Indian guide.

At successive viewpoints along the rim drives, one notes that the rock walls become higher eastward as the once flat-lying strata rise onto the great upwarp of the Defiance anticline. The hard, resistant **caprock** of the Defiance Plateau is the Shinarump Conglomerate, stream-deposited lowest member of the Chinle Formation. Both rim drives remain on this Triassic rock, and it is well exposed at numerous overlooks. Its short, very gently sloping cross-bedding, formed by the streams that carried its coarse pebbles, shows up in the across-the-canyon views. The wide variety of sand grain and pebble sizes is evident close at hand.

In places on the rock surface wind-scoured potholes may hold rainwater, forming minilakes in which tiny plants and animals are born, grow to maturity, and reproduce — living out their life cycles in the few short days before their little ponds dry up. These organisms secrete weak acids, products of their metabolism, which gradually attack the calcium carbonate cement that holds Shinarump sand grains together. When the ponds dry, desert winds blow away the loosened grains and deepen the potholes. Wind winnows the rest of the surface, too, carrying away sand grains but leaving the heavier Shinarump pebbles as close-packed desert pavement.

At many overlooks, the contact between the Shinarump Conglomerate and the pale, fine-grained De Chelly Sandstone below it is clearly exposed — a contact that represents 60 million years of non-deposition and erosion. Here and there along the rim, below the conglomerate, pre-Shinarump stream channels cut into otherwise

Massive cross-bedded De Chelly Sandstone (Permian) walls Canyon de Chelly, sheltering the homes of the Anasazi – the "ancient ones" – as well as summer hogans of present-day Navajos.

flat surfaces of the De Chelly Sandstone. These features show up best if you look for them across the width of the canyon.

Canyon walls expose the De Chelly Sandstone well — a pale peach-colored rock deposited in Permian time in a vast desert of the supercontinent Pangaea. Undermined by stream erosion, the rock weakens along vertical joints widened by repeated frost action and, in places, by tree roots. Ultimately giant slabs crash to the floor of the canyon. Such erosion was doubtless much more forceful in the past, when Ice Age climates brought increases in precipitation and long, frigid winters to northern Arizona.

What can we learn from the De Chelly Sandstone? What does it tell us about itself? Cross-bedding in this rock is on a grandiose scale: steep swooping laminae that in combination with rounded, pitted or frosted sand grains tell us that this is an **eolian** or wind-deposited sandstone. Horizontal laminae bevel the sloping ones, marking transient erosion surfaces, interdune valleys in which flat-lying layers of sand and dust accumulated. They may contain sediments, thin layers of silty sandstone that appear and disappear over relatively short distances. Similar interdune deposits are known today in the sand seas of the Sahara and other deserts, temporary stable surfaces soon to be covered as migrating dunes drift across them once more.

The great cross-beds form the sloping ceilings of many recesses and alcoves in Canyon de Chelly and Canyon del Muerto. Localized concentration of moisture in the porous sandstone probably accounts for the weakening of wall rock that ultimately, as slabs and sheets of rock fell away, deepened some of these recesses until they took shape as suitable living quarters for "the ancient ones," the Anasazi who inhabited this region 700 to 1000 years ago.

Cross-beds in this sandstone almost all slope southwestward, showing that the prevailing winds that formed them blew predominantly from today's northeast. The greatest present-day deserts, the Sahara and Arabian deserts and the dry interior of Australia, lie 20° to 30° north and south of the equator in zones where the easterlies of the tropics brush against the westerly winds of temperate belts. It is likely that the immense sand seas of Permian time, which extended far north and east of Arizona, lay in a similar belt a similar distance from an equator that ran from California to Newfoundland.

The thousand-foot walls of the two canyons and their tributaries are marked also with dark, shiny stains of manganese and iron oxides — desert varnish. Early inhabitants of this area appreciated the smooth veneer of some rock walls, pecking through the dark sheen into unstained sandstone (perhaps for religio-superstitious purposes, perhaps just to amuse themselves) and creating petroglyphs that can be seen from the canyon floor today. Duller streaks are carbonaceous material washed from overlying soils or formed in place as algae, lichens, and mosses grew on frequently dampened surfaces.

These petroglyphs were pecked through desert varnish and into the underlying rock sometime after Spanish explorers and settlers brought horses to this continent.

Chiricahua National Monument

Arizona Highway 186, branching from I-10 at Willcox, skirts the east side of Willcox Playa and weaves through some of the stabilized sand dunes around its border (see I-10 New Mexico — Willcox). In Pleistocene time, when the climate was a good deal rainier, a much larger lake filled the valley, a freshwater lake that drained southward into Mexico; the highway crosses part of its flat floor, and the horizontal terraces of its old shoreline can be seen still on some of the low mountain slopes. At present the entire valley drains toward Willcox Playa, which after heavy rains may be wet with a few inches of water.

Along the southwest side of the Dos Cabezas Mountains (though only *una cabeza* can be seen from here), the route runs through Precambrian granite and silvery schist, as well as through many steeply tilted layers of Paleozoic and Cretaceous sedimentary rocks that lie in a long fault sliver along this side of the range, lending their varied colors to foothills and mountains. At mile 343-344 the coarse granite, with crystals up to 3 inches across, appears at the side of the road. In this eroded mountain pediment the granite is slowly disintegrating, turning to "rotten granite" and then to coarse sand. A resistant Cambrian quartzite and conglomerate layer forms a particularly prominent ridge at mile 346 to 347, with the road cutting through it before emerging onto the gravel-colored bajada of the northern Chiricahua Mountains.

As seen from this southwest side, the Chiricahuas are almost entirely volcanic. The oldest rocks in the range, the same Precambrian schist and granite exposed in the Dos Cabezas Range, appear on the eastern side, however, with more slivers of Paleozoic and Cretaceous sedimentary rocks. All these older rocks were involved in Laramide, mid-Tertiary, and Basin and Range mountain-building, subjected to complex folding and faulting and sliding comparable to that in many other southeastern Arizona ranges.

Concealing these sedimentary rocks in the monument area are three rock units virtually confined to the Chiricahua Mountains, but which do have counterparts — close age equivalents — in some neighboring ranges. All three units bear evidence of a wild orgy of volcanic eruptions 30 to 25 million years ago. One of the units consists of small isolated patches of bright red and reddish brown lake-

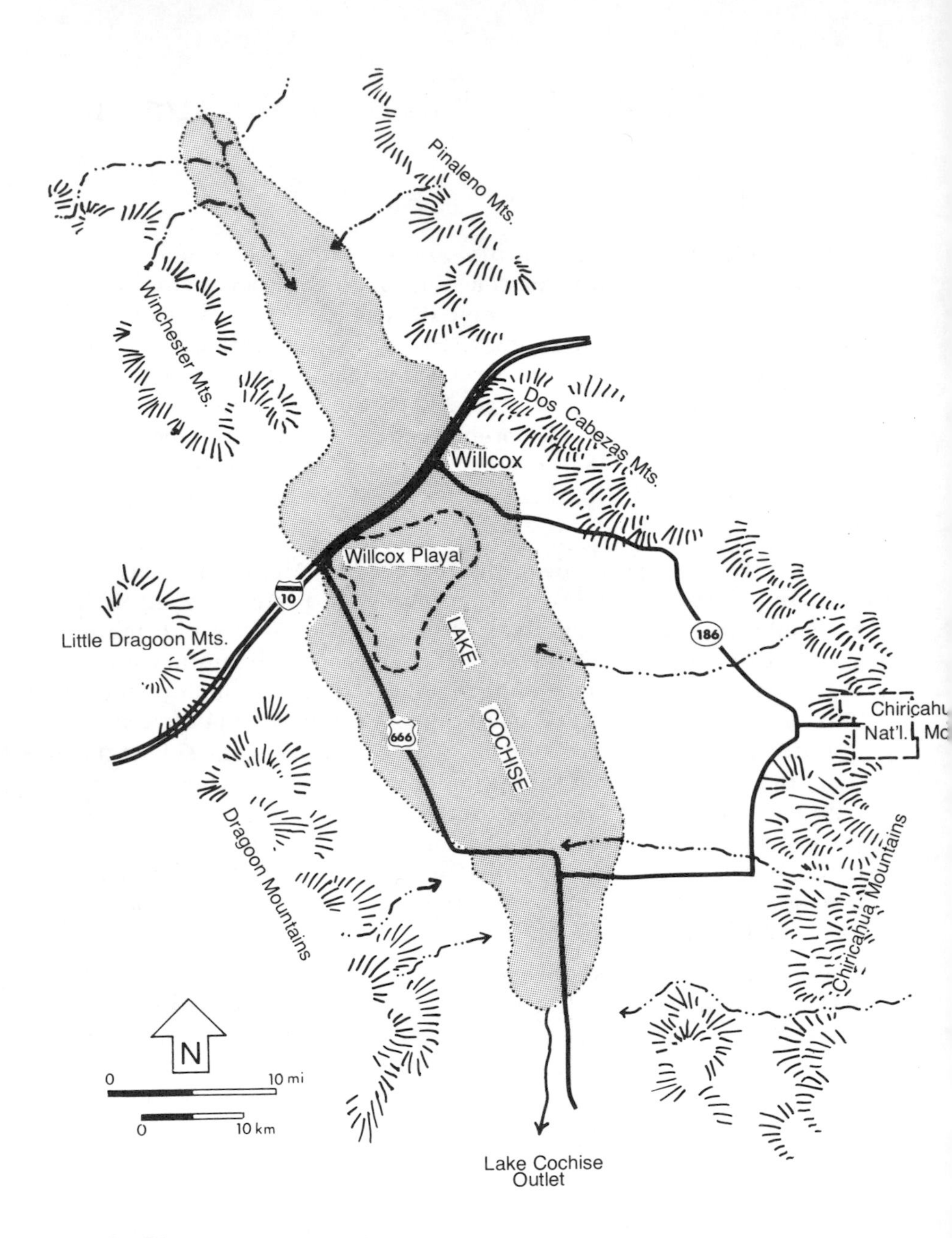

In Pleistocene time, when the climate was rainier than at present, Sulphur Springs Valley harbored Cochise Lake, a freshwater lake draining south into Mexico. Deep wells have encountered sediments of an older, larger lake.

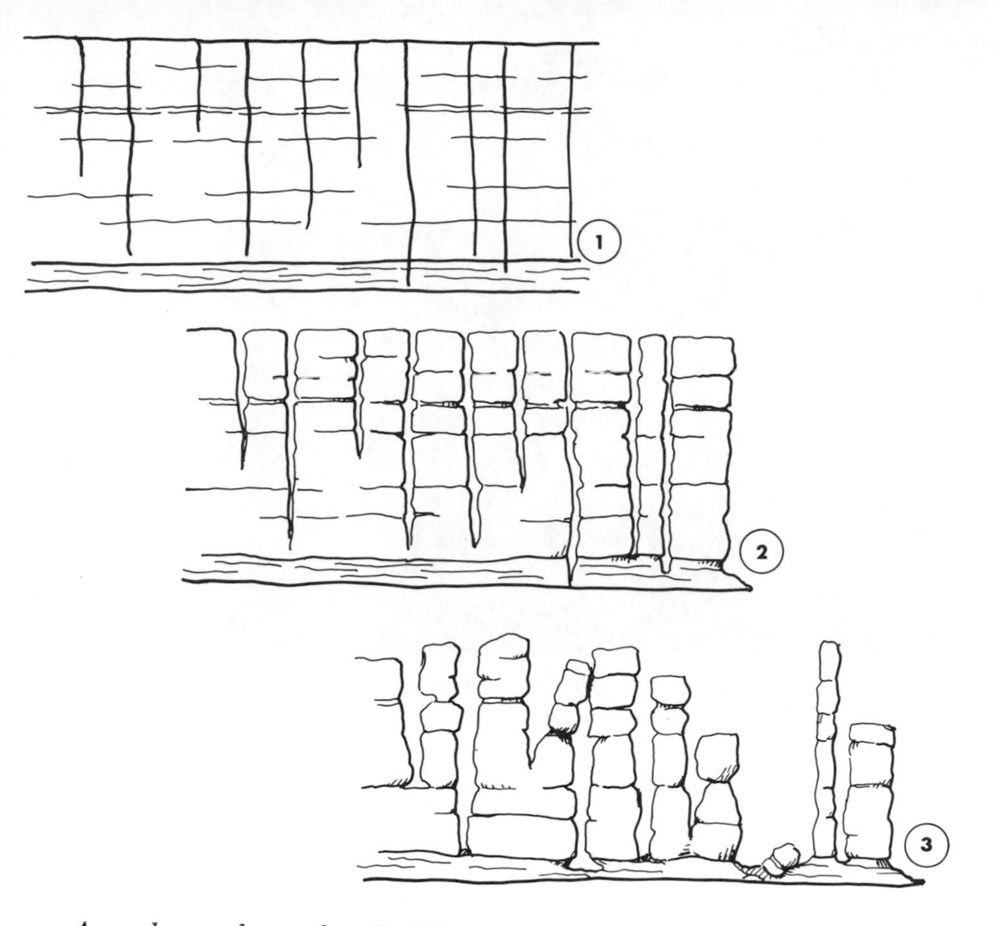

As rocks weather and erode, joints widen and alcoves form along zones of weakness. Bizarre columns, "totem poles," and balanced rocks result from continuing erosion.

deposited siltstone and sandstone, which apparently accumulated in small, short-lived lakes dammed by lava flows or perhaps by faulting. Of nearly the same age are much thicker units of sandstone and conglomerate containing fragments of volcanic rock, deposited along with volcanic ash or tuff and lava flows — the rocks exposed near and downstream from the visitor center. The conglomerates, deposited originally in alluvial fans, indicate the existence of sizeable volcanoes in this area in Oligocene time, probably 28 to 30 million years ago.

The weird spires and columns, the balanced rocks and hoodoos of Chiricahua Natiional Monument are eroded in yet younger rocks, products of fiery volcanic holocausts that about 25 million years ago flung gas-charged incandescent ash and molten pellets of pumice over this area. During the eruptions, which were many times greater

The rock spires of Chiricahua National Monument are carved in welded tuff expelled by an exploding volcano. Tad Nichols photo.

than any that historical man has seen, hot volcanic clouds expanding with hurricane force shot down mountain slopes and spread out over low surrounding areas, while others rose high into the sky in black billowing clouds like those emitted in 1980 by Mt. St. Helens, raining down fine debris as they spread and cooled. The ash flows that shot down-slope in incandescent avalanches came to rest while still extremely hot, and immediately fused into rock called **welded tuff**. The ash from high-rising clouds, cooler when it fell, did not fuse immediately.

Many of the tuff layers can be seen along the Bonita Canyon Road beyond the visitor center. They are also well exposed below Massai Point and along Echo Canyon, Rhyolite Canyon, and other park trails. They are composed of rhyolite, a silicic volcanic rock that, as magma, is stiff and sticky, likely to burst forth explosively rather than flowing smoothly as basalt lava does. Within the national monument eight individual ash flows or cooling units — produced by eight eruptions — have been recognized in the tuff, totaling nearly 2000 feet. Individual units vary in thickness from 2 to 880 feet, and each shows a definite and logical bottom-to-top sequence formed as an explosive eruption progressed: a thin layer of non-welded tuff formed from cool ash, overlain by welded tuff from hot, glowing avalanches, in turn overlain by more non-welded tuff that fell from ash clouds. The degree of welding, dependent on both the speed with

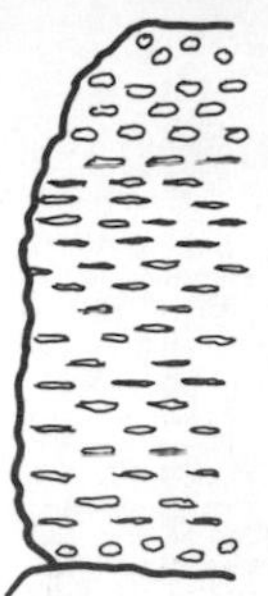

A cooling unit of volcanic tuff contains dense welded tuff sandwiched between layers of non-welded tuff. Thicknesses of cooling units vary.

which the material fell and cooled and the amount of pressure from overlying ash, is reflected in the shape of small lumps of pumice thrown out with the ash: lumps in very hot ash created at the height of the eruption were squashed and flattened by rapidly accumulating overburdens.

The tuff layers are capped with a single lava flow, of which only a remnant remains as the top of Sugarloaf Mountain. The fine-grained, dark gray lava contains scattered crystals of white feldspar, and in its upper part are numerous bubble holes, or **vesicles**, many of them filled with chalcedony.

Final carving of the armies of spires, totem poles, and other figures that march up Chiricahua's mountainsides came after Basin and Range mountain-building. Most of the tall columns are developed in a single 880-foot-thick cooling unit. Weathering and erosion concen-

Balanced rocks are created by differential erosion of soft and hard rock layers, often with the help of wind erosion.

Lichens now populate many rock surfaces in the Chiricahua Mountains. Their growth is measured in inches per century.

trated along intersecting sets of vertical joints, joints that developed partly by shrinking of the ash as it cooled, partly as a result of later earth movements. The spires are particularly abundant and well developed in canyons on the west side of the mountains, where the joints are best developed; the eastern slope bears few vertical fractures and as a result displays few columns.

Interesting details of weathering are revealed on these rocks. Even within single joint blocks the welded tuff does not weather uniformly, but erodes more easily along somewhat wavy, more or less horizontal layers that represent changes in grain size and degree of welding of the ash grains within the flowing ash clouds. In places the welded tuff appears more resistant at the surface than inside the rock — a result of surface hardening as moisture that soaked into the rock during rains absorbed minute quantities of rock minerals, which it later gave up near the surface as the rock dried out. Balanced rocks are shaped by wind erosion, which is greater near the surface where the wind uses blown sand grains as tools. **Volcanic hailstones**, which occur at Massai Point and along the Rhyolite Canyon trail, form as little mudballs in volcanic clouds of ash and rain, in the turbulent air currents that accompany volcanic eruptions. The greenish crust on the rocks is formed of lichens, primitive associations of algae and fungi.

Northwest of the national monument, visible from the rest of

Unsurpassed for beauty, Grand Canyon displays in its orderly rock layers a record of 2 billion years of earth history.

Bonita Canyon Road near Massai Point (as well as from I-10 near the New Mexico line) is the readily recognized landmark of Cochise Head, named for one of the southwest's most brilliant Indian warriors. It is formed of a thick layer of welded tuff older than that within the national monument, separated from the monument ridge crest by a deep valley eroded in faulted Precambrian and Paleozoic rocks.

Grand Canyon National Park

The Grand Canyon — Nature's masterpiece — is a textbook of geology laid open before our eyes. In its dark Inner Gorge are relics of ancient sedimentary and volcanic rocks crushed, broken, and distorted through time, intruded by magma, yet still able to tell us of early conditions on our planet, to hint of the origins of life, of towering mountain ranges, and of long millenia of erosion that beveled and smoothed whole continents and leveled ancient shores. The canyon's colorful upper walls — cliffs and slopes and ledges above the Inner Gorge — present orderly pages of Paleozoic sedimentary strata that tell a tale of seas and deltas and deserts, of unfolding life in a world not too unlike our own. And the mere presence of the canyon here, a mile deep, cutting across a 9000-foot plateau, challenges the reader of

In the dark walls of the Inner Gorge are Precambrian rocks that have been subjected to long eons of crushing, folding, and partial melting. Flat-lying Cambrian sandstone lies on their beveled upper surface.

the rocks with further history, the story of rising land, of faraway glaciers, and of a pirate river whose stolen riches far surpassed its own.

I've made no effort to present a map with this section. A full-color geologic map of eastern Grand Canyon (the ultimate in souvenirs!), published by the Grand Canyon Natural History Association and the Museum of Northern Arizona, is sold at visitor centers. It includes a number of cross sections keyed to the map.

The walls of Grand Canyon furnish an unequaled example of the premise on which much of geologic knowledge is based: In sedimentary sequences the oldest rocks are at the bottom, and those above them are sequentially younger. Here the oldest rocks are near the river — hard, crystalline rocks of the Vishnu Group. These dark schists, intruded by the Zoroaster Granite, stand on edge and are exposed along the river in almost all of eastern Grand Canyon — the part most visitors see. They originated more than 2 billion years ago as sedimentary rocks, sandstone and shale and limestone, interlayered with lava flows. They seem to have been deeply buried under perhaps as much as 12 vertical miles of other rock (a layer 12 times as thick as Grand Canyon is deep!), crushed and folded and even partly melted by heat and pressure. Then — erosion. All the 12 miles above were worn away, and the old rocks were smoothly beveled, before another round of sedimentation.

Below Desert View on the South Rim and Cape Royal on the North Rim, and in Bright Angel Canyon along the Kaibab Trail, are tilted wedges of younger Precambrian rocks, not as distorted, clearly showing the layered characteristics of sedimentary rock — conglomerate, sandstone, shale, and limestone, interlayered again with lava flows. Some of these rocks contain fossils, circular impressions that may be cysts of floating algae, as well as tracks, trails, and large cabbagelike "heads" called **stromatolites**, showing that organisms of various kinds did live in Precambrian seas.

The most easily recognized of these Precambrian strata are the bright orange-red Hakatai Shale and the dark Cardenas Lavas. The lava flows have been dated radiometrically as about 1 billion years old. The entire sequence, of course, was many million years in the making. Then, these layered rocks were neatly broken and tilted along north-south faults, creating a basin and range topography not too unlike that of western and southern Arizona. And finally, along with the older Precambrian crystalline rocks, they were worn down, except for a series of hard quartzite ridges, during a tremendously long period of erosion at the close of Precambrian time. The resulting Great Unconformity can be traced the length of eastern Grand Canyon, wherever Precambrian rocks are exposed. The nearly horizontal surface between these Precambrian rocks and the overlying Paleozoic rocks has been described as "one of the most striking exposures of an unconformity in the world" — an outcrop hundreds of miles long, winding in and out of the scalloped walls of the canyon and its tributaries.

Above all these old rocks are flat-lying Paleozoic sedimentary rocks — Cambrian, Devonian, Mississippian, Pennsylvanian, and Permian. Ordovician and Silurian are missing. The successive layers are shown in the accompanying diagram and on the geologic map, and are easily recognized from either rim by their color and cliff-forming or slope-forming tendencies. This is the story they tell:

About 570 million years ago the long erosion ended, and a Cambrian sea crept from west to east across the flattened land. Coarse near-shore Tapeats Sandstone, then greenish Bright Angel Shale, and finally the Muav Limestone were deposited in the deepening sea, where small marine animals grew and crawled and swam. Sponges, brachiopods, mollusks, corals, crinoids, and trilobites can be found and provide a record of flourishing life in a tropical sea.

Then the sea backed away and left the land to be scoured by erosion through Ordovician, Silurian, and part of Devonian time. Toward the

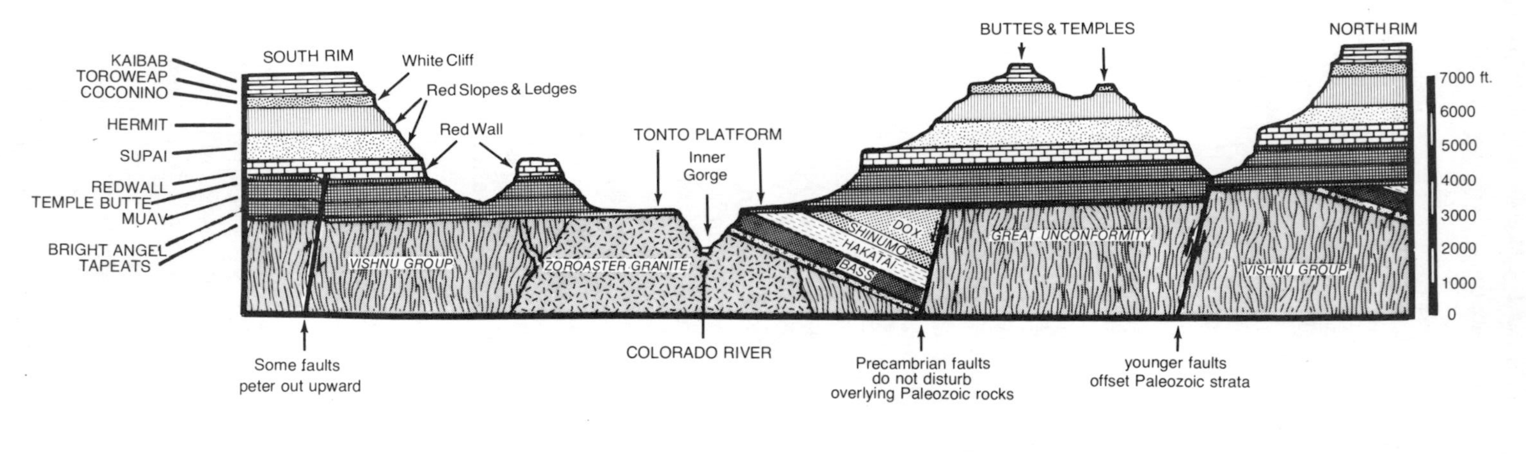

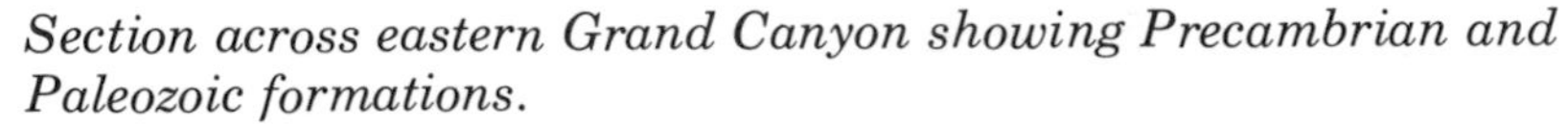

Section across eastern Grand Canyon showing Precambrian and Paleozoic formations.

end of the Devonian Period a little brackish-water or freshwater limestone — the Temple Butte Limestone — was deposited on a channeled surface. This is a thin layer, not everywhere present in the canyon, but where it occurs it contains a few fossils, including some strange armored Devonian fishes. Above it, and representing Mississippian time 365 to 330 million years ago, is the Redwall Limestone, a thick, fairly uniform gray limestone, stained red on the surface, that extends with other names over much of western North America. It displays a varied assemblage of fossils, especially of marine shellfish.

The top of the Redwall Limestone is marked by karst erosion, typical of many limestone areas today, particularly those in warm, humid regions such as Kentucky and Puerto Rico. Sinks, caves, and underground channels were dissolved in the limestone, and deep ravines and stream valleys developed as caverns collapsed. Many of the depressions filled with clayey red-orange soil similar to that known in the tropics today.

In Pennsylvanian time this irregular surface subsided, at times, to form a large bay in which repeated layers of silt, sand, and mud were deposited. These "redbeds", now the Supai Group and above it the Hermit Shale, make up the red slope-cliff-slope sequence above the Redwall Limestone. Although most of the red sandstones and shales do not contain fossils, in western Grand Canyon some limestones interlayered with them contain brachiopods, mollusks, trilobites, and other diagnostic fossils from which the age of the rock has been determined — Pennsylvanian for most of the Supai Group, Permian for uppermost Supai and for the Hermit Shale, which also contains reptile tracks and imprints of ferns.

Later in Permian time, a desert climate prevailed and wind-blown sand swept the Grand Canyon region, leaving a thick deposit of fine yellow-white sandstone marked with long, steep cross-bedding of

As seen from Desert View, tilted Precambrian sedimentary rocks are beveled and overlain by flat-lying Paleozoic strata.

Islands of Precambrian rock jut through the lowest Cambrian layers.
N. W. Carkhuff photo, courtesy of USGS

one-time sand dunes. The dune sands now form the white ribbon of Coconino Sandstone high up on the Grand Canyon's walls. Tracks and trails of insects and lizardlike reptiles mark some sloping dune surfaces.

Twice during the middle part of the Permian Period the western sea inundated the region again, leaving behind it the tan and buff limestones of the Toroweap and Kaibab Formations. Each of these rock units records a shallow sea that fairly teemed with life: brachiopods, trilobites, snails and clams, corals, and other animals that crept or swam or drifted with the currents. Many of the hard chert nodules so common in the formation developed around the silica skeletons of sponges.

At the end of Permian deposition the sea again retreated westward. All the younger sedimentary rocks deposited here since the end of the Paleozoic Era — thousands of feet of them — have since been washed off the Kaibab and Coconino Plateaus. They do remain east and southeast and north of the Grand Canyon: in the Painted Desert, Echo Cliffs, the Navajo country of northeastern Arizona, and southern Utah.

Many nearly vertical normal faults cut across Grand Canyon, and some of them can be seen easily from the rim. Just west of Grand Canyon Village, for instance, the rim is higher than at the village, lifted along Bright Angel Fault. The Bright Angel Trail zigzags through broken rock along the fault, one of the few places where a trail can scale the Coconino and Redwall cliffs, 300 and 500 feet high. The same fault extends across the canyon, controlling the long, straight course of Bright Angel Creek and offsetting the North Rim, too.

This brings us to the big "How come?" "How come" a river (the Colorado) cut a canyon (Grand Canyon) across a mountain (the Kaibab Upwarp)? Geologists have bandied this question about ever since explorer-geologist John Wesley Powell led a daring boat trip through the unexplored canyon in 1869. As more and more evidence comes to light, an interesting history has developed, a history that starts with the Ancestral Colorado River of mid-Tertiary time flowing south into Arizona from Utah, as it does now, and then continuing southward along what is now the valley of the Little Colorado River, staying, in other words, *east* of the new-formed Kaibab Upwarp. The river at that time may have flowed into a large lake, called Hopi Lake or Lake Bidahochi, in eastern Arizona.

Meantime another drainage system had developed *west* of the Kaibab Upwarp. One stream of this drainage system cut its way headward into the Kaibab Upwarp, much as Bright Angel Creek and other tributary streams are doing now. As its headwaters cut farther and farther east into the upwarp, the divide between the two drainages narrowed. Finally, erosion broke through the divide that separated the drainages. Relatively suddenly, the small "pirate" stream captured the waters of the mighty Colorado, which turned westward through this newer, steeper route.

From there on, erosion simply went to town. The new Colorado River, augmented by vastly increased rainfall of the Ice Ages and meltwater from glaciers in the Rockies, stripped away remaining Mesozoic sedimentary rocks, broke through the hard Kaibab Formation where the western drainage had not done so, and chiseled out the Grand Canyon as we see it today. The former channel of the old southeast-flowing Ancestral Colorado became the route of the northwest-flowing Little Colorado River, which cut for itself a sheer-walled gorge spectacular in its own right.

Erosion still goes on, but at a slower pace. Before Glen Canyon Dam was built, it was not unusual to see the untamed waters roiled and furious, and to hear — even from the rim — the booming thunder of rock against rock. But the Inner Gorge in which the river is now

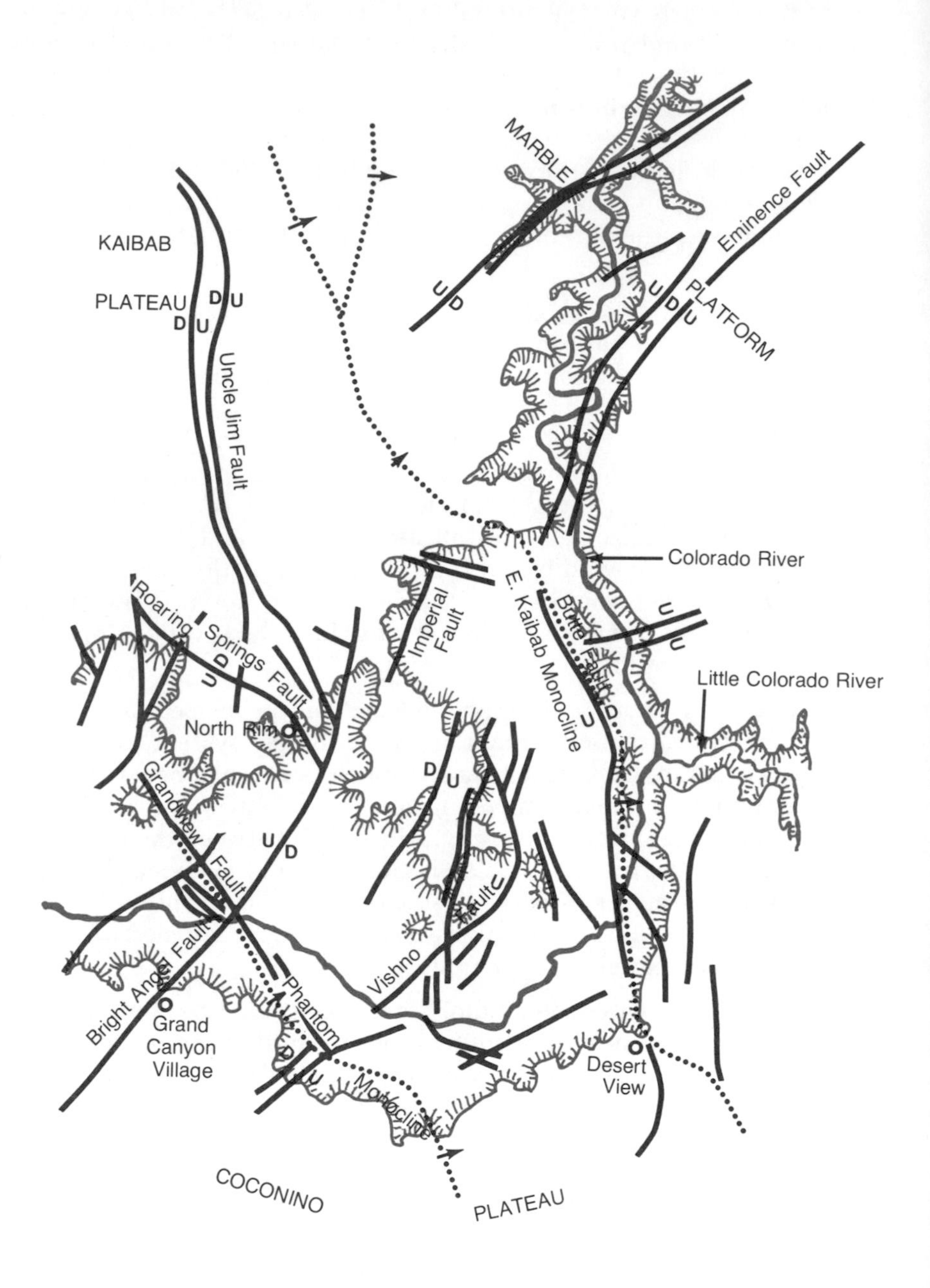

Faults of eastern Grand Canyon show a NW-SE trend superimposed on a NE-SW trend inherited from Precambrian time. In Grand Canyon these faults can be studied in three-dimensional detail.

confined is a fortress of hard crystalline rocks that do not yield as easily as the sedimentary rocks of the upper canyon walls, so even without upstream dams the pace of downward erosion has slowed. The canyon is widening — here a rockfall or a muddy, rain-fed cascade, there a flooded tributary carrying gravel and sand to the river in the endless battle between uplift and erosion.

Montezuma Castle National Monument

Prehistoric cliff dwellings cluster in a deep natural recess in horizontal layers of chalky limestone, sandstone, and volcanic ash of the Verde Formation. This rock unit was deposited in a series of shallow, probably intermittent lakes backed up, between 3 and 6 million years ago, by rise of the Central Arizona Highlands. The lake sediments spread along the Verde Valley for about 40 miles, probably the maximum extent of the lake, and are known to contain fossil mollusks and bones of many mammals. In places they contain salt, suggesting that the lakes in which they formed may have been as short-lived as those that occasionally cover the flat white playas of some of today's desert basins.

Montezuma Castle overlooks Beaver Creek, a small but permanent tributary of the Verde River. The stream draws part of its sustenance from Montezuma Well, a circular sinkhole some 400 feet across and about 135 feet deep, 6 miles upstream from Montezuma Castle. Groundwater in limestone regions, slightly acid from exposure to atmosphere and soil, slowly dissolves passages through the limestone. With changes in water level some of the underground solution caverns may collapse, leaving sinkholes such as this one, usually circular in plan, some with deep pools. Groundwater still enters and leaves Montezuma Well through submerged channels in the rubble that must lie beneath the water surface. Near Beaver Creek, whose downcutting caused the lowering of the water table, and thus collapse of the sinkhole, it gushes from rock crevices at a rate of more than 1000 gallons per minute. This reliable and generous water supply was used by prehistoric Sinagua people (emphatically *not* by Montezuma!), who cut channels near both Montezuma Castle and Montezuma Well to convey the water to their fields. The name **Sinagua** comes from the Spanish *sin agua*, without water; obviously Spanish explorers did not know about this water supply! Because the water is richly charged with calcium carbonate from its passage through limestones of the Verde Formation, it deposits calcium along the

Montezuma Castle is well concealed, surrounded by limy lakebeds of the Verde Formation.

edges of the ancient Sinagua ditches, automatically waterproofing the stone-lined channels with a natural cement.

The setting of this national monument, with the Black Hills rising like a fortress to the southwest and the lava-capped Coconino Plateau to the northeast, is discussed under I-17 Camp Verde — Flagstaff, in Chapter IV.

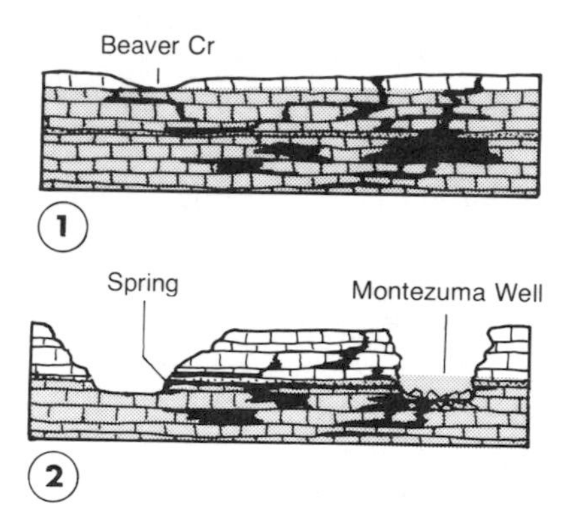

1. At Montezuma Well, caverns were dissolved in limestone in the saturated zone (shaded) below the water table. The water itself helped to support the cavern ceiling.
2. As Beaver Creek deepened its valley the water table dropped. Left unsupported, a cavern ceiling collapsed, creating a sinkhole that still drains through subterranean channelways to feed the large spring near Beaver Creek.

Monument Valley Tribal Park

Highway US 163 going north from Kayenta climbs through the tilted rocks of Comb Ridge, slipping between the "teeth" of the comb — pink, cross-bedded Navajo Sandstone. Beyond it are successively older rocks of the Glen Canyon Group. The Wingate Formation at the base of the group is prominent here, forming a single pinnacle with a Chinle Formation base at mile 398, and the cliffs of other plateaus and buttes to the west. Though also composed of dune sand, the Wingate Formation weathers into angular vertical cliffs that do not round off like those of the Navajo Sandstone.

Below the Wingate the Triassic Chinle Formation, of Painted Desert fame, splashes the slopes with color. Its layers of mudstone and volcanic ash weather into badlands with crusty surfaces creased by rill marks. At mile 403 there is an abrupt change to sandstone and conglomerate of the Shinarump Conglomerate, lowest part of the Chinle Formation. This unit forms a hard, sloping surface covered with pebbles concentrated as wind and water winnow the finer sand and silt particles. Wind is responsible for much of the erosion here.

Cross-bedded Navajo Sandstone occurs at roadside level near highway US 163. Irregular, distorted bedding seen in parts of the formation show where sand avalanched down steep lee faces of sand dunes. Tad Nichols photo.

The striking monuments of Monument Valley are created as soft shales of the Cutler Formation erode away, leaving massive, vertically jointed slabs of De Chelly Sandstone without support. Both formations are Permian. Tad Nichols photo.

East of the highway in this area are the Navajo Buttes, old volcanic conduits that now jut up as dark, resistant, rough-hewn volcanic necks. Agathla Peak, 1225 feet high, is the largest of the necks. They and the dikes that radiate from them are of Tertiary age, probably 15 to 20 million years old. The volcanoes once associated with them are, of course, completely gone now — eroded away.

Notice how quickly the sedimentary rock layers, steeply tilted at Comb Ridge, level out northward. It is this abrupt change that leads geologists to believe that the Comb Monocline lies above a fault that breaks and offsets Precambrian rocks far below, but bends the more supple sedimentary layers that overlie them.

The Shinarump Conglomerate is separated by a thin layer of red Moenkopi shale from the De Chelly Sandstone, the monument-forming formation of Monument Valley. These three units — the Shinarump, Moenkopi, and De Chelly — can be seen at roadside between mileposts 409 and 412. Near the highway they may be darkened with desert pavement.

The upwarp that brings Permian rock — the De Chelly Sandstone — to the surface here in this part of Arizona, where Jurassic and Cretaceous rocks predominate, is an oval-shaped dome that extends from US 160 northward about 125 miles, to Canyonlands National Park in Utah. The south and east flanks of the uplift are sharply defined by Comb Ridge; the north and west edges are less distinct.

Dramatic spires of Monument Valley, islands in a desert sea, soon come into view. Sweeping **eolian** cross-bedding of the De Chelly Sandstone can be seen on the flanks of the monuments themselves. The towers form as wind and rain etch their soft mudstone and siltstone bases, leaving unsupported the massive, vertically fractured rock slabs above.

There are modern dunes in the valley also. Much of their sand is derived directly from the Permian dune sandstone; some comes also from the Jurassic dunes of the Navajo Sandstone. Within some of the modern dunes buried soil layers indicate periods when the sand stabilized and plants were able to grow. The soil dates back into Pleistocene time, when major climate changes brought increased precipitation to these deserts. For much of its history, from Permian time to the present, this region has alternately been wet and dry, floodplain and sand-swept desert.

Navajo National Monument

Hidden in a high-arched recess, accessible only through a narrow, spring-fed canyon, the ruins at Betatakin went unnoticed until 1909. This national monument demonstrates well the nature of the Navajo Sandstone — its pale salmon color, its wide-sweeping cross-bedding inherited from ancient sand dunes, its horizontal lenses of pink clay and light gray freshwater limestone, and its usefulness as an aquifer even in prehistoric time.

The entrance drive to the national monument climbs Organ Rock Monocline, rising with the Navajo Sandstone onto the Shonto Plateau. Much of the surface is completely barren. Sloping rocks, marked by arcs and swirls of cross-bedding, are fully exposed to view. Along joints that parallel the rock surfaces, the thin sheets of sandstone spall off to give the rounded surfaces so characteristic of this formation. The rocks also show several intersecting sets of vertical joints, as one might expect, products of strains and stresses of 7000 feet of uplift. Deeply incised vertical joints serve as watercourses down which tiny streams flow during rainstorms. Many of these incisions are marked by rows of miniature pools that fill with rainwater, pools in which small plants and animals, many of them microscopic, live out their life cycles in a few hours or days. Since plant and animal by-products include acids that dissolve the limy cement hold-

Betatakin Ruin clings to a rock recess walled and roofed with Navajo Sandstone. Springs emerge at the base of this porous sandstone, nourishing the greenery of the valley floor.

ing the sand grains together, with every pool-forming cycle the little basins grow larger. Lichens attached to dry rock surfaces secrete acids as well, loosening grains that are blown away by wind. The "soil" that surrounds the barren patches is hardly more than wind-blown sand held together by roots of trees, shrubs, and grass.

Betatakin Canyon, visible from the visitor center, again exposes this salmon-colored rock. The canyon is a tributary to Tsegi Wash, which in turn flows into Laguna Wash near Highway US 160. Ultimately these streams drain into the San Juan River at Mexican Hat.

Sandal Trail, starting near the visitor center, provides views of Betatakin Canyon and its remarkable ruins. The ruins occupy a cave nearly 500 feet high, formed as the Navajo Sandstone broke away in arching slabs where it was undermined by erosion of less resistant rocks (the Kayenta Formation) below. Near the overhang the Kayenta Formation is pretty well hidden by fallen rock and tangled vegetation, but it can be seen clearly farther down the canyon. In the floor of the overhang, at the Navajo-Kayenta interface, greenery surrounds a small spring. Since the Navajo Sandstone is porous and permeable and the Kayenta Formation is not, water sinking into the sandstone on the Shonto Plateau moves laterally at this interface, to appear in springs along the canyon walls. The presence of the water here almost certainly was a factor in localizing the excavation process. It was also of course a factor in making this site attractive to prehistoric peoples.

Near the visitor center the highest hillocks are capped with thin beds of freshwater limestone probably deposited in ephemeral ponds in interdune areas. A similar limetone layer, light gray and finely veined with red, and several horizontal beds of silty red claystone are intercepted by the trail leading down to Betatakin. Many small alcoves can be seen along the switchbacking, stairstepped trail, their sandy floors marked by tracks and strewn with dropping of small animals. Some alcoves have arched ceilings like that at Betatakin; others display sloping ceilings regulated by the cross-bedding of the sandstone. Springs in the floor of the canyon, again at the contact of the Navajo and Kayenta Formations, nourish a lush growth of aspen, box elder, Douglas fir, scrub oak, and other plants. Black, gray, and green lichens mat rock surfaces. In places the blue-purple stains of desert varnish darken the canyon walls.

Navajo National Monument includes two other cave ruins, Keet Seel — the largest cliff-dwelling in America — and Inscription House, but since they are not readily accessible by road they are not discussed here. They share many geologic features with Betatakin.

In the Ajo Mountains, irregular lava flows alternate with bands of light-colored volcanic ash, now tuff.

Organ Pipe Cactus National Monument

Well off the beaten track, this national monument offers a glimpse into desert processes, desert landforms, and the varied and eventful Tertiary history of this part of southern Arizona. Two loop drives give easy access to geologic features, and several good trails invite exploration on foot.

As elsewhere in Southern Arizona, steep, linear mountain ranges here are separated by sloping desert plains. Mountain pediments grade smoothly into the surface of the sloping valley fill. Both pediments and valley fill wear armor of desert pavement that partially protects the desert soils from ravages of wind and rain.

All the mountains in the national monument are fault-block ranges, but they differ in topography because of differences in type and age of the rocks. They can be classed into four groups:

• Flat-topped, cliff-edged mesas topped with Quaternary basalt lava flows, as in the Bates Mountains of the northwest part of the monument, which are visible about 10 miles west of Arizona 85 as it enters the monument from the north. Slight faulting and tilting of the basalt lavas shows that mountain-building forces were active here in comparatively recent time.

• Quite rugged, deeply eroded ranges of Tertiary volcanic rocks, with tilted layers of lava, tuff, and breccia faulted upward, as in the Ajo Range and on the northwest slope of the Puerto Blanco Mountains.

• Rounded hills of Mesozoic granite, such as those near Senita Basin in the southern Puerto Blanco Mountains.

• Rougher hills of light-colored Mesozoic metamorphic rock — gneiss and schist — as in the rugged central part of the Puerto Blanco Mountains.

Of the rock types exposed in these ranges, the gneiss and schist are the oldest, the basalt lava flows the youngest. Mesozoic granite intruded the gneiss and schist. Volcanism came early enough in Tertiary time that most Tertiary volcanic rocks were bent and broken during mid-Tertiary mountain-building and later disrupted again by Basin and Range faulting.

As block faulting subsided, erosion took over, erosion that may have become particularly severe in Pleistocene time when rainfall was much greater than today's frugal 4 to 14 inches. As the mountains were stripped back, the basins filled with their debris. Eventually a thin veneer of gravel was spread across the mountain pediments to merge smoothly with the valley gravels and to hide the faults that separate mountain and basin. Torrential summer rains continue to whittle at the mountains and to deposit debris on the wide slopes below them.

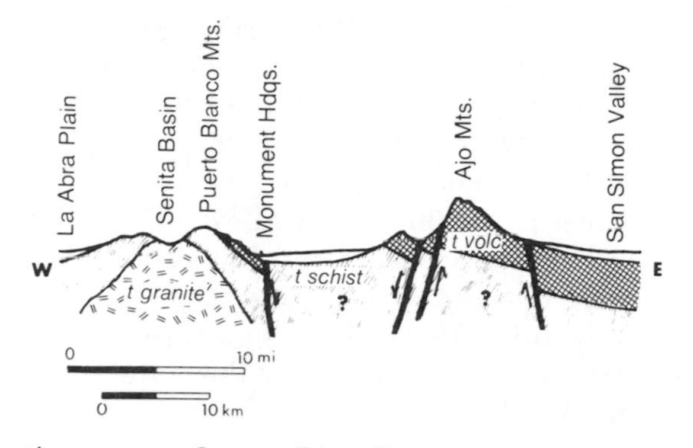

Section across Organ Pipe Cactus National Monument.

Ajo Mountain Drive (a guide leaflet details geologic and biologic features) crosses the south-draining part of the alluvial apron on the west side of the Ajo Range, a slope which stretches north to Arizona Highway 86 and south into Mexico. About 80 per cent of the drainage on this side of the Ajo Range funnels northwestward toward steep-walled Growler Canyon between the Bates and Growler Mountains. The other 20 per cent drains southwest toward the Sonoyta Valley in Mexico.

The drive winds through Tertiary volcanic rocks of some little fault block hills, crossing typical dry desert washes and broad expanses of desert pavement as it approaches the larger fault block of the Ajo Range. Note the concentration of vegetation along the washes. Even after summer downpours, which may cause flash floods along the washes, water sinks rapidly into the loose gravel and sand.

The dark bastion of the Ajo Range reveals irregular layers of brown lava alternating with tan or light yellow tuff, in a mountain face that exposes the cut edges of faulted, east-tilting layers. A number of easily recognized dikes cut across the face of the range. The faulted block is now deeply eroded, with stark cliffs, deep canyons, and turreted mountain ridges. Trails lead into several canyons and to a natural arch that is also visible from the road.

A longer loop, Puerto Blanco Drive, circles the Puerto Blanco Mountains. From the first part of this drive there are good views of the Ajo Range and its pediment. The washes that radiate from the range converge at the west side of the valley to flow through Growler Canyon between the Bates Mountains and the Growler Mountains farther north.

On along the road, well to the north, are views of lava-capped mesas of the Bates Mountains. The uppermost lavas are Quaternary and have suffered only slightly from faulting and tilting. West of the drainage divide between the Bates Mountains and the Puerto Blancos to the south, look north into the heart of the Bates Mountains, with their somber purple-brown lava flows. The deep, fairly straight valley up which one looks lies along a major north-south fault. Both Kino Peak in the Bates Mountains and the Cipriano Hills northwest of the loop route are on the uplifted side of this fault, and their now eroded fault faces reveal layers of lava flows and volcanic ash similar to those in the Ajo Range.

In the Puerto Blanco Mountains, on the other hand, rocks are lighter in color: gneiss, schist, and granite, all Tertiary in age. In places they contain **pegmatite** dikes with large crystals of quartz, feldspar, and mica. At Golden Bell Mine they are enriched with mineralized quartz veins.

On the west side of the Puerto Blancos a bedrock pediment extends well west of the base of the range. Along here, the road crosses one wash after another, their sandy floors favorite sunning places for rattlesnakes. All drain southwest toward Aguajita Wash. Ground-

Cracked by heat of sun and cold of nigh a single cobble continues the disintegration that will end in soil.

water seems to be near the surface along Aguajita Wash, where vegetation is denser than elsewhere. The stream lies along the trend of the fault described above. Differences in permeability of rocks on opposite sides of the fault govern the position of the wash, the flow of groundwater, and the position of Aguajita Spring near the Mexican border. Quitobaquito Springs, half a mile west, lie along another fault. There, water bubbles from terraced hillsides to collect in ponds constructed years ago for use by cattle and now maintained as a watering place for wild animals and birds. Domes of white, powdery **siliceous sinter** surround the hillside springs; they were deposited over many thousands of years as groundwater dissolved mineral matter, in this case silica, SiO_2, from surrounding rocks and redeposited it at the surface.

Agricultural groundwater use is increasing in Sonoyta Valley in Mexico, and there is some fear that increased pumping from wells there will lower the water table, destroying the flow at Quitobaquita Springs.

Petrified Forest National Park

The Petrified Forest is not really a forest at all, but a region where in Triassic time tree trunks rafted by flooded streams were buried quickly with stream sediments and volcanic ash, and later impregnated with silica and preserved. The great logs are almost all lying horizontally. Most are battered, with limbs and roots broken off and bark stripped away. The fossil logs occur throughout the Chinle Formation of the Painted Desert, but are most abundant within the area of the national park.

Volcanic ash is made of tiny bubbly particles of silica glass, and it is this characteristic that preserved the trees so remarkably. Because the bubbles and the small size of the particles provided large surface areas relative to volume, groundwater seeping through the ash soon became highly charged with dissolved silica. Some of this mineral material came out of solution when it met up with organic material such as the wood of buried tree trunks, so that it gradually replaced the wood, cell for cell, in a type of fossilization called **silicification**. The process is fairly common, occurring also with buried animal bones, and silicified wood is found the world over. But here in the Painted Desert the wood is exceptionally well preserved, and small amounts of other minerals create decorator colors. Since silicified wood is much harder than the volcanic ash in which it occurs, tree trunks commonly weather out and are now exposed above ground.

Logs of the Petrified Forest may have been uprooted by catastrophic volcanic explosions and transported by mudflows similar to those that followed the Mt. St. Helens explosion in 1980.

The saga of the Petrified Forest and Painted Desert began in Triassic time, 240 to 205 million years ago, in marshes and channels of a broad floodplain. In high mountain ranges that had risen in central Arizona, tall forests grew, with pine-like trees related to the araucarian pines of the modern Southern Hemisphere — tall and gracefully symmetrical. On floodplains below the mountains, shifting streams and marshes surrounded a scattering of tree-covered "islands." Occasionally, volcanoes well to windward belched and exploded, releasing clouds of fine volcanic ash. Trees toppled or were flattened by explosive shock waves, and as the torrential rains that often accompany volcanic cataclysms sent mudflows and floods surging down the mountain flanks fallen trees were washed out onto the plain.

Quickly buried where they came to rest, the tree trunks were gradually silicified, their pore spaces filled with silica, their woody tissues being replaced so slowly that the basic cellular structure of the wood can still be recognized. With iron and manganese the percolating groundwater painted the petrified wood with hues of yellow and white, blue, red, and black. In some cavities, quartz crystals grew.

The Painted Desert of today is a far cry from the lush, watery floodplain of Triassic time. The present desert exists not just because of a shortage of rain, but also because the soft, poorly consolidated silt and volcanic ash of the Chinle Formation erodes extremely easily. The ash has gradually altered into bentonite, a type of clay that swells when it is wet into gummy, tacky mud, but when dry is cracked and puffy, readily slaking into dust. Erosion by wind and water is too fast to permit plants to take hold, and no soil develops. No roots hold the loose sediment; no leafy canopies protect it. In places, it is true, the surface is covered with a desert pavement of pebbles winnowed by wind from conglomerate layers above the Chinle Formation. (In some areas, the winnowed pebbles are fragments of petrified wood.) But elsewhere the barren surfaces are fully exposed to the vagaries of wind and weather.

The petrified logs of this region are unusually large, abundant, and well preserved. And they are unusually colorful — truly agatized rather than just silicified. As a result, rock and mineral collectors find them exceptionally attractive. The park was established with the express purpose of protecting them, so rules against collecting are strictly enforced. Specimens for sale in curio shops come from outside the national park.

Other Triassic fossils occur in the Chinle Formation: more than 40 other types of plant fossils, from fern fronds to leaves of deciduous trees; vertebrate fossils such as armored amphibians and crocodile-like phytosaurs; and fossil dragonflies as large as swallows. Some of these fossils are displayed at the Rainbow Forest Visitor Center.

The drive through the park will take you right among the blue, lavender, and red badlands of the Painted Desert, where there are petrified logs aplenty, as well as examples of desert pavement, desert varnish (with Indian petroglyphs), small shield volcanoes, cross-bedded sandstone, and a great many fascinating badland erosional shapes and forms. Short trails lead to points of interest, all well explained with trailside signs or leaflets.

Turned to agate by slow infiltration of silica, the well preserved fossil wood is some of the most colorful in the world.

Saguaro National Monument

The Rincon Mountain Unit of Saguaro National Monument lies east of Tucson at the foot of the Rincon Mountains, part of an immense 3-humped metamorphic core complex. The granite dome of these mountains is surfaced with a metamorphic shell or carapace which has had its atomic calendar reset to a Tertiary age. Tanque Verde Ridge, plunging down the Rincon flank toward the monument headquarters, extends westward like an anticline from the three-humped dome.

Along Cactus Forest Loop Drive, many of the metamorphic carapace rocks are exposed. One can easily recognize silvery, silky schist or **phyllite** made up almost entirely of parallel mica grains, and strongly and intricately banded light and dark gray gneiss. Paleozoic sedimentary rocks along the flank of the metamorphic complex appear as a ridge encircled by the southern part of the loop drive and as another small mass southwest of the picnic area. The layers of Paleozoic limestone are torn and somewhat crumpled from their long-ago movement. Much larger hills of these sedimentary rock layers, with clearcut folding, appear near Colossal Cave, which you may want to visit while you are nearby.

Along the loop drive and most park trails there are many good exposures of both gneiss and schist. In some places the schist appears rusty gray, the rust or iron oxide forming from iron-bearing biotite or black mica. In a few places both gneiss and schist are cut by dark gray diabase dikes; almost everywhere the rock is finely laced with white quartz veins and veinlets. Gneiss, granite, schist, and quartz veins are intermixed as if they were taffy swirled around in some giant melting pot, making it evident that the immense heat and pressure that created the metamorphic core complex were great enough to partially melt the outer part of the granite core.

Sand and gravel made up of fragments of gneiss, schist, and granite surround the Rincon Mountains. In the pebbles of desert pavement and in the rounded cobbles of dry washes are found most of the rock types of the Rincons: granite, gneiss, dark fine-grained diabase, silvery schist, and occasional bits of sedimentary rocks. Close to the mountains several dry washes have cut down into the mountain pediment to lay bare the gneiss and schist of the carapace.

The unit of the national monument west of Tucson reveals an equally complicated but quite different geologic picture. A small range, the Tucson Mountains consist of a faulted, east-tilted wedge of Paleozoic and Mesozoic sedimentary rocks overlain and in part concealed by Tertiary volcanic rocks.

Sloping lava flows edge parts of the Tucson Mountains. Foreground buildings are a now-abandoned mine. P.B. King photo, courtesy of USGS.

The red-brown layers near the visitor center and less colorful layers above them are Cretaceous floodplain, delta, and marine deposits. Although well exposed here, farther east in the Tucson Mountains they are covered with an enigmatic rock unit that is a medley of large blocks of Precambrian schist, Paleozoic limestone, Mesozoic sedimentary rock, and Tertiary volcanic rock, all tossed together in a matrix of either sandstone or tuff. For obvious reasons the unit is called the Tucson Mountain Chaos. Its exact mode of origin is not known, but it may be related to the volcanic rocks immediately above it.

Overlying these units, forming the east slope and some of the highest parts of the Tucson Mountains, are irregular layers of light-colored tuff stained brown by desert varnish. This tuff is firmly welded like the rocks that result from explosive volcanic eruptions that shoot out clouds of volcanic ash mixed with very hot, very rapidly expanding volcanic gases. Propelled horizontally or down-slope with something like hurricane force and speed, the ash clouds maintain the internal heat that welds the ash particles together as they reach the ground. Similar rock occurs in Chiricahua and Organ Pipe Cactus National Monuments and in several other southern Arizona ranges. Welded tuff in the Tucson Mountains has been dated at 70 to 60 million years old. Several lava flows occur within the national monument as well.

A coarse-grained, light-colored granite intrusion completes the geologic picture here. It forms both Wasson and Amole Peaks and the westward-jutting hills below them.

Several lime kilns built by Spanish settlers still exist within the monument area. Lime was obtained by roasting limestone dug from nearby slivers of Paleozoic and Mesozoic sedimentary rocks. Because copper, silver, and other metallic ores commonly occur at contacts between shallow igneous intrusions, like the Tertiary stock described above, and limestones such as this, the monument area as well as other parts of the Tucson Mountains are dotted with prospect holes and old mining claims.

From various places in the western unit one can look out over the alluvial fans and bajada of the Tucson Mountains into Avra Valley, a typical broad desert basin. In the distance to the southwest is sharp-pointed Baboquivari Peak, in the Baboquivari Mountains, a metamorphic core complex range. At their north end these mountains merge with the Quinlan Range, easily identified by the white towers of Kitt Peak National Observatory. To the northwest are the Silver Bell Mountains, with tailings from an open pit copper mine at their base. Just to the right of the Silver Bells, and farther away, is sharp-spired Picacho Peak, part of an old volcano sliced in two by Basin and Range faulting. Closer at hand, Twin Peaks, two small hills of steeply tilted Paleozoic sedimentary rock, are now being quarried for limestone to be made into cement.

Sunset Crater and Wupatki National Monuments

A classic cinder cone only 900 years old, Sunset Crater is one of the newest, youngest geologic features in Arizona. The sunset tints of red and yellow that give the crater its name are due to oxidation by steaming hot gases, apparently enduring for some time after other activity ceased.

A blanket of cinders surrounds the crater and extends far east and north, covering an area of about 120 square miles. Be sure to feel and look closely at a handful of this lightweight, frothy rock. The bubble-filled material has not weathered significantly in this cool, dry, climate — note its black color and its crunchiness underfoot. Rain and snowmelt sink right in without gullying or otherwise remodeling the surface.

Bonita Flow, western of two lava flows associated with this cinder cone, is still fresh and dark, as if it oozed only yesterday from fissures at the base of the cone. But if you look closely you will see that the

Sunset Crater is about 1000 feet high. The lopsidedness of its rim was caused by leeward drift of cinders during its eruption.
Tad Nichols photo.

rough, jagged lava surfaces are spotted with lichens, pioneers of the plant world. Already grass and shrubs are moving in from the edges of the flow. In places, trees have taken root in cinder patches enriched by fallen leaves and winter-killed grass.

The lava shows a variety of flow patterns. Aa lava, rough and quite impossible to walk on, is the most common. In places the lava is blocky, with more or less rectangular chunks. The rock itself is very fine-grained basalt; scattered through it are individual crystals of green olivine and white or glassy feldspar. In places it contains fragments of other rock torn from the conduit walls. A self-guiding trail near the base of Sunset Crater leads past squeeze-ups, lava tunnels, spatter cones, and lava blisters.

Drive or walk to the top of O'Leary Peak for a view of Sunset Crater and its lava flows. O'Leary Peak itself, outside the national monument, is another expression of volcanism — a volcanic dome much like those that have developed in Mt. St. Helen's crater. The vista also includes San Francisco Mountain and its Inner Basin, which seems to have been roughed out by explosion or collapse and then smoothed by glacial erosion.

Sunset Crater is one of 400 or so cinder cones formed late in the history of the San Francisco Volcanic Field, long after San Francisco Mountain itself had taken shape. It is the most recent expression of volcanic activity in this field, or for that matter in southwestern United States. To reconstruct its birth and growth we can draw on eye-witness reports of a modern cinder cone's birth and growth. Paricutin erupted in central Mexico in 1943, born before the as-

Squeeze-ups *form as gummy, partly cooled lava squeezes up through cracks in a hardened lava surface.*
Tad Nichols photo

tonished eyes of a Mexican farmer as a tiny fissure in his cornfield began to emit strange sounds and thin wisps of steam. Then the fissure belched bits of spongy cinders — and more, and more — slowly building a miniature volcanic mountain. Soon, cinders and ash and spinning football-sized bombs of hot lava rained down on cornfield, house, and surrounding country, driving local inhabitants from their homes. Vegetation was killed, roads were buried, walls collapsed beneath cinder-weighted roofs. Soon the volcano grew to a hundred feet — 200 — 300 — . As more and more of the frothy, gas-rich foam at the top of its underground magma column was blown from the summit crater, less gaseous, more fluid magma rose in the conduit and broke out through fissures in the base of the cone, flooding across the already Dali-esque countryside, engulfing the entire village of Paricutin in a choppy sea of black basalt.

Sunset Crater, too, excited human wonderment and fear. It, too, deluged fields of corn, a staple crop of the early Sinagua people who inhabited this region. The weight of its cinders collapsed the roofs of prehistoric pit houses as the owners fled from trembling earth and fiery rain. From tree ring studies of charred pit-house roof timbers come dates for the initial eruptions: autumn of 1064 or winter of 1065.

The prehistoric people who fled Sunset Crater's bombardment returned later to find that the cinders curbed evaporation and conserved soil moisture, adding to the productivity of their fields. In an early population explosion, they built clustered multi-storied pueblos along the fringes of the cinder fields. From Sunset Crater a highway loop continues to Wupatki National Monument, where several of these skillfully built stone-slab apartment houses remain. The route leads past other cinder cones in various stages of erosion. Through broad, open pine forest carpeted with black cinders from Sunset Crater, and among long, finger-like lava flows reaching out toward the Little Colorado River. Leaving the forest, it approaches the edge of the Painted Desert, where flat-lying strata of the Coconino Plateau — buff Kaibab Limestone and abovc it deep red Moenkopi Formation sandstone and siltstone — tilt eastward. Here where the San Francisco volcanic field, the Coconino Plateau, and the Painted Desert come together, are the ruins of Wukoki, Wupatki, Citadel, Lomaki, and other pueblos, built of the rocks that surround and support them.

Where molten lava splashes from a small volcanic vent, a splatter cone *develops.*
Tad Nichols photo.

Walnut Canyon National Monument

Less than a thousand years ago prehistoric people of the Sinagua culture found protection from weather and prhaps enemies in shallow, low-ceilinged alcoves along the walls of Walnut Canyon. The alcoves are in the Kaibab Limestone, the Permian marine limestone that surfaces most of the Coconino Plateau and rims both Walnut Canyon and Grand Canyon. In the Walnut Canyon area, not far from the ancient shoreline of the Kaibab sea, the limestone contains increasing amounts of **dolomite** and fine quartz sand. Dolomite, like limestone, is a carbonate rock, but it differs from limestone in that it contains magnesium as well as calcium carbonate. It weathers a little differently than does limestone, too, and forms a rough, rather prickly surface that is unpleasant to scrape against or sit on. Both limestone and dolomite occur along the stairs and ramps descending to the Island Trail.

The Kaibab Limestone in the upper canyon walls is exposed as a series of thick, resistant ledges separated by slopes or recesses of soft, thinly layered silty limestone or limy sandstone. Below the ledges and slopes and recesses, in the lower canyon walls, are cross-bedded sandstones of the Toroweap Formation. Like the Kaibab, these rocks reflect a near-shore environment, but the Toroweap shore was even nearer than that of the Kaibab sea. The rocks reflect this: almost to the exclusion of limestone, they contain stream-deposited and wind-deposited sand, and would be difficult to delineate precisely from Coconino Sandstone, the windblown dune formation that underlies them in Grand Canyon.

Walnut Creek has carved a steep-walled canyon about 350 feet deep. There is little water in the stream now except after heavy rains, largely because it is held back in Lake Mary as part of Flagstaff's water supply. Cave dwellings are under two massive overhanging ledges, low in the Kaibab Formation but still well above the floor of the canyon. Along the trail to the cliff houses watch for fossils in this formation, most of them just hollow molds left when the shells were dissolved away by groundwater. The most common fossils are the rather plump brachiopod *Dictyoclostus*; spiral impressions of snails and molds of small clams can be seen, too. Hard, resistant nodules of chert, abundant along some of the limestone layers, formed around certain kinds of sponges whose "skeletons" were made of little needles of silica, SiO_2, of the same composition as chert or quartz.

Many parallel vertical joints slice through the rock ledges. Some form natural drainage routes and have eroded into steep ravines. Others guided Walnut Creek's detour around the so-called Island.

Where horizontal layers are slightly offset along a break, the structure should of course be termed a fault.

Exhibits in the visitor center tell the story of the prehistoric Sinagua people, and show how the science of tree-ring dating, or **dendrochronology**, documents the development of the cave dwellings and their later abandonment. The same technique is used by geologists to date the eruption of Sunset Crater.

Glossary

Agate: a translucent, very finely crystalline variety of quartz, usually with color bands or other patterns.

Alabaster: a very fine snow-white or pale pink variety of gypsum.

Alkali: a mixture of calcium, potassium, and sodium carbonates common in dried-up lake (playa) deposits.

Alluvial: deposited by running water.

Alluvial fan: a sloping, spreading mass of gravel and sand deposited by a stream as it issues from a narrow mountain valley.

Ammonite: an extinct shell-forming mollusk related to the living chambered nautilus.

Andesite: a medium-colored volcanic rock containing a high proportion of feldspar.

Anhydrite: an evaporite (formed by evaporation) mineral, $CaSO_4$, closely related to gypsum.

Anticline: a fold that is convex upward. An anticline has the oldest rocks near the center.

Aquifer: a porous rock layer from which water may be obtained.

Arroyo: a desert gully with near-vertical banks.

Artesian: water under pressure in an aquifer beneath an impermeable rock layer. Artesian water rises in a well without pumping.

Asbestos: a group of fibrous minerals that separate into long, thin, spinable fibers that are heat-resistant and chemically and electrically inert.

Azurite: a deep blue copper mineral, copper carbonate, $Cu_3(CO_3)_2(OH)_2$.

Badlands: barren, rough, steeply gullied topography in arid areas with soft sedimentary rocks that contain swelling clays such as bentonite.

Bajada (baHAda): coalescing alluvial fans extending from a mountain base into the surrounding valley.

Basalt: a dark gray or black volcanic rock, very fine-grained and often with gas bubbles or vesicles.

Basement: undifferentiated rocks (usually igneous and metamorphic) that underlie the oldest sedimentary or volcanic rocks.

Batholith: a very large mass of granite intruded as molten magma.

Bed: a thin layer or stratum of sedimentary rock.

Bedrock: solid rock exposed at the surface.

Bentonite: clay formed from decomposition of volcanic ash.

Biotite: black mica.

Braided: branching and rejoining of stream channels, characteristic of overloaded streams.

Breccia: volcanic rock consisting of broken fragments ejected from a volcano, cemented together with lava or volcanic ash.

Butte: an isolated steep-walled hill, often capped with a resistant horizontal layer of basalt or sandstone.

Calcite: a mineral, calcium carbonate ($CaCO_3$), the principal mineral in limestone.

Caldera: a large basin-shaped volcanic depression formed when a volcano collapses into its partly emptied magma chamber.

Caliche (caLEEchee): a hard, crusty, whitish rock that accumulates in place as calcium carbonate and other minerals are precipitated in pore spaces in gravel, especially in arid regions.

Caprock: a comparatively resistant rock layer (either sedimentary or volcanic) forming the top of a mesa, butte, or plateau.

Chalcocite: a black or dark lead-colored ore of copper, Cu_2S.

Chalcopyrite: a brass-colored mineral, $CuFeS_2$, the most important ore of copper.

Chert: a very hard, dense, opaque variety of silica (SiO_2), white or colored, occurring as nodules or layers in limestone.

Cinder cone: a small, conical volcano formed by accumulation of volcanic ash and cinders around a volcanic vent.

Clay: a very fine-grained, pasty sediment, plastic when wet, consolidating into claystone or shale.

Columnar jointing: vertical jointing arranged in polygons, due to shrinkage of cooling lava and volcanic ash flows.

Conglomerate: sedimentary rock composed of rounded, waterworn pebbles of older rock, usually in combination with sand.

Cross-bedding: obliquely slanting laminae between the main horizontal layers of sedimentary rock, usually sandstone.

Cuesta: a hill with a long, gentle slope formed by a resistant caprock, and a short steep slope on cut edges of underlying rock.

Dacite: a silicic volcanic rock with a high proportion of quartz and feldspar.

Delta: low, nearly flat land at or near a river's mouth, consisting of clay, sand, and gravel deposited by the river.

Dendritic drainage: a drainage pattern with streams branching irregularly in all directions.

Desert pavement: a surface veneer of closely spaced pebbles resulting when finer material is blown away.

Desert varnish: a dark, shiny surface of manganese and iron oxides that characterizes many exposed rock surfaces in deserts.

Diabase: a dark-colored intrusive rock commonly occurring as dikes and sills.

Diatreme: a volcanic pipe or conduit filled with breccia (some of which may be derived from the wall rock), resulting from a volcanic explosion.

Differential erosion (or weathering): erosion (or weathering) at different rates as governed by difference in resistance or hardness of rocks.

Dike: a thin body of igneous rock resulting when magma intrudes and cools in a vertical crack or joint.

Dip: downward slope of a rock layer.

Dripstone: calcium carbonate deposited by dripping water, usually in caves.

Eolian: created or acted upon by wind.

Epoch: a geologic time unit, subdivision of a period.

Era: the largest geologic time unit.

Evaporite: a mineral deposited as mineralized water evaporates. Principal evaporite minerals are salt, gypsum, and anhydrite.

Exfoliation: peeling off of thin concentric rock layers from the bare outer surface of rock: also called **sheeting**.

Extrusive rock: igneous rock that cooled on the surface; also called **volcanic rock**.

Fanglomerate: conglomerate deposited as an alluvial fan.

Fault: a break in solid rock along which rocks on either side have moved relative to each other.

Feldspar: the most abundant group of light-colored rock-forming minerals.

Flowstone: calcium carbonate deposited by flowing water, usually in caves.

Foliation: layered texture in rock commonly produced by alignment of crystals, as in metamorphic rock such as schist.

Formation: a named, recognizable, mappable unit of rock.

Fossil: any remains or traces of plants or animals preserved in rock, including hard parts, tracks or trails, and impressions.

Garnet: a group of hard red-brown minerals occurring in igneous or metamorphic rock, occasionally of gem quality.

Glacier: a slowly flowing mass of ice.

Gneiss: a coarse-grained metamorphic rock with alternating light and dark minerals.

Graben: a long down-dropped valley bounded by two parallel faults.

Granite: a coarse-grained intrusive igneous rock with feldspar and quartz as principal minerals.

Groundwater: subsurface water completely filling rock pore spaces, joints, and solution channels.

Group: a stratigraphic unit that includes several related formations.

Gypsum: a common evaporite mineral, $CaSO_4 \cdot 2H_20$.

Halite: common salt, NaCl.

Hematite: a dark red ore of iron (Fe_2O_3) that in low concentrations colors rocks in various shades of red.

Honeycomb weathering: weathering of sandstone into small, deep pits.

Hoodoos: bizarre pillars of rock developed by differential weathering and erosion.

Hornblende: a common, black, dark green, or brown mineral occurring as rod-like crystals in igneous and metamorphic rocks.

Igneous rocks: rocks formed by cooling of molten magma.

Intrusive rock: igneous rock that cools and hardens slowly below the surface.

Jasper: red chert.

Joint: a fracture in rock along which no appreciable movement has occurred.

Karst: a distinctive type of landscape where solution of limestone has created caves, sink holes, and solution valleys.

Laccolith: a sill-like igneous intrusion that domes up overlying rock layers.

Lava: magma that reaches the surface, or rock hardened from it.

Limonite: a group of brown iron minerals that color rocks pale yellow to brown; yellow rust.

Listric fault: a concave-upward fault that flattens out with depth.

Lithosphere: the rigid outer layer of the earth, made up of the crust and the outer part of the mantle.

Maar volcano: a low, shallow volcanic crater formed by a volcanic explosion as molten magma comes in contact with water.

Magma: molten rock in or from the earth's interior.

Malachite: a bright green ore of copper, $Cu_2CO_3(OH)_2$.

Member: a subdivision of a formation.

Mesa: a flat-topped mountain or hill capped with a resistant rock layer and edged with steep cliffs.

Metaconglomerate: rock altered by heat and pressure but still recognizable as having been conglomerate.

Metamorphic core complex: a dome of ancient igneous or metamorphic rock with a shell or carapace of mylonite — intensely deformed metamorphic rock.

Metamorphic rocks: rocks formed from older rocks that have been subjected to great heat and pressure or to chemical changes.

Metasedimentary rock: rock altered by heat and pressure but still recognizable as having been sedimentary in origin.

Metavolcanic rock: rock altered by heat and pressure but still recognizable as having been volcanic in origin.

Mica: a group of minerals characterized by the way they separate into thin, shiny plates or flakes.

Monocline: a fold or flexure in stratified rock in which all the strata dip in the same direction.

Muscovite: white or light brown mica.

Mylonite: compact streaky or banded rock produced by shearing movements and intense pressures associated with certain types of faults.

Obsidian: black, dark gray, or reddish volcanic glass, formed by extremely rapid cooling of lava.

Olivine: an olive green mineral, $(Mg,Fe)_2SiO_4$, common in basalt magma whose source is in the earth's mantle.

Ore: economically mineable rock containing valuable minerals.

Orogeny: mountain-building.

Outcrop: surface exposure of geologic materials.

Overburden: useless rock material lying above an ore deposit.

Overthrust: the upper plate of a thrust fault, or the fault itself.

Pediment: a gently inclined erosion surface carved in bedrock at the base of a mountain range.

Pegmatite: very coarse-grained igneous rock, similar to granite in composition, usually occurring in veins in igneous or metamorphic rock.

Period: a geologic time unit longer than an epoch, a subdivision of an era.

Petroglyph: a drawing carved or pecked on rock; usually prehistoric.

Phenocryst: a large, conspicuous crystal in a finer matrix in an igneous rock that cooled moderately slowly at shallow depth.

Placer: a gravel or sand deposit containing particles of gold or other valuable minerals.

Playa: a flat-floored dry lake bottom in an undrained desert basin.

Porphyry: an igneous rock containing conspicuous large crystals in a fine-grained matrix.

Quartz: a hard, glassy mineral composed of crystalline silica, SiO_2; one of the commonest rock-forming minerals.

Quartzite: a metamorphic rock formed from sandstone cemented by silica.

Redbeds: red, pink, and purple sedimentary rocks, usually sandstone and shale colored by hematite.

Rhyolite: light-colored silicic volcanic rock containing very small or microscopic crystals of quartz, feldspar, and mica.

Rotten granite: granite weakened by decomposition of mica grains so that it is in effect turning to sand.

Schist: metamorphic rock with parallel orientation of abundant mica flakes, so that it breaks easily along more or less parallel planes.

Sedimentary rocks: rocks resulting from consolidation of loose sediment that has accumulated in layers.

Shale: a fine-grained sedimentary rock formed by consolidation of clay, silt, or mud, characterized by fine stratification so that it breaks into flat sheets.

Shield volcano: a low, broad volcano built of flows of very fluid basaltic lava.

Silica: silicon dioxide, SiO_2, occurring as quartz and as a major part of many other minerals.

Siliceous sinter: hotspring deposits composed largely of silica.

Silicic: igneous rock or magma of 65% or more silica.

Silicified wood: petrified or fossil wood replaced, commonly cell by cell, with silica.

Sill: a thin body of igneous rock intruded between horizontal rock layers.

Sink: a depression caused when ground collapses into an underlying limestone solution cavern.

Solution cavern: a cave resulting from solution of limestone.

Spheroidal weathering: weathering in which rectangular blocks of rock are attacked from all sides, so that corners are rounded off.

Stalactite: a dripstone "icicle" hanging from the roof of a limestone cave.

Stalagmite: a dripstone pedestal rising from the floor of a limestone cave.

Stock: an igneous intrusion smaller than a batholith, commonly rising like a finger from a batholith.

Strata: layers or beds of rock. Singular is stratum.

Stratovolcano: a cone-shaped volcano built of alternating layers of silicic lava and volcanic ash.

Stromatolite: dome-shaped, cabbagelike fossil algae.

Syncline: a fold that is concave upward.

Tailings: waste debris from ore-processing mills.

Talus: broken rock that collects at the foot of a hill or cliff.

Thrust fault: a fault in which one side is pushed horizontally or nearly horizontally over the other side.

Travertine: hotspring deposits composed largely of calcite.

Tuff: a rock composed of compacted volcanic ash.

Unconformity: a surface of erosion that separates younger strata from older rock.

Vein: a thin, sheetlike igneous intrusion into a crevice, or an ore deposit of similar shape.

Vesicle: a bubble cavity in lava.

Volcanic ash: fine rock material ejected from a volcano.

Volcanic dome: a steep-sided, rounded extrusion of very viscous lava, forming a dome-shaped mass, often within a caldera.

Volcanic neck: an erosional remnant of volcanic rock that formerly filled a volcano's conduit.

Water table: the upper surface of groundwater, below which soil and rock are saturated with water.

Welded tuff: volcanic ash hardened by the original heat of the ash and its gases.

Index